Fundamentals of
Diophantine Geometry

T0211100

Serge Lang

Fundamentals of
Diophantine Geometry

Springer-Verlag
New York Berlin Heidelberg Tokyo

Serge Lang
Department of Mathematics
Yale University
New Haven, CT 06520
U.S.A.

AMS Subject Classifications: 10B99, 14GXX

Library of Congress Cataloging in Publication Data
Lang, Serge. 1927–
    Fundamentals of diophantine geometry.
    1. Diophantine analysis.   2. Geometry, Algebraic.
I. Title.
QA242.L235   1983   512'.74   83-361

An earlier version of this book, *Diophantine Geometry*, was published by Wiley-Interscience.

Published in 2010 by Springer-Verlag New York Inc.

ISBN 978-1-4419-2818-4

9 8 7 6 5 4 3 2 1

Foreword

Diophantine problems represent some of the strongest aesthetic attractions to algebraic geometry. They consist in giving criteria for the existence of solutions of algebraic equations in rings and fields, and eventually for the number of such solutions.

The fundamental ring of interest is the ring of ordinary integers $\mathbf{Z}$, and the fundamental field of interest is the field $\mathbf{Q}$ of rational numbers. One discovers rapidly that to have all the technical freedom needed in handling general problems, one must consider rings and fields of finite type over the integers and rationals. Furthermore, one is led to consider also finite fields, p-adic fields (including the real and complex numbers) as representing a localization of the problems under consideration.

We shall deal with global problems, all of which will be of a qualitative nature. On the one hand we have curves defined over say the rational numbers. If the curve is affine one may ask for its points in $\mathbf{Z}$, and thanks to Siegel, one can classify all curves which have infinitely many integral points. This problem is treated in Chapter VII. One may ask also for those which have infinitely many rational points, and for this, there is only Mordell's conjecture that if the genus is ≥ 2, then there is only a finite number of rational points.

On the other hand, we have abelian varieties. If A is an abelian variety defined over $\mathbf{Q}$, then its group of rational points is finitely generated (Mordell–Weil theorem). The proofs do not give a constructive method to get hold of the generators. Abelian varieties include of course curves of genus 1, which have been the fundamental testing ground for conjectures, theorems, and proofs in the theory.

As a curve of genus ≥ 1 is characterized by the fact that it is embeddable in an abelian variety, one is led to study subvarieties of abelian varieties, and especially Mordell's conjecture, in this light. A curve of genus ≥ 2 is characterized by the fact that it is unequal to any non-zero translation of itself in its Jacobian. This is a very difficult hypothesis to use.

It is also interesting to look at subvarieties of other group varieties. For us, the natural ones to consider are the toruses, i.e. products of multiplicative groups. It will be shown that if G is such a group, V a curve contained in G, Γ a finitely generated subgroup of G, and $V \cap \Gamma$ is infinite then V is the translation of a subtorus (in characteristic 0).

For various generalizations, and comments on the above results and problems, see Chapters VI and VII and the Historical Notes at the end of these chapters.

The prototypes of the diophantine results and methods given in this book were developed by Mordell, Weil, and Siegel between 1920 and 1930. Considering recent developments in the techniques of algebraic geometry, and Roth's theorem, it seemed quite worth while to give a systematic exposition of the theory as it stands, i.e. essentially a qualitative theory. One of the directions of research at present lies in making theorems quantitative, that is, getting estimates for the number of rational or integral points, or the generators of the Mordell–Weil group. It is a striking feature of the proofs that they are highly non-constructive. There do exist still some unsolved qualitative problems. Aside from Mordell's conjecture, for instance, I would conjecture that Siegel's result on integral points remains true for an affine open subset of an abelian variety, defined over $\mathbf{Z}$ (or a ring of finite type over $\mathbf{Z}$). In general, the extension of Roth's theorem and the theory of integral points as given in Chapter VII to varieties of dimension > 1 is still lacking.

New York, 1961   S. LANG

[*The above is reproduced verbatim from* Diophantine Geometry. *Chapters VI and VII are now included in Chapters 7 and 8 respectively.*]

Preface

Diophantine Geometry has been out of print for a while. Advances in algebraic geometry, especially in some of the problems of diophantine geometry, have motivated me to bring out a new book, and I thank Springer-Verlag for publishing it.

Three years after *Diophantine Geometry* appeared, Néron published his fundamental paper giving his canonical heights and explaining how they could be decomposed as sums of local functions. Tate also gave his simple global existence proof for the canonical height. These constitute the main new topics which I treat here. I have also included the Chevalley–Weil theorem; Mumford's theorem concerning heights on curves; Liardet's theorem on the intersection of curves with division points; Schanuel's counting of points in projective space; and theorems of Silverman and Tate on heights in algebraic families.

I still emphasize the Mordell–Weil theorem, Thue–Siegel–Roth theorem, Siegel's theorem on integral points and related results, and the Hilbert irreducibility theorem. These basically depend on heights rather than Néron functions, and so the material on Néron functions (as distinguished from the canonical heights) has been put at the end.

Twenty years ago, the modern period of algebraic geometry was in its infancy. Today the situation is different. Several major advances have not been included although they would be equally worthy. They deserve books to themselves, and I list some of them. Néron's paper proving the existence of a minimal model for abelian varieties was already a book, and only a full new book could do justice to a new exposition of his result and some of its applications; these include Spencer Bloch's interpretation of the Birch–Swinnerton-Dyer conjecture together with its relation to the Néron pairing. I have not included Parsin's and Arakelov's rigidity theorems. I have not included Manin's proof of the Mordell conjecture over function fields; nor Grauert's subsequent other proof, depending on differential geometric considerations on the curve; nor Parsin's proof of Manin's theorem based on entirely different considerations. I have not included Raynaud's proof that the intersection of a curve with the torsion in an abelian variety is finite, nor Bogomolov's results in this same direction. The methods of proof depend on group schemes and p-adic representations. I have not included Arakelov's intersection theory on arithmetic surfaces. Such results and others would fill several volumes, as a continuation of the present book, which merely lays down the basic concepts of diophantine geometry. And the results mentioned above are some of those with a unity of style relative to the topics chosen here. This leaves out transcendental methods, the theory of equations in many variables (quasi-algebraic closure, circle methods), etc.

When *Diophantine Geometry* first came out in 1961, some people regarded methods of algebraic geometry in number theory as something rather far out, and some mathematicians objected to the general style, as when Mordell wrote (*Bull. AMS* (1964), p. 497): "... When proof of an extension makes it exceedingly difficult to understand the simpler cases, it might sometimes be better if the generalizations were left in the Journals." (Mordell's review and my review of his book are reproduced as an appendix.) Of course, a question here is: "exceedingly difficult" to whom? When?

Mordell wrote that he felt like Rip van Winkle. But great progress in mathematics is often accompanied by such feelings. An algebraic geometer who went to sleep in 1961 and woke up in 1981 also might feel like Rip van Winkle. At the time of the first edition, the foundations of algebraic geometry were about to shift from the van der Waerden–Chevalley–Weil–Zariski methods to the Grothendieck methods. Today we have a more extensive perspective. On the whole, neither will bury the other. They serve different purposes, and even complementary purposes. Just as it is essential to reduce mod p^2 (as in Raynaud's proof), and thereby rely on the full power of schemes and commutative algebra, it is also essential to compute estimates on the coefficients and degrees of algebraic polynomial operations, as in the theory of transcendental numbers and algebraic independence several years ago, and in recent work of Wustholz and others. For these applications, one needs the older coordinates (I was about to write of van der Waerden–Chow–Chevalley–Weil–Zariski) of Lasker, Gordan and the nineteenth century algebraic geometers. Anyone who rejects any part of these contributions does so at their own peril. Furthermore, the fact that estimates of coefficients in algebraic operations occurred during recent times first in the direction of transcendental numbers does not mean that this direction is the only application for them. It is a legitimate, and to many people an interesting point of view, to ask that the theorems of algebraic geometry from the Hilbert Nullstellensatz to the most advanced results should carry with them estimates on the coefficients occurring in these theorems. Although some of the estimates are routine, serious and interesting problems arise in this context.

As in *Diophantine Geometry* the content of the present volume is still essentially qualitative. My conjecture that there is only a finite number of integral points on an affine open subset of an abelian variety is still unproved. Roth's theorem is still not effective, and the main advance has been due to Schmidt's generalizations to higher dimensions.

My introduction to *Elliptic Curves: Diophantine Analysis* makes it unnecessary to discuss here once more the relative roles of general theorems and concrete examples or special cases, which are complementary. The present book is addressed to those whose taste lies with abelian varieties.

New Haven, 1983   S. LANG

As this book goes to press, Faltings has announced his proof of the Mordell conjecture.

Arbeitstagung, Bonn
June 1983

Contents

Acknowledgment

I thank Michel Laurent and Michel Waldschmidt for useful comments. I am especially indebted to Joe Silverman for his thorough going over of the manuscript and a long list of valuable suggestions and corrections. I am also indebted to Silverman and Tate for their manuscripts which formed the basis for the next to last chapter.

Some Standard Notation

The following notation is used in a standard way throughout.

Rings are assumed commutative and without divisors of 0, unless otherwise specified.

$\boldsymbol{\mu}$	group of all roots of unity.
$\boldsymbol{\mu}(K)$	subgroup of roots of unity in a field K.
R^*, K^*	invertible elements in a ring R (resp. in a field K).
K^a	algebraic closure of a field K.
$K(P)$	field obtained by adjoining to K a set of affine coordinates for a point P (equal to $K(x_0/x_i, \ldots, x_n/x_i)$ if $(x_0, \ldots, x_n)$ are projective coordinates for P).
$R(\mathfrak{a})$	$R/\mathfrak{a}$ for any ideal $\mathfrak{a}$.
$\mathbf{Z}(N)$	$\mathbf{Z}/N\mathbf{Z}$.
$A^{(p)}$	p-primary part of an abelian group A (that is, the subgroup of elements whose order is a power of p).
A_m	subgroup of elements x in an abelian group A such that $mx = 0$.
$[n]$	multiplication by an integer n on an abelian group.
$V(K)$	set of K-valued points of a variety or scheme V.
$D \sim D'$	for divisors D, D', linear equivalence.
$D \approx D'$	for divisors D, D', algebraic equivalence.
$\mathrm{Div}(V)$	group of divisors on a variety V.
$\mathrm{Div}_a(V)$	subgroup of divisors algebraically equivalent to 0.
$\mathrm{Div}_l(V)$	subgroup of divisors linearly equivalent to 0.

Pic(V) $\mathrm{Div}(V)/\mathrm{Div}_l(V)$.

$\mathrm{Pic}_0(V)$ $\mathrm{Div}_a(V)/\mathrm{Div}_l(V)$.

NS(V) $\mathrm{Div}(V)/\mathrm{Div}_a(V)$ (the **Néron–Severi** group of V).

$h \sim h'$ equivalence for functions, $|h - h'|$ is bounded.

$h \approx h'$ quasi-equivalence for functions: for each $\varepsilon > 0$,

$$-C_1 + (1 - \varepsilon)h \leqq h' \leqq (1 + \varepsilon)h + C_2.$$

$h \ll h'$ for functions, with h' positive, there exists a constant $C > 0$
                 such that $|h| \leqq Ch'$. Same as $h = O(h')$.

$h \gg\ll h'$ both h, h' positive, $h \ll h'$ and $h' \ll h$.

Usually h, H denote **heights** with $h = \log H$. These are indexed to specify
qualifications:

h_φ height determined by a morphism φ into projective space.

k_K height relative to a field K.

h_X height determined by a morphism derived from the linear
              system $\mathscr{L}(X)$, well defined up to $O(1)$.

h_c on an arbitrary variety, the height associated with a divisor
              class c, determined only up to $O(1)$; on an abelian variety, the
              canonical height.

$\hat{h}_c$ canonical height if one needs to distinguish it from an equiva-
              lence class of heights.

Standard references:

IAG $=$ *Introduction to Algebraic Geometry* [L 2].

AV $=$ *Abelian Varieties* [L 3].

Weil's *Foundations* is still quoted in canonical style, F^2-X_y, Theorem Z,
which refers to the second edition.

For schemes, see Hartshorne's *Algebraic Geometry*, and also Mumford's
Abelian Varieties.

CHAPTER 1

Absolute Values

What distinguishes an arithmetic problem from a geometric one is the nature of the ground ring or ground field and any prime structure it may have. In addition to the existence of primes (prime numbers, prime ideals, etc.) one must also, or even principally, take into account archimedean absolute values. It turns out that all the foundational material we need can be stated in terms of absolute values, and we have included in this chapter everything we need in the sequel.

It is convenient to localize certain questions by taking the completion of our field with respect to an absolute value. In fact, our insight concerning the extensions of an absolute value on K to a finite extension E will come from an embedding of E into the algebraic closure of the completion of K.

§1. Definitions, Dependence and Independence

Let K be a field. An **absolute value** on K is a real valued function $x \mapsto |x|_v = |x|$ on K satisfying the following three properties:

AV 1. We have $|x| \geq 0$, and $|x| = 0$ if and only if $x = 0$.

AV 2. $|xy| = |x||y|$ for all $x, y \in K$.

AV 3. $|x + y| \leq |x| + |y|$.

If instead of **AV 3**, the absolute value satisfies the stronger condition

AV 4. $|x + y| \leq \max\{|x|, |y|\}$

then we shall say that it is a **valuation**, or that it is **non-archimedean**. In this chapter we emphasize absolute values, taking more or less for granted the elementary properties of valuations as given, for instance, in IAG [16], Chapter I.

The absolute value which is such that $|x| = 1$ for all $x \neq 0$ is called **trivial**.

We define

$$v(x) = -\log|x|_v,$$

so v is a homomorphism $v: K^* \to \mathbf{R}$ (the additive group of $\mathbf{R}$).

An absolute value on K defines a metric, and thus a topology. Two absolute values are called **dependent** if they define the same topology. If they do not, they are called **independent**.

Proposition 1.1. *Two non-trivial absolute values,* $|\ |_1$ *and* $|\ |_2$, *on a field K, are dependent if and only if the relation*

$$|x|_1 < 1$$

implies $|x|_2 < 1$. *If they are dependent, then there exists a real number $\lambda > 0$ such that* $|x|_1 = |x|_2^\lambda$ *for all $x \in K$.*

Proof. If the two absolute values are dependent, then our condition is satisfied, because the set of $x \in K$ such that $|x|_1 < 1$ is the same as the set such that $\lim x^n = 0$ for $n \to \infty$. Conversely, assume the condition satisfied. Then $|x|_1 > 1$ implies $|x|_2 > 1$ since $|x^{-1}|_1 < 1$. By hypothesis, there exists an element $x_0 \in K$ such that $|x_0|_1 > 1$. Put $a = |x_0|_1$ and $b = |x_0|_2$, and let $\lambda = \log(b)/\log(a)$. Let $x \in K$, and $x \neq 0$. Say $|x|_1 \geq 1$. Then $|x|_1 = |x_0|_1^\alpha$ for some $\alpha \geq 0$. If m, n are integers > 0 such that $m/n > \alpha$ we have $|x|_1 < |x_0|_1^{m/n}$ whence $|x^n/x_0^m|_1 < 1$, and thus $|x^n/x_0^m|_2 < 1$, or in other words, $|x|_2 < |x_0|_2^{m/n}$. Similarly if m, n are integers such that $m/n < \alpha$, then $|x|_2 > |x_0|_2^{m/n}$. Hence $|x|_2 = |x_0|_2^\alpha$. The assertion of our proposition is now obvious.

One of the main applications of the theory of absolute values will be to the rational numbers $\mathbf{Q}$ and its finite extensions (which are called **number fields**). It is easy to show that a non-trivial absolute value on $\mathbf{Q}$ is dependent on one of the following:

The ordinary absolute value.

The p-**adic absolute value** $|\ |_p$, defined for each prime number p by the formula

$$|p^r m/n|_p = 1/p^r,$$

where r is an integer, and m, n are integers $\neq 0$ and not divisible by p.

In fact, if the absolute value on $\mathbf{Q}$ is a valuation, then looking at its valuation ring $\mathfrak{o}$ and maximal ideal $\mathfrak{m}$ one sees immediately that $\mathfrak{m} \cap \mathbf{Z}$ is a prime ideal, and thus is generated by a prime number p. Thus by Proposition 1.1 our absolute value is a power of the p-adic absolute value (cf. IAG, Chapter I).

Suppose that we have an absolute value on a field which is bounded on the prime ring (i.e. the integers $\mathbf{Z}$ in characteristic 0, or the integers mod p

in characteristic p). Then it is non-archimedean. One sees this immediately by looking at the expansion of $(x + y)^n$ and $|(x + y)^n)|^{1/n}$ as $n \to \infty$. This is always the case in characteristic > 0.

If the absolute value is archimedean, then we refer the reader to Bourbaki or Van der Waerden for a proof that it is dependent on the ordinary absolute value. We do not need this fact in our applications, for we always will start with a concretely given set of absolute values on fields which will interest us.

The relation of dependence on absolute values is obviously an equivalence relation (symmetric, reflexive, transitive). An equivalence class of non-trivial absolute values on a field K will be called a **prime** of K, and will sometimes be denoted by a German letter $\mathfrak{p}$, or if we deal with an extension of K, by $\mathfrak{P}$. If E is an extension of K, and $\mathfrak{p}$ a prime of K, and $\mathfrak{P}$ a prime of E, we say that $\mathfrak{P}$ **extends** $\mathfrak{p}$ and write $\mathfrak{P}|\mathfrak{p}$ if the absolute values in $\mathfrak{P}$ when restricted to K give absolute values of $\mathfrak{p}$. Similarly, if w is an absolute value on E extending an absolute value on K, we write $w|v$. We shall study in §4 the manner in which an absolute value on K can be extended to algebraic extensions of K.

For the convenience of the reader, we recall some terminology. A **prime** $\mathfrak{p}$ is called **non-archimedean** if its absolute values are non-archimedean. If this is the case, then it belongs to a valuation ring $\mathfrak{o} = \mathfrak{o}_\mathfrak{p}$ with maximal ideal $\mathfrak{m} = \mathfrak{m}_\mathfrak{p}$. Let E be a finite extension of K, and let $\mathfrak{P}$ be a prime extending $\mathfrak{p}$. It corresponds to a valuation ring lying above $\mathfrak{o}$. Let $\mathfrak{o}_E$ be this valuation ring, and $\mathfrak{m}_E$ its maximal ideal. Then the residue class field $\mathfrak{o}_E/\mathfrak{m}_E$ is an extension of $\mathfrak{o}/\mathfrak{m}$, and its degree is called the **residue class degree**, denoted by $d_\mathfrak{P}$, or d_w if w is one of the valuations of $\mathfrak{P}$. The value group $w(K^*)$ is a subgroup of $w(E^*)$, whose index is called the **ramification index** and denoted by $e_\mathfrak{P}$ or e_w. They are both finite (cf. *Algebra*, Chapter XII).

In our applications we shall deal mostly with archimedean absolute values, or with discrete valuations. Let us say a few words about the latter.

Suppose we have a field K with a valuation ring $\mathfrak{o}$ and maximal ideal $\mathfrak{p}$. In general, we obtain a valuation whose canonical value group is the factor group K^*/U where U are the units of $\mathfrak{o}$ (cf. *Algebra*, Chapter XII). Suppose this value group is infinite cyclic. It can then be embedded into the positive reals, preserving the ordering. We select a canonical generator for it (from two possible ones) namely, $|\pi|$ where π lies in $\mathfrak{p}$. Then every element of K^* can be written in the form

$$x = u\pi^r$$

where u is a unit, r an integer, which is called the **order** of x at the valuation (or prime). In particular, $\mathfrak{p}$ is principal, generated by π. Using the same letter $\mathfrak{p}$ for the maximal ideal of $\mathfrak{o}$, or for the equivalence class of absolute values (valuations) which correspond to it, we write

$$r = \text{ord } x = \text{ord}_\mathfrak{p} x = \text{ord}_v x.$$

Using Proposition 1.1, we see that all the absolute values corresponding to o can be described as follows. We choose a number c such that $0 < c < 1$, and put

$$|x| = c^{\operatorname{ord} x}.$$

Such absolute values, or valuations, will be called **discrete**.

There is of course considerable arbitrariness in the choice of the constant c, and there is usually no way of choosing it in a natural manner. However, in some applications, the residue class field o/p is a finite field. In that case, denoting the number of elements in it by **Np**, we choose $c = 1/\mathbf{Np}$, and our absolute value becomes

$$\|x\|_{\mathfrak{p}} = (1/\mathbf{Np})^{\operatorname{ord}_{\mathfrak{p}} x}.$$

This will be the case in one of the most important applications, namely, number fields, whose non-trivial valuation rings all have a finite residue class field. We shall return to this in greater detail in Chapter 2.

In Proposition 1.1 we derived a strong condition on dependent absolute values. We shall now derive a condition on independent ones.

Theorem 1.2. *Let K be a field, and $|\ |_1, \ldots, |\ |_s$ non-trivial pairwise independent absolute values on K. Let $x_1, \ldots, x_s$ be elements of K, and $\varepsilon > 0$. Then there exists $x \in K$ such that*

$$|x - x_i|_i < \varepsilon$$

for all i.

Proof. Consider first two of our absolute values, say v_1 and v_s. By hypothesis, we can find $\alpha \in K$ such that $|\alpha|_1 < 1$ and $|\alpha|_s \geq 1$. Similarly, we can find $\beta \in K$ such that $|\beta|_1 \geq 1$ and $|\beta|_s < 1$. Put $y = \beta/\alpha$. Then $|y|_1 > 1$ and $|y|_s < 1$.

We shall now prove that there exists $z \in K$ such that $|z|_1 > 1$ and $|z|_j < 1$ for $j = 2, \ldots, s$. We prove this by induction, the case $s = 2$ having just be proved. Suppose we have found $z \in K$ satisfying

$$|z|_1 > 1 \qquad \text{and} \qquad |z|_j < 1 \quad \text{for } j = 2, \ldots, s - 1.$$

If $|z|_s \leq 1$, then the element $z^n y$ for large n will satisfy our requirements.

If $|z|_s > 1$, then the sequence

$$t_n = z^n/(1 + z^n)$$

tends to 1 at v_1 and v_s, and tends to 0 at v_j ($j = 2, \ldots, s - 1$). For large n, it is then clear that $t_n y$ satisfies our requirements.

Using the element z that we have just constructed, we see that the sequence $z^n/(1 + z^n)$ tends to 1 at v_1 and to 0 at v_j for $j = 2, \ldots, s$. For each $i = 1, \ldots, s$ we can therefore construct an element z_i which is very close to 1 at v_i and very close to 0 at v_j $(j \neq i)$. The element

$$x = z_1 x_1 + \cdots + z_s x_s$$

then satisfies the requirement of the theorem.

§2. Completions

Let K be a field with a non-trivial absolute value v, which will remain fixed throughout this section. One can then define in the usual manner the notion of Cauchy sequence, and K is said to be complete if every Cauchy sequence converges.

Proposition 2.1. *There exists a pair (K_v, i) consisting of a field K_v complete under an absolute value, and an embedding $i: K \to K_v$ such that the absolute value on K is induced by that of K_v (i.e. $|x|_v = |ix|$), and such that iK is dense in K_v. If (K_v', i') is another such pair, then there exists a unique isomorphism $\varphi: K_v \to K_v'$ preserving the absolute values, and making the following diagram commutative:*

The uniqueness is obvious. One proves the existence in the well known manner: One considers the Cauchy sequences in K, which form a ring. The null sequences form a maximal ideal, and the residue class ring is a field K_v. Our field K is naturally embedded in K_v (by means of the sequences whose elements consist of a fixed element of K), and the absolute value on K can be extended to K_v by continuity.

One usually identifies K inside K_v, and calls K_v the **completion** of K.

If the absolute value is archimedean (and thus K has characteristic 0), then it is well known that K_v is either the field of real numbers or of complex numbers, and that the absolute value is dependent on the ordinary one. For a proof, the reader may see Bourbaki. We shall not need this result in this book, and hence we shall always *assume* that the archimedean absolute values we deal with have this property.

As the completion depends only on the prime determined by our absolute value, we also denote it by $K_{\mathfrak{p}}$.

If the absolute value is a valuation, then we refer the reader to *Algebra*, Chapter XII, for the proof that it can be extended in some way to the algebraic closure of K.

Proposition 2.2. *Let K be complete with respect to a non-trivial absolute value v. If E is any algebraic extension of K, then v has a unique extension to E. If E is finite over K, then E is complete.*

Proof. We give the proof only in the non-archimedean case, the other being trivial once one knows that the only complete archimedean fields are the reals or complex. Without loss of generality, we may assume E finite over K. Let us prove the uniqueness. In view of Proposition 1.1, it suffices to prove that two extensions define the same topology. Furthermore, If $\omega_1, \ldots, \omega_n$ are linearly independent elements of E over K, and

$$(1) \qquad x^{(v)} = \xi_1^{(v)}\omega_1 + \cdots + \xi_n^{(v)}\omega_n, \qquad \xi_n^{(v)} \in K,$$

is a Cauchy sequence in E, then it will suffice to prove that the n sequences $\xi_j^{(v)}$ converge in K. We do this by induction on n. It is obvious for $n = 1$. Assume $n \geq 2$. We consider a sequence as above and, without loss of generality, we may assume that it converges to 0. (If necessary, consider $x^{(v)} - x^{(\mu)}$ for $v, \mu \to \infty$.) We must then show that the sequences of the coefficients converge to 0 also. If this is not the case, then there exists a number $a > 0$ such that we have for some j, say $j = 1$,

$$|\xi_1^{(v)}| > a$$

for all v sufficiently large. Then $x^{(v)}/\xi_1^{(v)}$ converges to 0, and we can write

$$\frac{x^{(v)}}{\xi_1^{(v)}} - \omega_1 = \frac{\xi_2^{(v)}}{\xi_1^{(v)}}\omega_2 + \cdots + \frac{\xi_n^{(v)}}{\xi_1^{(v)}}\omega_n.$$

We let $y^{(v)}$ be the right-hand side of this equation. Then $y^{(v)}$ yields a sequence which converges (according to the left-hand side of the equation). By induction, we conclude that its coefficients in terms of $\omega_2, \ldots, \omega_n$ also converge in K, say to $\eta_2, \ldots, \eta_n$. Taking the limit, we get

$$\omega_1 = \eta_2\omega_2 + \cdots + \eta_n\omega_n$$

contradicting the independence of the ω_j.

We have thus proved that the extension of our valuation is unique. Furthermore, we have seen that if E is finite over K, of degree n, then it has the topology of Cartesian n-space over K: in particular, it is complete. The Cauchy sequences of the coefficients in (1) above converge in K, say to $\xi_1, \ldots, \xi_n$; then our sequence $x^{(v)}$ converges to $\xi_1\omega_1 + \cdots + \xi_n\omega_n$.

From the uniqueness, we can get an explicit determination of the absolute value on an algebraic extension. Indeed, observe first that if E is a normal extension of K, and σ is an automorphism of E over K, then the function

$$x \mapsto |\sigma x|$$

is an absolute value on E extending that of K. Hence we get

$$|\sigma x| = |x|.$$

If E is algebraic over K, and σ is an isomorphism of E over K, then the same conclusion remains valid, as one sees immediately by embedding E in a normal extension of K. (We assume, without loss of generality, that all our algebraic extensions are contained in a fixed algebraic closure of K.) In particular, if α is algebraic over K, of degree n, and if $\alpha_1, \ldots, \alpha_n$ are its conjugates (counting multiplicities), then all the absolute values $|\alpha_i|$ are equal. Denoting by N the norm from $K(\alpha)$ to K, we see that

$$|N(\alpha)| = |\alpha|^n,$$

and taking the n-th root, we get:

Proposition 2.3. *Let K be complete with respect to a non-trivial absolute value. Let α be algebraic over K, and let N be the norm from $K(\alpha)$ to K. Let $n = [K(\alpha):K]$. Then*

$$|\alpha| = |N(\alpha)|^{1/n}.$$

Of course, the arguments we gave above hold when we deal with a valuation, but in the archimedean case the proposition is also true, taking into account that we are dealing with the complex and real numbers.

Let us now consider the case of a discrete valuation. We assume that K is complete under a discrete valuation, with valuation ring $\mathfrak{o}$, and maximal ideal $\mathfrak{m}$, and we let π be a prime element of $\mathfrak{o}$. If $x, y \in K^*$ are such that $|x| = |y|$, we shall write $x \sim y$. Let $\pi_i\,(i = 1, 2, \ldots)$ be a sequence of elements of $\mathfrak{o}$ such that $\pi_i \sim \pi^i$. If $x \in \mathfrak{o}$ we denote by x' its canonical image in $\mathfrak{o}/\mathfrak{m}$. Let R be a set of representatives of $\mathfrak{o}/\mathfrak{m}$ in $\mathfrak{o}$. This means that the map $x \mapsto x'$ of R into $\mathfrak{o}/\mathfrak{m}$ is bijective. Then every element of $\mathfrak{o}$ can be written as a convergent series

$$x = a_0 + a_1\pi_1 + a_2\pi_2 + \cdots$$

with $a_i \in R$, and the a_i uniquely determined. This is easily seen by a recursive argument, namely, suppose we have written

$$x \equiv a_0 + \cdots + a_n\pi_n \pmod{\mathfrak{m}^{n+1}}$$

then $x - (a_0 + \cdots + a_n\pi_n) = \pi_{n+1}y$ with some $y \in \mathfrak{o}$. But by hypothesis, we can write $y = a_{n+1} + \pi z$ with some $a_{n+1} \in R$. From this we get

$$x \equiv a_0 + \cdots + a_{n+1}\pi_{n+1} \pmod{\mathfrak{m}^{n+2}},$$

and it is clear that the n-th term in our series tends to 0. Thus our series converges. The fact that R contains precisely one representative for each residue class mod $\mathfrak{m}$ implies that the a_i are uniquely determined.

Let E be a finite extension of K, let $\mathfrak{o}_E$, $\mathfrak{m}_E$ be the valuation ring and maximal ideal in E lying above $\mathfrak{o}$, $\mathfrak{m}$ in K, and let Π be a prime element in E. If Γ_E and Γ_K are the value groups of the valuation in E and K respectively, let $e = (\Gamma_E : \Gamma_K)$. It is the **ramification index**. We then have

$$|\Pi^e| = |\pi|,$$

and the elements

$$\Pi^i\pi^j, \qquad \begin{matrix} 0 \leq i \leq e - 1, \\ j = 0, 1, 2, \ldots \end{matrix}$$

have order $je + i$ in E.

Let $\omega_1, \ldots, \omega_d$ be elements of E such that their residue classes $\omega'_1, \ldots, \omega'_d$ in $E' = \mathfrak{o}_E/\mathfrak{m}_E$ form a basis for E' over K'. If R is as above a set of representatives of $\mathfrak{o}/\mathfrak{m}$ in $\mathfrak{o}$, then

$$a_1\omega_1 + \cdots + a_d\omega_d,$$

as the a_ν range over R, form a system of representatives of $\mathfrak{o}_E/\mathfrak{m}_E$ by definition. From this, one sees that every element of $\mathfrak{o}_E$ admits a convergent expansion

$$\sum_{i=0}^{e-1} \sum_{\nu=1}^{d} \sum_{j=0}^{\infty} \alpha_{\nu,i,j}\pi^j\omega_\nu\Pi^i,$$

which shows that the ed elements $\omega_\nu\Pi^i$ form a set of generators of $\mathfrak{o}_E$ over $\mathfrak{o}$. On the other hand, one sees easily that they are linearly independent over K, and hence we get:

Proposition 2.4. *Let K be complete with respect to a discrete valuation. Let E be a finite extension of K, with ramification index e and residue class degree d. Then*

$$ed = [E : K].$$

If we now combine the interpretation of the degree given in Proposition 2.4 for discrete valuations with the expression given in Proposition 2.3 we get:

Proposition 2.5. *Let K be complete with respect to a discrete valuation and let $\mathfrak{p}$ be its prime. Let E be a finite extension of K, $\mathfrak{P}$ the prime extending $\mathfrak{p}$, and $d_\mathfrak{P}$ the degree of the residue class field extension. Then for $\alpha \in E^*$, we have*

$$\operatorname{ord}_\mathfrak{p} N_K^E(\alpha) = d_\mathfrak{P} \operatorname{ord}_\mathfrak{P} \alpha.$$

Proof. This is immediate from the formula

$$|N_K^E(\alpha)| = |\alpha|^{e_\mathfrak{P} d_\mathfrak{P}}$$

and the definitions.

§3. Unramified Extensions

Let K be a field, $\mathfrak{o}$ a valuation ring of K, and $\mathfrak{m}$ its maximal ideal. We let $K' = \mathfrak{o}/\mathfrak{m}$ be the residue class field, and we let φ_0 be the canonical K'-valued placed on K. We assume that φ_0 is extended in a fixed way to the algebraic closure K^a of K, and call φ this extension.

Let E be an algebraic extension of K, and $\mathfrak{o}_E$ one of the valuation rings of E lying above $\mathfrak{o}$, with $\mathfrak{m}_E$ as maximal ideal. If E is finite over K, then $\mathfrak{o}_E/\mathfrak{m}_E$ is finite over $\mathfrak{o}/\mathfrak{m}$, and thus φ is K'^a-valued. We frequently identify $\varphi(\mathfrak{o}_E)$ with $\mathfrak{o}_E/\mathfrak{m}_E$, and denote it by E'.

Let v be a valuation associated with $\mathfrak{o}$, and w an extension of v to E associated with $\mathfrak{o}_E$. We shall say that w is **unramified** over v, or simply unramified, if the residue class field extension E' over K' is separable, and if the ramification index is equal to 1, i.e. if

$$(w(E^*) : v(K^*)) = 1.$$

If every w extending v is unramified we say that v is **unramified**.

The purpose of this section is to study unramified extensions when K is complete. However, we may just as well make a weaker assumption:

Throughout the rest of this section, we shall assume that for each algebraic extension E of K, there exists precisely one valuation ring of E lying above $\mathfrak{o}$.

In view of our assumption, we know that if ψ is another place of K^a having the same restriction as φ on K, then ψ and φ are equivalent, i.e. there exists an automorphism σ' of K'^a over K' such that

$$\psi = \sigma'\varphi.$$

If $f(X)$ is a polynomial in $K^a[X]$ whose coefficients are finite under φ, then we let f^φ be the polynomial obtained by applying φ to its coefficients. Suppose f has leading coefficient 1 and that its coefficients are integral over $\mathfrak{o}$. If

$$f(X) = \prod_{i=1}^{d} (X - \alpha_i)$$

is a factorization in K^a, then

$$f^\varphi(X) = \prod_{i=1}^{d} (X - \varphi\alpha_i)$$

is a factorization of f^φ in the algebraic closure of K'.

If $f'(X)$ is a polynomial in $E'[X]$ then any polynomial f in $\mathfrak{o}_E[X]$ of the same degree as f', with leading coefficient 1 if f' has leading coefficient 1, and such that $f^\varphi = f'$ will be said to be **lifted back from** f' **into** E. It is clearly always possible to lift back a polynomial.

The above remark concerning the factorization of f shows that φ maps K^a onto the algebraic closure of K'.

Our assumption concerning the uniqueness of the extension of the valuation ring yields the following result.

Proposition 3.1. *Let f be a polynomial in $\mathfrak{o}[X]$ such that f is irreducible in $K[X]$ and has leading coefficient 1. Then f^φ is the power of an irreducible polynomial in $K'[X]$.*

Proof. Suppose $f^\varphi = g'h'$ where g', h' are relatively prime of degrees ≥ 1. Suppose we have a factorization as above, and say $\varphi\alpha_1$ is a root of g' and $\varphi\alpha_2$ is a root of h'. There exists an isomorphism σ of $K(\alpha_1)$ over K such that $\sigma\alpha_1 = \alpha_2$ because f is irreducible, and we can extend σ to K^a. Thus $\varphi\sigma\alpha_1 = \varphi\alpha_2$. But $\varphi\sigma$ is a place coinciding with φ on K, and thus must be conjugate to φ, i.e. there exists an isomorphism σ' of K'^a over K' such that $\varphi\sigma = \sigma'\varphi$, and we must have $\sigma'\varphi\alpha_1 = \varphi\alpha_2$. This contradicts the fact that g' and h' are relatively prime, and proves that f^φ is the power of an irreducible polynomial.

Our purpose is now to establish a bijective correspondence between certain extensions of K and separable extensions of K'. We shall say that a finite extension E of K is **unramified** if it satisfies the following properties:

UN 1. The residue class field E' is separable over K'.

UN 2. We have $[E:K] = [E':K']$.

Another way of formulating these conditions is:

Proposition 3.2. *Let E be a finite extension of K, and let φ_0 be the canonical place of K. Let $[E' : K'] = n$. Then E is unramified over K if and only if φ_0 has at least n distinct extensions to places of E (in the given algebraic closure of K'), and in that case, it has exactly n.*

Proof. By uniqueness, all extensions of φ_0 to E are conjugate, and the number of conjugates is equal to the separable degree of E' over K'. As $[E' : K'] \leq [E : K]$ our assertion is immediate.

We observe that if E is unramified over K, then the ramification index is equal to 1. In the case of greatest interest to us, this is characteristic.

Proposition 3.3. *Assume K complete under a discrete valuation. Then the finite extension E of K is unramified if and only if E' is separable over K' and the ramification index is equal to 1.*

Proof. This follows immediately from Proposition 2.4.

An arbitrary algebraic extension of K is unramified if every finite sub-extension is unramified.
We shall now obtain an algebraic criterion for non-ramification.

Proposition 3.4. *Let E be finite over K. If E is unramified, let $\alpha' \in E'$ be such that $E' = K'(\alpha')$, and let $a \in E$ be such that $\varphi\alpha = \alpha'$. Then $E = K(\alpha)$, and the irreducible polynomial $f(X)$ of α over K is such that f^φ is irreducible. Conversely, if $K = K(\alpha)$ for some integer α satisfying a polynomial $f(X)$ in $o[X]$ having leading coefficient 1 and such that f^φ has no multiple root, then E is unramified over K and $E' = K'(\varphi\alpha)$.*

Proof. First assume E unramified. Let $f'(X)$ be the irreducible polynomial of α' over K'. Let α be an element of E such that $\varphi\alpha = \alpha'$, and let $f(X)$ be its irreducible polynomial over K. Then α is integral over o, and α' is a root of f^φ, and hence f' divides f^φ. On the other hand,

$$\deg f' = [E' : K'] = [E : K] \geq \deg f$$

and so $f' = f^\varphi$. This proves the first statement.
Conversely, if α satisfies the stated condition, then we may assume without loss of generality that its irreducible polynomial $g(X)$ is such that g^φ has no multiple roots, because g divides f. We can now simply apply Proposition 3.1, to conclude that g^φ is irreducible. Using the inequalities

$$[K'(\varphi\alpha) : K'] \leq [E' : K'] \leq [E : K],$$

we now conclude that we must have an equality everywhere, and that $E' = K'(\varphi\alpha)$. This proves our proposition.

Next, we give the formalism for unramified extensions under various operations.

Proposition 3.5. *Let E be a finite extension of K.*

(i) *If $E \supset F \supset K$, then E is unramified over K if and only if E is unramified over F and F unramified over K.*

(ii) *If E is unramified over K, and K_1 is a finite extension of K, then EK_1 is unramified over K_1.*

(iii) *If E_1 and E_2 are finite unramified over K, then so is $E_1 E_2$.*

Proof. The first assertion comes from the inequalities

$$[E' : K'] \leqq [E : K], \qquad [E' : F'] \leqq [E : F], \qquad [F' : K'] \leqq [F : K]$$

together with their multiplicativity property in towers. One must also use the fact that assertion (i) holds when "unramified" is replaced by "a finite separable extension." The second assertion is an immediate consequence of our criterion in Proposition 3.4. The third comes formally from the first and second.

Let us denote by φE the image under φ of the valuation ring $\mathfrak{o}_E$, for any finite extension E of K.

Proposition 3.6. *The map $E \mapsto \varphi E$ gives a bijection of the finite unramified extensions of K and the finite separable extensions of $K' = \varphi K$.*

Proof. We have shown in Proposition 3.4 that every finite separable extension of K' is obtainable as an image φE, where E is unramified. We now must show uniqueness. If $E_1 \subset E_2$ are unramified, then clearly $\varphi E_1 \subset \varphi E_2$. It will therefore suffice to prove that if E_1, E_2 are two unramified extensions of K then $\varphi(E_1 E_2) = \varphi E_1 \cdot \varphi E_2$. To do this, we can write $E_1 = K(\alpha_1)$ and $E_2 = K(\alpha_2)$ where α_1 and α_2 respectively satisfy the properties expressed in Proposition 3.4. Then $E_1 E_2 = K(\alpha_1, \alpha_2)$, and we observe that with respect to the intermediate field E_1, α_2 satisfies a polynomial $f(X)$ such that f^φ has no multiple root. Hence $(E_1 E_2)' = E_1'(\varphi\alpha_1) = K'(\varphi\alpha_1, \varphi\alpha_2)$ which proves Proposition 3.6.

§4. Finite Extensions

Throughout this section, we shall deal with a field K having a nontrivial absolute value v. We wish to describe how this absolute value extends to finite extensions of K.

If we let K_v be the completion, we know that v can be extended to K_v, and then uniquely to its algebraic closure K_v^a. If E is a finite extension of K, or even an algebraic one, then we can extend v to E by embedding E in K_v^a by an isomorphism over K, and taking the induced absolute value on E. We shall now prove that every extension of v can be obtained in this manner.

Proposition 4.1. *Let E be a finite extension of K. Let w be an absolute value on E extending v, and let E_w be the completion. Let K_w be the closure of K in E_w and identify E in E_w. Then $E_w = EK_w$ (the composite field).*

Proof. We observe that K_w is a completion of K, and that the composite field EK_w is algebraic over K_w and therefore complete by Proposition 2.2. Since it contains E, it follows that E is dense in it, and hence that $E_w = EK_w$.

If we start with an embedding $\sigma: E \to K_v^a$ (always assumed to be over K), then we know again by Proposition 2.2 that $\sigma E \cdot K_v$ is complete. Thus this construction and the construction of the proposition are essentially the same, up to an isomorphism. In the future, we take the embedding point of view. We must now determine when two embeddings give us the same absolute value on E.

Given two embeddings $\sigma, \tau: E \to K_v^a$, we shall say that they are **conjugate over K_v** if there exists an automorphism λ of K_v^a over K_v such that $\sigma = \lambda\tau$. We see that actually λ is determined by its effect on τE, or $\tau E \cdot K_v$.

Proposition 4.2. *Let E be an algebraic extension of K. Two embeddings $\sigma, \tau: E \to K_v^a$ give rise to the same absolute value on E if and only if they are conjugate over K_v.*

Proof. Suppose they are conjugate over K_v. Then the uniqueness of the extension of the absolute value from K_v to K_v^a guarantees that the induced absolute values on E are equal. Conversely, suppose this is the case. Let $\lambda: \tau E \to \sigma E$ be an isomorphism over K. We shall prove that λ extends to an isomorphism of $\tau E \cdot K_v$ onto $\sigma E \cdot K_v$ over K_v. Since τE is dense in $\tau E \cdot K_v$, an element $x \in \tau E \cdot K_v$ can be written

$$x = \lim \tau x_n$$

with $x_n \in E$. Since the absolute values induced by σ and τ on E coincide, it follows that the sequence $\lambda\tau x_n = \sigma x_n$ converges to an element of $\sigma E \cdot K_v$ which we denote by λx. One then verifies immediately that λx is independent of the particular sequence τx_n used, and that the map $\lambda: \tau E \cdot K_v \to \sigma E \cdot K_v$ is an isomorphism, which clearly leaves K_v fixed. This proves our proposition.

In view of the previous two propositions, if w is an extension of v to a finite extension E of K, then we may identify E_w and a composite extension

EK_v of E and K_v. If $N = [E:K]$ is finite, then we shall call

$$N_w = [E_w : K_v]$$

the **local degree**. It is clear that $\sum_{w|v} [E_w : K_v] \leqq [E:K]$.

Proposition 4.3. *Let E be a finite separable extension of K, of degree N. Then*

$$N = \sum_{w|v} N_w.$$

Proof. We can write $E = K(\alpha)$ for a single element α. Let $f(X)$ be its irreducible polynomial over K. Then over K_v, we have a decomposition

$$f(X) = f_1(X) \cdots f_r(X)$$

into irreducible factors $f_i(X)$. They all appear with multiplicity 1 according to our hypothesis of separability. The embeddings of E into K_v^a correspond to the maps of α onto the roots of the f_i. Two embeddings are conjugate if and only if they map α onto roots of the same polynomial f_i. On the other hand, it is clear that the local degree in each case is precisely the degree of f_i. This proves our proposition.

There exist interesting cases in characteristic p for which the relation in the proposition holds without the hypothesis of separability. This is notably the case for the discrete valuation arising from a simple subvariety of co-dimension 1 on an algebraic variety.

Whenever v is an absolute value on K such that for any finite extension E of K we have

$$[E:K] = \sum_{w|v} [E_w : K_v],$$

we shall say that v is **well behaved**. Suppose we have a tower of finite extensions, $L \supset E \supset K$. Let w range over the absolute values of E extending v, and u over those of L extending v. If $u|w$ then L_u contains E_w. Thus we have:

$$\sum_{u|v} [L_u : K_v] = \sum_{w|v} \sum_{u|w} [L_u : E_w][E_w : K_v]$$

$$= \sum_{w|v} [E_w : K_v] \sum_{u|w} [L_u : E_w]$$

$$\leqq \sum_{w|v} [E_w : K_v][L : E]$$

$$\leqq [E:K][L:E].$$

From this we immediately see that if v is well behaved, E finite over K, and w extends v on E, then w is well behaved (we must have an equality everywhere).

Let E be a finite extension of K. Let p^r be its inseparable degree. We recall that the norm of an element $\alpha \in K$ is given by the formula

$$N^E_K(\alpha) = \prod_\sigma \sigma\alpha^{p^r},$$

where σ ranges over all distinct isomorphisms of E over K (into a given algebraic closure).

If w is an absolute value extending v on E, then the norm from E_w to K_v will be called the **local norm**.

Replacing the above product by a sum, we get the trace, and the local trace. We abbreviate the trace by Tr.

Proposition 4.4. *Let E be a finite extension of K, and assume that v is well behaved. Let $\alpha \in E$. Then:*

$$N^E_K(\alpha) = \prod_{w|v} N^{E_w}_{K_v}(\alpha),$$

$$\mathrm{Tr}^E_K(\alpha) = \sum_{w|v} \mathrm{Tr}^{E_w}_{K_v}(\alpha).$$

Proof. Suppose first that $E = K(\alpha)$, and let $f(X)$ be the irreducible polynomial of α over K. If we factor $f(X)$ into irreducible terms over K_v, then

$$f(X) = f_1(X) \cdots f_r(X),$$

where each $f_i(X)$ is irreducible, and the f_i are distinct because of our hypothesis that v is well behaved. The norm $N^E_K(\alpha)$ is equal to $(-1)^{\deg f}$ times the constant term of f, and similarly for each f_i. Since the constant term of f is equal to the product of the constant terms of the f_i, we get the first part of the proposition. The statement for the trace follows by looking at the penultimate coefficient of f and each f_i.

If E is not equal to $K(\alpha)$, then we simply use the transitivity of the norm and trace. We leave the details to the reader.

Taking into account Proposition 2.3, we have:

Theorem 4.5. *Let K have a well behaved absolute value v. Let E be a finite extension of K, and $\alpha \in E$. Let*

$$N_w = [E_w : K_v]$$

for each absolute w on E extending v. Then

$$\prod_{w|v} |\alpha|_w^{N_w} = |N_K^E(\alpha)|_v.$$

We conclude this section with some remarks on valuations which show how the ramification index and residue class degree remain the same when we pass to the completions. This will be especially useful in our applications to discrete valuations, taking into account Proposition 2.4.

Suppose that we start with a field K with a non-trivial valuation v, corresponding valuation ring $\mathfrak{o}$, and maximal ideal $\mathfrak{m}$. Let us denote for a moment by $\hat{K}$ the completion of K, and let $\hat{\mathfrak{o}}$ resp. $\hat{\mathfrak{m}}$ be the closure of $\mathfrak{o}$ resp. $\mathfrak{m}$ in $\hat{K}$. Using the fact that if $|x| \neq |y|$ then $|x + y| = \max(|x|, |y|)$ we see that $\hat{\mathfrak{o}}$ is a valuation ring of $\hat{K}$, that $\hat{\mathfrak{m}}$ is its maximal ideal, and that

$$\hat{\mathfrak{o}} \cap K = \mathfrak{o}, \qquad \hat{\mathfrak{m}} \cap K = \mathfrak{m}.$$

Thus we have a canonical isomorphism

$$\mathfrak{o}/\mathfrak{m} \to \hat{\mathfrak{o}}/\hat{\mathfrak{m}}.$$

If E is a finite extension of K, and $\mathfrak{o}_E$, $\mathfrak{m}_E$ a valuation ring and its maximal ideal lying above $\mathfrak{o}$, $\mathfrak{m}$ and if we denote by a $\hat{\ }$ the completion, then we have a canonical commutative diagram:

$$
\begin{array}{ccc}
\mathfrak{o}_E/\mathfrak{m}_E & \xrightarrow{\approx} & \hat{\mathfrak{o}}_E/\hat{\mathfrak{m}}_E \\
\uparrow & & \uparrow \\
\mathfrak{o}/\mathfrak{m} & \xrightarrow{\approx} & \hat{\mathfrak{o}}/\hat{\mathfrak{m}}
\end{array}
$$

the vertical arrows being inclusions. Thus the residue class field extension can be studied either over the given field K or its completion.

Similarly for the ramification index. If $x \in \hat{K}$ and $y \in K$ are such that $|y - x| < |x|$, then $|y - x + x| = |y|$ and hence the inclusion map

$$v(K^*) \to v(\hat{K}^*)$$

is an isomorphism (the value groups are the same). If now E is a finite extension of K, then we have the commutative diagram

$$
\begin{array}{ccc}
w(E^*) & \xrightarrow{\approx} & w(\hat{E}^*) \\
\uparrow & & \uparrow \\
v(K^*) & \xrightarrow{\approx} & v(\hat{K}^*)
\end{array}
$$

for any valuation w extending v to E, the vertical arrows being inclusions. Thus the ramification index can be studied locally over the completion.

Proposition 4.6. *Let v be a discrete, well behaved valuation on K, and let E be a finite extension of K. For each extension w of v to E, let*

$$e_w = \big(w(E^*) : v(K^*)\big),$$
$$d_w = [\mathfrak{o}_w/\mathfrak{m}_w : \mathfrak{o}_v/\mathfrak{m}_v]$$

be the ramification index and residue class degree respectively. Then

$$\sum_{w|v} e_w d_w = [E : K].$$

If E is Galois over K, then all e_w are equal to the same number e, all d_w are equal to the same number d, and so

$$[E : K] = edr,$$

where r is the number of extensions of v to E.

Proof. Our first assertion comes from our assumption, and Proposition 2.4. If E is Galois over K, we know that given two extensions w_1, w_2 of v there exists an automorphism σ of E over K such that $w_1 = w_2\sigma$. This symmetry implies our second assertion.

Historical Note

The theory of absolute values is perfectly classical. We have, of course, not been exhaustive, giving only what we need most in the sequel. The reader wanting a more complete treatment may consult Bourbaki's *Commutative Algebra*, or Artin's *Algebraic Numbers and Algebraic Functions*. Actually, our treatment is much influenced by Artin's. Furthermore, although many special cases of the approximation theorem were known before (Chinese remainder theorems), it was first stated and proved in full generality for absolute values by Artin–Whaples [A–W].

CHAPTER 2

Proper Sets of Absolute Values.
Divisors and Units

In this chapter, we give the most important examples of the type of field which concerns us. They are essentially fields of finite type over the prime field or over some constant field, the latter giving rise to relative theories.

On these fields, when we take suitable models of them, we get families of absolute values. These absolute values are either discrete or archimedean. A proper set of absolute values will be defined so as to axiomatize what is common to all of them, and to give us what we shall need later for certain applications, for instance, our study of height functions, or divisors and divisor classes.

It is particularly important when our set of absolute values satisfies a product formula, the classical case being that of number fields. I limit myself mostly to qualitative results. Quantitative results for number fields can be found, for instance, in my *Algebraic Number Theory*. Here I repeat the very brief proofs that certain groups of units and ideal classes are finitely generated. Even in fields of finite type, the proofs are essentially elementary, except for one reference to the Mordell–Weil theorem of Chapter 6, which is used axiomatically in §7.

§1. Proper Sets of Absolute Values

Let K be a field. An absolute value v on K is said to be **proper** if it is non-trivial, well behaved, and if, K having characteristic 0, its restriction to $\mathbf{Q}$ is either trivial, the ordinary absolute value, or a p-adic absolute value v_p.

A set M_K of absolute values on K is said to be **proper** if every absolute value in it is proper, if any two distinct absolute values are independent, and if, given $x \in K$, $x \neq 0$, there exists only a finite number of $v \in M_K$ such that $|x|_v \neq 1$. In particular, if M_K is proper, there can be only a finite number of archimedean absolute values in M_K.

If E is an algebraic extension of K, we shall denote by M_E the set of absolute values on E extending some absolute value in M_K. If E is finite over K, then M_E is proper if M_K is proper. (Its absolute values are well behaved by Chapter 1, §4, and the other conditions are trivially verified.)

Let M_K be a proper set of absolute values on K. For each $v \in M_K$, let λ_v

be a real number > 0. We shall say that M_K satisfies the **product formula with multiplicities** λ_v if for each $x \in K$, $x \neq 0$, we have

$$\prod_{v \in M_K} |x|_v^{\lambda_v} = 1.$$

By assumption, there is only a finite number of terms in this product which are not equal to 1, so the product makes sense. We shall say that M_K satisfies the **product formula** if all $\lambda_v = 1$. When we deal with a fixed set of multiplicities λ_v, then we write for convenience

$$\|x\|_v = |x|_v^{\lambda_v}$$

so that our product formula reads

$$\prod_{v \in M_K} \|x\|_v = 1.$$

Suppose now that we have a field $\mathbf{F}$ with a proper set $M_\mathbf{F}$ of absolute values satisfying the product formula with multiplicities 1. Let K be a finite extension of $\mathbf{F}$, and let M_K be the set of absolute values on K which extend the absolute values of $M_\mathbf{F}$. Then M_K is also a proper set of absolute values on K. If $v_0 \in M_\mathbf{F}$ and $v \in M_K$, with $v|v_0$, set $N_v = [K_v : \mathbf{F}_{v_0}]$. Then for any element $\alpha \in K$, $\alpha \neq 0$, we get by Theorem 4.5 of Chapter 1:

$$\begin{aligned}
1 &= \prod_{v_0 \in M_\mathbf{F}} |N_\mathbf{F}^K(\alpha)|_{v_0} \\
&= \prod_{v_0 \in M_\mathbf{F}} \prod_{v|v_0} |\alpha|_v^{N_v} \\
&= \prod_{v \in M_K} |\alpha|_v^{N_v}.
\end{aligned}$$

This shows that M_K satisfies the product formula with multiplicities N_v.

It was Artin–Whaples [A–W] who first showed how the product formula can be taken as the basic axiom for algebraic number theory, if one makes one further assumption of finiteness. For us here, it is irrelevant to know that if one of the absolute values in M_K is archimedean or discrete and its residue class field is finite, or finite over a constant field, then this characterizes K as being either a number field or a function field of one variable over the constant field. In fact, there are product formulas which occur in higher dimensions; we shall give examples in §3.

§2. Number Fields

The classical example is that of the rational numbers $\mathbf{Q}$. For each prime number p we have the absolute value v_p described in Chapter 1, §1. The ordinary archimedean absolute value will be said to be **at infinity**.

If l is a prime number, then

$$|l|_p = \begin{cases} 1 & \text{if } p \text{ is a prime number} \neq l \\ 1/l & \text{if } p \text{ is a prime number} = l, \end{cases}$$

$$|l|_\infty = l.$$

Thus the set $M_{\mathbf{Q}}$ of all absolute values v_p and v_∞ satisfies the product formula, the above argument showing this for prime numbers of $\mathbf{Z}$, the general case following by multiplicativity. It is also clear that our set is proper (using Proposition 4.3 of Chapter 1).

If K is a number field, then the set M_K of absolute values extending those of $M_{\mathbf{Q}}$ will be called the **canonical set**.

If $\mathfrak{o} = \mathfrak{o}_K$ is the ring of integers of K (i.e. the integral closure of $\mathbf{Z}$ in K) then the valuation rings of K are in bijective correspondence with the prime ideals $\mathfrak{p}$ of $\mathfrak{o}$. We shall always exclude the trivial one. Each valuation ring is of type $\mathfrak{o}_{\mathfrak{p}}$, the local ring of $\mathfrak{o}$ at $\mathfrak{p}$. The set of archimedean absolute values in M_K is usually denoted by M_K^∞ or S_K^∞. It is called the set of **absolute values at infinity.**

Let K be a number field, and v one of the absolute values extending the ordinary absolute value on $\mathbf{Q}$. Then K_v is either the field of real numbers or the field of complex numbers. We then say that v is **real** or **complex**, accordingly. The multiplicity

$$N_v = [K_v : \mathbf{Q}_v]$$

is then 1 or 2 according as v is real or complex, and thus for $x \in K$, we have

$$\|x\|_v = |x|_v^{N_v}$$

is either $|x|$ or $|x|^2$.

Let us now consider a prime ideal $\mathfrak{p}$ in the ring of integers of K. Then there is a unique prime number p lying in $\mathfrak{p}$ and we denote by $v_{\mathfrak{p}}$ or $|\ \ |_{\mathfrak{p}}$ the absolute value belonging to $\mathfrak{p}$ extending the p-adic absolute value on $\mathbf{Q}$. Its multiplicity is

$$N_{\mathfrak{p}} = [K_{\mathfrak{p}} : \mathbf{Q}_p] = e_{\mathfrak{p}} d_{\mathfrak{p}},$$

where $e_{\mathfrak{p}}$ and $d_{\mathfrak{p}}$ are the ramification index and residue class degree, respectively. Letting $\mathfrak{o}_{\mathfrak{p}}$ and $\mathfrak{m}_{\mathfrak{p}}$ be the valuation ring and maximal ideal corresponding to $\mathfrak{p}$, we see that the number of elements in $\mathfrak{o}_{\mathfrak{p}}/\mathfrak{m}_{\mathfrak{p}}$, which we denote by $\mathbf{N}\mathfrak{p}$, is

$$\mathbf{N}\mathfrak{p} = p^{d_{\mathfrak{p}}},$$

since $\mathfrak{o}_\mathfrak{p}/\mathfrak{m}_\mathfrak{p}$ is of degree $d_\mathfrak{p}$ over $\mathbf{Z}/p\mathbf{Z}$. Furthermore, let π be a prime element at $\mathfrak{p}$. Then by definition,

$$|\pi^{e_\mathfrak{p}}|_\mathfrak{p} = |p|_\mathfrak{p} = 1/p.$$

Hence

$$|\pi|_\mathfrak{p}^{e_\mathfrak{p} d_\mathfrak{p}} = (1/\mathbf{N}\mathfrak{p}).$$

From this we get:

Proposition 2.1. *Let K be a number field, $\mathfrak{p}$ a prime ideal, $\mathbf{N}\mathfrak{p}$ the order of $\mathfrak{o}_\mathfrak{p}/\mathfrak{m}_\mathfrak{p}$, and $|\ \ |_\mathfrak{p}$ the absolute value extending the p-adic absolute value on $\mathbf{Q}$. Let $\|\ \|_\mathfrak{p} = |\ \ |_\mathfrak{p}^{N\mathfrak{p}}$. Then for $x \in K$, $x \neq 0$,*

$$\|x\|_\mathfrak{p} = (1/\mathbf{N}\mathfrak{p})^{\mathrm{ord}_\mathfrak{p} x}.$$

§3. Divisors on Varieties

We assume that the reader is acquainted with the notion of **variety** (especially affine and projective) over a field k, and with the Zariski topology where the closed sets are finite unions of subvarieties. We adopt the convention that if V is a **variety**, then by definition, its field of rational functions $k(V)$ is a regular extension of k, that is separably generated and such that k is algebraically closed in $k(V)$. This condition insures that nothing bad happens under extension of the ground field k, and in particular that V is **absolutely irreducible**.

Let V be a variety, defined over a field k. Every point lies in some affine open subset. Suppose V is affine, and has an affine coordinate ring

$$R = k[f_1, \ldots, f_n],$$

where $f_1, \ldots, f_n$ are rational functions. If $\mathfrak{p}$ is a prime ideal in R then we have the local ring $R_\mathfrak{p}$, contained in the field of rational functions

$$k(f_1, \ldots, f_n) = k(V).$$

Let W be a subvariety, defined over a finite extension k'. Associated with W we have its coordinate ring $k'[W] = k'[f'_1, \ldots, f'_n]$, where to each $f \in k[V]$ we associate its induced function $f' = f_W$ on W. Let $\mathfrak{p}$ be the kernel of the homomorphism $f \mapsto f_W$. Then $R_\mathfrak{p}$ is called the **local ring of** W in $k(V)$.

If $\sigma: k' \to \sigma k'$ is an isomorphism over k, and W is the set of zeros of polynomial equations with coefficients in k', then we may apply σ to these coefficients to get polynomial equations said to be conjugate by σ, and defining a **conjugate subvariety** denoted by W^σ. The local ring of W^σ in $k(V)$ is the same as the local ring of W.

An arbitrary variety is pieced together from affine open pieces, and thus the notion of local ring and conjugate variety extend to an arbitrary variety from the same notions for affine varieties. The group of r-**chains** on a variety is the free abelian group generated by the subvarieties of dimension r. The subgroup generated by the subvarieties which are non-singular on V is called the subgroup of r-**cycles**. By **non-singular on** V, one means that the subvariety contains at least one point which is simple on V.

Let $[k':k]_i$ be the degree of inseparability of k' over k. If k_1 is the maximal separable subextension, then $[k':k]_i$ is defined by

$$[k':k] = [k':k]_i[k_1:k].$$

Let us suppose that k' is the smallest field of definition for W. By a **prime rational chain** over k determined by W we mean the chain

$$[k':k]_i \sum_\sigma W^\sigma,$$

where the sum is taken over the distinct conjugates of W. If V is affine such chains are in bijection with the prime ideals of the coordinate ring $k[V]$. We use the letter $\mathfrak{p}$ to denote the above sum in general.

We shall now assume that V is non-singular in codimension 1, and is projective. A **prime rational divisor** $\mathfrak{p}$ of V over k is a cycle of codimension 1,

$$\mathfrak{p} = \mu \sum_\sigma W^\sigma$$

where W is subvariety of V, of codimension 1, defined over an algebraic extension of k, W^σ ranges over the distinct conjugates of W over k, and μ is its order of inseparability over k. Such a prime rational divisor (also called a **prime divisor**) or **prime** determines its local ring in the function field $k(V)$ of V over k.

Proposition 3.1. *If W is non-singular on V, and of codimension 1, then its local ring $\mathfrak{o}$ in $k(V)$ is a discrete valuation ring.*

We assume this basic fact from elementary algebraic geometry. In light of this fact, we have the notion of order at $\mathfrak{p}$ of a function $x \in k(V)$, $x \neq 0$.

On the other hand, from a projective embedding of V, we also have the **degree** of $\mathfrak{p}$, denoted $\deg \mathfrak{p}$, by which we mean the **projective degree**, i.e. the number of points of intersection with a generic linear variety of complementary dimension in the given projective embedding. Similarly, define

the group of **divisors** to be the free abelian group generated by the subvarieties of codimension 1, and the subgroup of **divisors rational over** k
to be generated by the prime rational divisors over k. If

$$D = \sum n(\mathfrak{p})\mathfrak{p}$$

is a divisor rational over k, then its **degree** is defined by

$$\deg D = \sum n(\mathfrak{p}) \deg \mathfrak{p}.$$

Again we may say that it is the number of points of intersection of D with a
generic linear variety of complementary dimension.

Let $x \in k(V)$, $x \neq 0$. We can associate with x its divisor

$$(x) = \sum_{\mathfrak{p}} \mathrm{ord}_{\mathfrak{p}}(x)\mathfrak{p},$$

where $\mathrm{ord}_{\mathfrak{p}}(x)$ is the order of x in the discrete valuation ring of $\mathfrak{p}$. We let
$(x)_0$ be the sum taken only over those $\mathfrak{p}$ such that $\mathrm{ord}_{\mathfrak{p}}(x) > 0$, and $(x)_\infty$ be
such that $(x) = (x)_0 - (x)_\infty$. We call $(x)_0$ the divisor of **zeros**, and $(x)_\infty$
the divisor of **poles**. We have

$$\deg(x)_0 = \deg(x)_\infty$$

and hence $\deg(x) = 0$.

Now let c be a number, $0 < c < 1$, and for each prime divisor $\mathfrak{p}$ of V
over k, let, for $x \neq 0$,

$$|x|_{\mathfrak{p}} = c^{(\mathrm{ord}_{\mathfrak{p}} x)\,\deg(\mathfrak{p})}.$$

To each prime divisor we have therefore associated an absolute value
$|\ |_{\mathfrak{p}}$. We let $K = k(V)$ and let $M_K = M_{k(V)}$ be the set of such absolute
values. It is an elementary fact that this set is a proper set of absolute values,
and it satisfies the product formula because of the relation

$$\deg(x)_0 = \deg(x)_\infty,$$

in other words, because a function has as many zeros as it has poles, roughly
speaking. Once our number c is chosen, we have a bijective correspondence
between prime divisors and absolute values in $M_{k(V)}$, and thus we shall also
write our product formula

$$\prod_{\mathfrak{p} \in M_{k(V)}} |x|_{\mathfrak{p}} = 1, \quad \text{or} \quad \sum_{\mathfrak{p}} v_{\mathfrak{p}}(x) = 0.$$

As we have multiplicities $= 1$, we have $\|x\|_{\mathfrak{p}} = |x|_{\mathfrak{p}}$.

By a **function field** K over the constant field k we shall always mean a finitely generated, regular extension of k. Each model V of K over k which is projective and non-singular in codimension 1 gives rise to absolute values as above. This slightly restricts the notion of function fields as used in the classical theory of function fields in one variable, i.e. finitely generated extensions K of a field k which are of transcendence degree 1 and such that k is algebraically closed in K. For such fields, one also has a product formula for the set of discrete valuations, trivial on k, such valuations being well behaved and hence proper.

Finally we observe that there are product formulas arising from more general situations. Indeed, let us take a complete, non-singular surface V and a curve C on V which has the following property: for every positive 1-cycle X on V, the intersection number $I(X.C)$ of the numerical equivalence classes of X and C is $\geqq 0$. For every prime rational cycle $\mathfrak{p}$ on V we define the multiplicity $d(\mathfrak{p}) = I(\mathfrak{p}.C)$. If φ is a function on V, then we define

$$v_{\mathfrak{p}}(\varphi) = d(\mathfrak{p})\,\mathrm{ord}_{\mathfrak{p}}(\varphi).$$

The set of such absolute values does satisfy the product formula. If the curve is a generic hyperplane section, then we have our standard example. However, as Mumford pointed out to me, there exist examples of such curves which do not lie in a pencil, and thus give rise to other types of product formulas, to which Roth's theorem would also apply. As far as I know, this line of thought has not been pursued by algebraic geometers.

In the next section, we give a generalization of the notion of divisor when we do not work over a constant field. We shall also recall some useful lemmas which apply to both situations, and are useful in dealing with divisor classes, as explained below.

§4. Divisors on Schemes

To a large extent, the constant field in the last section was irrelevant. We used it mostly in connection with the projective embedding, and to get the product formula. For the general theory of divisors, it played no role, and so we go through once again the general discussion of divisors in the context of schemes.

We first start in the analogue of the affine case. Let R be a Noetherian ring which we assume **normal**, meaning that it has no divisors of zero and is integrally closed in its quotient field, which we denote by K. If $\mathfrak{p}$ is a minimal prime ideal of R not equal to 0, then the local ring $R_{\mathfrak{p}}$ is a discrete valuation ring in K. Given $x \in R$, $x \neq 0$, there is only a finite number of such primes $\mathfrak{p}$ where x has a zero at $\mathfrak{p}$, that is x lies in the maximal ideal $\mathfrak{m}_{\mathfrak{p}}$ of

$R_{\mathfrak{p}}$. Each such prime $\mathfrak{p}$ therefore gives rise to a discrete valuation on K. Suitably normalizing the absolute values, from such a $\mathfrak{p}$ in case they induce a non-trivial absolute value on the prime ring, we get a proper set of absolute values on K, which will be denoted by M_R or $M_{R.K}$.

Now let V be a scheme (which, throughout this book, is assumed separated). For the rest of this section, we assume that V is covered by a finite number of open affine subschemes, each of which is isomorphic to $\mathrm{spec}(R)$ for a Noetherian normal ring R as above. We also assume that V is irreducible.

Then the quotient field K is "the same" for all such rings R. We call K the **function field** of V. In this case, a **prime divisor** or **prime** for short, is an irreducible subscheme of codimension 1, corresponding to a point on V having a representative $\mathfrak{p}$ on some affine open subset $\mathrm{spec}(R)$ as above. Such a prime divisor W is then covered by affine open subsets $\mathrm{spec}(R/\mathfrak{p})$. We let M_V be the union of $M_{R.V}$ for all affine open subschemes $\mathrm{spec}(R)$ as above. Then M_V is a set of discrete absolute values on K. In general, M_V does not satisfy the product formula.

The following property is often useful when dealing with finitely generated rings over $\mathbf{Z}$ (which is one of the most important cases of interest to us).

Proposition 4.1. *Let R be a ring finitely generated over the prime ring. Then there exists an element $a \neq 0$ in the prime ring such that the integral closure of $R[1/a]$ (in its quotient field) is a finite module over $R[1/a]$.*

Proof. The case of characteristic > 0 is standard, since the prime ring is a field (Noether's Normalization Theorem). The case of interest to us here is that of characteristic 0. Denote by $R_\mathbf{Q}$ the algebra generated over $\mathbf{Q}$ by R. By the normalization theorem, there exist elements $t_1, \ldots, t_r \in R$ algebraically independent over $\mathbf{Q}$ such that $R_\mathbf{Q}$ is integral over $\mathbf{Q}[t_1, \ldots, t_r]$. Since R is finitely generated over $\mathbf{Z}$, there exists an element $a \in \mathbf{Z}$, $a \neq 0$ such that R is integral over $\mathbf{Z}[1/a, t_1, \ldots, t_r]$. Since this latter ring is integrally closed, its integral closure in the quotient field of R is a finite module over it, say by IAG, Theorem 2 of Chapter V, §1.

Remark. It is in fact true that the integral closure of R itself is a finite module over R, but this theorem, due to Nagata, is very much more difficult to prove.

Suppose R is finitely generated over $\mathbf{Z}$ or over the prime field $\mathbf{F}_p$. In the second case, the situation is essentially that of the last section, that is there is a constant field. Over $\mathbf{Z}$, however, there is no constant field until we tensor with $\mathbf{Q}$ or some finite extension of $\mathbf{Q}$. Then a prime divisor may be of two types. First, suppose that $\mathfrak{p}$ does not contain any prime number. Then $\mathfrak{p} \cap \mathbf{Z} = \{0\}$, and in that case, $\mathfrak{p}$ extends uniquely to a prime ideal in

$\mathbf{Q} \otimes R$, giving rise essentially to a prime divisor as in the preceding section. On the other hand, if $\mathfrak{p}$ contains the prime number p, then $\mathrm{Spec}(R/\mathfrak{p})$ is a scheme over the prime field $\mathbf{F}_p$ in characteristic p, and is one of the components of the reduction of V modulo $\mathfrak{p}$.

More generally, let $\mathfrak{o}$ be a Noetherian normal ring, and suppose we have a morphism $V \to \mathrm{spec}(\mathfrak{o})$. Locally in the Zariski topology, this means that V is covered by affine open subsets $\mathrm{spec}(R)$ where R is finitely generated as an algebra over $\mathfrak{o}$. Then again prime divisors on V may be of two types. The first type corresponds to a minimal prime ideal $\mathfrak{p}$ of R whose intersection with $\mathfrak{o}$ is 0. We shall call such a prime divisor **geometric over** $\mathfrak{o}$. Let k be the fraction field of $\mathfrak{o}$. Then $\mathfrak{p}$ extends to a unique prime ideal in $k \otimes R$, defining a prime rational divisor over k essentially in the sense of the preceding section. The second type corresponds to a minimal prime ideal $\mathfrak{p}$ whose intersection with $\mathfrak{o}$ is a non-zero prime ideal $\mathfrak{p}_0$. Such prime divisors may be called **arithmetic over** $\mathfrak{o}$.

We qualified our use of the term prime rational divisor over k by the word "essentially" because it may be that the function field K of V is not a regular extension of k. If K is a regular extension of k, we shall say that V is a **variety scheme** over $\mathfrak{o}$. In other words, the scheme $k \otimes V$ obtained by extension of scalars from $\mathfrak{o}$ to k is a variety in the precise sense of §3.

We let $M_{k,V}$ be the set of discrete valuations in M_V which are trivial on k. We let $M_{\mathfrak{o},V}$ be the set of discrete valuations in M_V which are non-trivial on $\mathfrak{o}$ (and hence on k). Thus we have the disjoint union

$$M_V = M_{k,V} \cup M_{\mathfrak{o},V}.$$

Of course, $M_{\mathfrak{o}}$ is the set of prime divisors of $\mathfrak{o}$ itself. It is an elementary fact that all but a finite number of primes of $M_{\mathfrak{o}}$ actually extend to primes of $M_{\mathfrak{o},V}$.

Now suppose that V is a variety scheme over $\mathfrak{o}$ as defined above, so $k(V)$ is a regular extension of k. Then it is also an elementary fact that for all but a finite number of primes of $\mathfrak{o}$, a prime $\mathfrak{p}_0$ in $M_{\mathfrak{o}}$ has a *unique extension* to V, and corresponds to a "non-degenerate" reduction of V mod $\mathfrak{p}_0$. Locally, this means that if $\mathrm{spec}(R)$ is an affine open subset of V, then for all but a finite number of primes $\mathfrak{p}_0$ in $M_{\mathfrak{o}}$ the ideal $\mathfrak{p}_0 R$ is prime in R, and $R/\mathfrak{p}_0 R$ is the affine coordinate ring of a variety scheme over $\mathfrak{o}/\mathfrak{p}_0$. We shall denote by $M'_{\mathfrak{o},V}$ the subset of $M_{\mathfrak{o},V}$ lying above such $\mathfrak{p}_0$ in $M_{\mathfrak{o}}$. Then $M'_{\mathfrak{o},V}$ is in natural bijection with a subset $M'_{\mathfrak{o}}$ differing from $M_{\mathfrak{o}}$ by only a finite set of elements. The bijection is given by

$$\mathfrak{p}_0 \mapsto \mathfrak{p}_0 R.$$

In the exceptional cases, when $\mathfrak{p}_0 R$ is not prime, there is only a finite number of minimal primes $\mathfrak{p}$ in R containing $\mathfrak{p}_0 R$. In geometric language, there is only a finite number of prime divisors on V inducing the same non-trivial prime divisor on $\mathrm{spec}(\mathfrak{o})$.

The above decomposition of M_V into geometric and arithmetic parts will be used in §7. For the rest of this section, we leave aside these relative considerations (relative to some base ring o), and return to an arbitrary scheme, irreducible, Noetherian and normal.

The scheme V being as above, one can define the group of **divisors** on V to be the free abelian group generated by the irreducible subschemes of codimension 1. We call elements $x \in K$ **rational functions** on V. To each rational function we can associate its divisor.

$$(x) = \sum \operatorname{ord}_W(x) . W.$$

If a prime divisor W is represented on an affine open subset $\operatorname{spec}(R)$ for R as above, then by definition $\operatorname{ord}_W(x)$ is the order of x at the discrete valuation whose valuation ring is $R_\mathfrak{p}$. We let:

Div(V) = group of divisors on V;

Div$_l(V)$ = subgroup of divisors of functions.

We say that two divisors D, D' are **linearly equivalent** if $D - D'$ is the divisor of a function. The factor group is called the **Picard group**, or the group of **divisor classes**

$$\operatorname{Pic}(V) = \operatorname{Div}(V)/\operatorname{Div}_l(V).$$

In the next section, we shall give a definition of divisors which also takes into account archimedean absolute values.

In the affine case, $V = \operatorname{spec}(R)$ where R is Noetherian and normal, we also write

$$\operatorname{Div}(V) = \operatorname{Div}(R) \qquad \text{and} \qquad \operatorname{Pic}(V) = \operatorname{Pic}(R).$$

If R is a Dedekind ring, for instance the ring of algebraic integers in a number field, then there is a natural isomorphism between the group of divisors Div(V) (or Div(R)) and the group of fractional ideals of R. The group Pic(V) (or Pic(R)) is precisely the group of ideal classes in the classical sense.

We now recall some properties relating to divisors passing through points.

Let P be a point of V. If P lies in $\operatorname{spec}(R)$, then P corresponds to a prime ideal in R, and thus to the maximal ideal in the local ring R_P. The point P is **closed** (in the Zariski topology) if and only if its corresponding ideal in R is maximal. If D is a divisor, we can represent D as a sum

$$D = \sum n(W)W,$$

with coefficients $n(W)$ which are 0 for all but a finite number of prime divisors W. If $n(W) \neq 0$ we say that W is a **component** of D. The **support** of D, denoted $|D|$, or supp(D), is the set of points of V lying in some component of D. We say that a point P is **regular** if the local ring R_P is regular. We say that V is **regular** if every point is regular. It is a fact from basic commutative algebra (due to Zariski [Za]) that a regular local ring has unique factorization. If t is a prime element in R_P, then t generates a principal ideal $\mathfrak{p}$ in R_P which is a minimal prime ideal. Then $\mathfrak{p}$ defines a prime divisor, which is said to **pass through** P. We also say that t **represents** the prime divisor locally at P. A rational function f on V is said to **represent** D at P if P does not lie in the support of $D - (f)$.

A divisor $D = \sum n(W)W$ is said to be **positive** if $n(W) \geqq 0$ for all prime divisors W.

Proposition 4.2. *Let S be a finite set of regular points on V, all contained in the same affine open subset* spec(R), *and let D be a divisor on V. Then there exists a rational function $\varphi \in K$ such that no point of S lies in the support of $D + (\varphi)$.*

Proof. Without loss of generality, we may assume that D is positive. We may also assume that the points in S are closed. Let I be the ideal in R generated by all the functions f such that $(f) \geqq D$. Since R is Noetherian, I has a finite number of generators $f_1, \ldots, f_n$. Let P be a closed regular point of Spec(R) and R_P its local ring, with maximal ideal M. Then there is a function t in R_P which represents D locally at P. We can write

$$t = \sum_{i=1}^{n} z_i f_i,$$

with $z_i \in R_P$. If $w_i \in R$ is such that $w_i \equiv z_i \bmod M$, then

$$\sum_{i=1}^{n} w_i f_i = tu$$

for some unit u in R_P. Indeed,

$$\sum w_i f_i = \sum (w_i - z_i)f_i + t.$$

Since $(f_i) \geqq D$, we can write $f_i = tg_i$ with $g_i \in R$. Then

$$\sum w_i f_i \in t(1 + M),$$

thus showing that $\sum w_i f_i = tu$ with some unit u in R_P.

Finally, let $P_1, \ldots, P_r$ be a finite number of distinct regular closed points. Let $R_j = R_{P_j}$ be their local rings with maximal ideals M_j. Given elements

$x_j \in R_j$ by the Chinese remainder theorem we can find $x \in R$ such that $x \equiv x_j \bmod M_j \cap R$ because for $j \neq j'$ the maximal ideals $M_j \cap R$ and $M_{j'} \cap R$ in R are distinct. We now apply the approximation in the first part of the proof to each point P_j to find the desired element in R which represents D at each point $P_1, \ldots, P_r$.

Remark. The proposition applies to a finite number of simple points on a projective normal variety, since any finite set of such points lies on some affine open subset.

As usual, if $f \in K$ and $\mathrm{ord}_W(f) < 0$ for some prime divisor W, then we say that f has a **pole** at W. We say that f is **defined at** P if $f \in R_P$.

Proposition 4.3. *If f is not defined at some point P of V, then f has a pole passing through P, that is there exists a prime divisor W passing through P such that f has a pole at W.*

Proof. By our convention on V, P is a point in $\mathrm{Spec}(R)$ for some Noetherian normal ring R. The statement of the proposition is then equivalent with the basic statement of algebra that R is equal to the intersection of all the local rings $R_\mathfrak{p}$ taken over all the minimal prime ideals $\mathfrak{p}$ in R. Such $\mathfrak{p}$ correspond precisely to the prime divisors of $\mathrm{Spec}(R)$.

§5. M_K-divisors and Divisor Classes

Let K be a field with a proper set of absolute values M_K. By a **multiplicative M_K-divisor** one means a real valued function

$$v \mapsto \mathfrak{d}(v)$$

of the absolute values in M_K with the following properties:

(i) $\mathfrak{d}(v) > 0$ for all $v \in M_K$, and $\mathfrak{d}(v) = 1$ for all but a finite number of the v's.

(ii) For each non-archimedean $v \in M_K$, there exists $\alpha \in K^*$ such that $\mathfrak{d}(v) = |\alpha|_v$.

By an **additive M_K-divisor** one means a real valued function

$$v \mapsto \gamma(v)$$

of the absolute values in M_K with the following properties:

(i) $\gamma(v) = 0$ for all but a finite number of the v's.

(ii) For each non-archimedean $v \in M_K$, there exists $\alpha \in K^*$ such that $\gamma(v) = v(\alpha) = -\log|\alpha|_v$.

Thus an additive M_K-divisor is just the log of a multiplicative M_K-divisor. When dealing with number fields, it is often more convenient to deal with multiplicative M_K-divisors because one counts the number of elements, say as in Theorems 6.6 or 6.7. When there are no archimedean absolute values, and one deals in an algebraic geometric context, it is more convenient to deal with the additive M_K-divisors. *For the rest of this section, we keep the multiplicative notation.*

Remark. The second condition in each case is made to allow for accurate counting of elements of K in parallelotopes, as in Theorem 6.7, or as in the Riemann–Roch theorem. In Chapter 10 we consider real valued functions $v \mapsto \gamma(v)$ satisfying only the first condition, and use such functions merely as bounds. We then think of the values $\gamma(v)$ as constants, parametrized by the elements of M_K.

We note that M_K-divisors, including possibly archimedean absolute values, were defined by Weil [We 1], with the idea of making his decomposition theorem valid for them also. We shall reproduce Weil's result in the form given to it by Néron in Chapter 10.

If $\alpha \in K^*$ then the function $|\alpha|_v$ on M_K is an M_K-divisor, denoted by $\mathfrak{d}_\alpha$, and is called **principal**.

We observe that the M_K-divisors are closed under the operations of sup and inf, and that they form a group under multiplication (componentwise).

If E is an algebraic extension of K, then for each $w \in M_E$ we extend an M_K-divisor $\mathfrak{d}$ to M_E by the formula

$$\mathfrak{d}(w) = \mathfrak{d}(v)$$

if w extends $v \in M_K$. If E is finite over K, then this is an M_E-divisor, and our extension maps the M_K-divisors injectively into the M_E-divisors. By abuse of language, even if E is infinite, we sometimes say that our extension is an M_E-divisor (although condition (i) might be violated).

Suppose that v is discrete. If π is a prime element at v, then

$$\mathfrak{d}(v) = |\pi|_v^n$$

for some integer n, which is called the **order** of $\mathfrak{d}$ at v, and abbreviated $\text{ord}_v \mathfrak{d}$, or $\text{ord}_\mathfrak{p} \mathfrak{d}$ if $\mathfrak{p}$ is the prime associated with v.

Let us denote by M_K^0 the subset of M_K consisting of the nonarchimedean absolute values.

Suppose that all the absolute values M_K^0 are discrete, and that for each M_K-divisor we take its restriction to these non-archimedean absolute values. These restrictions will be called **finite divisors**, and they form a group which is clearly isomorphic in the natural manner with the free abelian group generated by the primes corresponding to the absolute values in M_K^0. This latter group will be called the group of M_K-**ideals**. Each element $x \in K, x \neq 0$ determines an M_K-ideal denoted by (x), and called a **principal** M_K-**ideal**. The principal M_K-ideals form a group, and the factor group is called the group of M_K-**ideal classes**

If $\mathfrak{a}$ is an M_K-ideal, then we can write

$$\mathfrak{a} = \sum_{\mathfrak{p}} n_{\mathfrak{p}} \mathfrak{p}$$

with $n_{\mathfrak{p}} \in \mathbf{Z}$, and $\mathfrak{p}$ ranging over the primes of M_K^0, all but a finite number of $n_{\mathfrak{p}}$ being 0. We write

$$n_{\mathfrak{p}} = \text{ord}_{\mathfrak{p}}\, \mathfrak{a}.$$

If $\alpha \in K^*$, then $\text{ord}_{\mathfrak{p}}\, \alpha = \text{ord}_{\mathfrak{p}}(\alpha)$, i.e. it is the order of its divisor at $\mathfrak{p}$.

If two M_K-ideals $\mathfrak{a}$ and $\mathfrak{b}$ lie in the same ideal class, we write

$$\mathfrak{a} \sim \mathfrak{b}.$$

This is the notion of linear equivalence in §3. An M_K-ideal is said to be **integral** if $\text{ord}_{\mathfrak{p}}\, \mathfrak{a} \geq 0$ for all $\mathfrak{p}$ in M_K^0.

By an M_K-**unit** we shall mean an element $\alpha \in K^*$ such that

$$|\alpha|_v = 1$$

for all non-archimedean absolute values v of M_K. The M_K-units obviously form a multiplicative group, which we denote by U_{M_K}, U_K, or simply U.

In §2 our group of M_K-ideals is isomorphic in the obvious manner with the classical group of ideals of the ring of integers (including the fractional ideals, of course). The M_K-units are simply the units of the number field in the classical sense.

In §3, the group of M_K-ideals is simply the group of divisors on the variety, rational over the constant field k. The units are the non-zero constants (since we assume the variety complete and non-singular in codimension 1).

In §4, we have:

Proposition 5.1. *Let R be a Noetherian, integrally closed ring. Let M_R be its set of primes as in §4. Then R^* is the set of M_R-units of the field K of fractions of R.*

Proof. It is well known from commutative algebra that $R = \bigcap R_{\mathfrak{p}}$, the intersection being taken over all $\mathfrak{p} \in M_R$. If $u \in K$ is an M_R-unit, then u lies in each $R_{\mathfrak{p}}$ since, in particular, it has no pole among the $\mathfrak{p}$'s. Hence it is in R, and since it has no zero, it must be invertible (Krull's principal ideal theorem). The converse is clear.

Starting with a proper set of absolute values M_K on a field K, we shall frequently let S be a finite subset, containing the archimedean absolute values which may occur in M_K. It is customary to call S-**units** the $(M_K - S)$-units, i.e. those elements of K having absolute value 1 for all $v \notin S$. These form a multiplicative group K_S. We shall study in §7 the qualitative effect of removing such a finite set from consideration.

§6. Ideal Classes and Units in Number Fields

Let K be a number field. Let

$$\mathfrak{a} = \sum_\mathfrak{p} (\operatorname{ord}_\mathfrak{p} \mathfrak{a}) \mathfrak{p}$$

be an integral M_K-ideal. We define

$$\mathbf{N}\mathfrak{a} = \prod (\mathbf{N}\mathfrak{p})^{\operatorname{ord}_\mathfrak{p} \mathfrak{a}}.$$

It is clear that there is only a finite number of $\mathfrak{p}$'s such that $\mathbf{N}\mathfrak{p}$ lies below a given number, and hence that there is only a finite number of integral ideals $\mathfrak{a}$ such that $\mathbf{N}\mathfrak{a}$ lies below a given number.

We define two ideals $\mathfrak{a}$, $\mathfrak{b}$ to be **linearly equivalent**, and write

$$\mathfrak{a} \sim \mathfrak{b}$$

if there exists an element $\alpha \in K$ such that $\mathfrak{a} = \mathfrak{b} + (\alpha)$. Also as usual, we denote by (α) the principal M_K-ideal of α.

We shall now prove that the M_K-ideal classes form a finite group. We need a lemma.

Lemma 6.1. *Let $\mathfrak{o}$ be the ring of integers of the number field K. Let $\mathfrak{a}$ be an integral ideal, $\mathfrak{a} = \sum n_\mathfrak{p} \mathfrak{p}$. Then $\mathbf{N}\mathfrak{a}$ is the number of elements of the module*

$$\prod \mathfrak{o}_\mathfrak{p} / \mathfrak{m}_\mathfrak{p}^{n_\mathfrak{p}}$$

(the product being taken over the finite primes, and all but a finite number of terms being trivial).

Proof. The valuation ring $\mathfrak{o}_\mathfrak{p}$ is a discrete valuation ring, and its maximal ideal $\mathfrak{m}_\mathfrak{p}$, generated by π, is principal. As a module over $\mathfrak{o}_\mathfrak{p}/\mathfrak{m}_\mathfrak{p}$,

$$\pi^r \mathfrak{o}_\mathfrak{p} / \pi^{r+1} \mathfrak{o}_\mathfrak{p}$$

is isomorphic to $\mathfrak{o}_\mathfrak{p}/\mathfrak{m}_\mathfrak{p}$ itself, under multiplication by π^r, r being an integer $\geqq 0$. We have the chain of ideals

$$\mathfrak{o}_\mathfrak{p} \supset \pi\mathfrak{o}_\mathfrak{p} \supset \pi^2\mathfrak{o}_\mathfrak{p} \supset \cdots \supset \pi^r\mathfrak{o}_\mathfrak{p}$$

and from this we see that the number of elements in the module $\mathfrak{o}_\mathfrak{p}/\pi^r\mathfrak{o}_\mathfrak{p}$ is equal to $(N\mathfrak{p})^r$. By definition, this is equal to $N(r\mathfrak{p})$ for the M_K-ideal $r\mathfrak{p}$. Our assertion is now obvious by linearity.

Proposition 6.2. *Let K be a number field, $\alpha \in K^*$, and $\mathfrak{a} = (\alpha)$ its M_K-ideal. Then for each prime number p, we have*

$$\operatorname{ord}_p N\mathfrak{a} = \operatorname{ord}_p N_\mathbf{Q}^K(\alpha)$$

and consequently

$$|N\mathfrak{a}| = |N_\mathbf{Q}^K(\alpha)| = N\mathfrak{a},$$

where $|\ \ |$ is the ordinary absolute value.

Proof. For each $\mathfrak{p}|p$, let $d_\mathfrak{p}$ be the residue class degree. By Proposition 4.4 of Chapter 1 we have

$$N_\mathbf{Q}^K(\alpha) = \prod_{\mathfrak{p}|p} N_{\mathbf{Q}_p}^{K_\mathfrak{p}}(\alpha)$$

and hence, using Proposition 2.5 of Chapter 1,

$$\operatorname{ord}_p N_\mathbf{Q}^K(\alpha) = \sum_{\mathfrak{p}|p} \operatorname{ord}_p N_{\mathbf{Q}_p}^{K_\mathfrak{p}}(\alpha)$$

$$= \sum_{\mathfrak{p}|p} d_\mathfrak{p} \operatorname{ord}_\mathfrak{p} \alpha.$$

Our first assertion now follows by definition, and the second is obvious from the first.

Theorem 6.3. *Let K be a number field. Then there is a constant C depending on K only, such that for any integral M_K-ideal $\mathfrak{a}$ there exists an integral ideal $\mathfrak{b} \sim \mathfrak{a}$ such that $N\mathfrak{b} \leqq C$. The group of M_K-ideal classes is finite.*

Proof. Our second assertion is an immediate consequence of the first and of the fact that there are only a finite number of integral M_K-ideals $\mathfrak{b}$ with $N\mathfrak{b} < C$. Let us prove our first assertion.

Let I be the integers of K. It is a finite module over $\mathbf{Z}$, which is a principal ideal ring, and thus has a basis $\omega_1, \ldots, \omega_N$ over $\mathbf{Z}$. Let S be the set of elements of I of type

$$a_1\omega_1 + \cdots + a_N\omega_N$$

with a_i integers, such that

$$0 \leqq a_i \leqq (\mathbf{N}\mathfrak{a})^{1/N} + 1.$$

Then there are more than $\mathbf{N}\mathfrak{a}$ elements in S, and thus there are two distinct elements α, β in S such that $\alpha - \beta = \xi$ will map into 0 in the homomorphism

$$I \to \prod I_\mathfrak{p}/\mathfrak{m}_\mathfrak{p}^{\mathrm{ord}_\mathfrak{p}\,\mathfrak{a}}$$

(referring to Lemma 6.1). It follows from this that $(\xi) \geqq \mathfrak{a}$, and that we can write

$$(\xi) = \mathfrak{a} + \mathfrak{b}$$

with an M_K-integral ideal $\mathfrak{b}$. On the other hand, we estimate

$$|N_\mathbf{Q}^K(\xi)| = \prod_\sigma |c_1\omega_1^\sigma + \cdots + c_N\omega_N^\sigma|$$

where $0 \leqq c_i \leqq (\mathbf{N}\mathfrak{a})^{1/N} + 1$, and we see that there is a constant C (depending on the maximum of the archimedean absolute values of the ω_i, and on N) such that

$$|N_\mathbf{Q}^K(\xi)| \leqq C\mathbf{N}\mathfrak{a}.$$

Using Proposition 6.2, we get $\mathbf{N}\mathfrak{b} \leqq C$, and $\mathfrak{b} \sim \mathfrak{a}$ by definition. This proves our theorem.

To get a better explicit bound, one has to use not only the geometry of numbers and the Minkowski constant (cf. [L 5], Chapter 5, §4); but also deeper methods, see below in connection with the regulator and analytic number theory.

Let K be a number field, and S a finite subset of M_K containing the archimedean absolute values. Let s be the number of elements of S. Let K_S be the S-units, i.e. the elements $x \in K^*$ such that

$$\|x\|_v = 1 \quad \text{for } v \notin S.$$

We map K_S into Euclidean s-space as follows. Let $v_1, \ldots, v_s$ be the absolute values of S. Map

$$x \mapsto (\log\|x\|_1, \ldots, \log\|x\|_s),$$

and call this map

$$\log: K_S \to \mathbf{R}^s.$$

By the product formula, the image of K_S is contained in the hyperplane defined by the equation

$$\xi_1 + \cdots + \xi_s = 0,$$

so that this image is at most $(s - 1)$-dimensional.

Theorem 6.4. *The group K_S is finitely generated.*

Proof. The kernel of the log mapping consists of all those elements of K which have absolute value 1 at all $v \in M_K$. It is therefore a group, and is contained in the units of $\mathfrak{o}_K$. The elementary symmetric functions of the conjugates of any such element lie in $\mathbf{Z}$ and have bounded absolute value. Hence such elements satisfy only a finite number of equations with co-efficients in $\mathbf{Z}$, so they form a finite group, namely the group of roots of unity in K.

Let L be the image of K_S under the log mapping. By an elementary theorem of linear algebra, it will suffice to prove that L is discrete, or in other words that in any bounded region of space there is only a finite number of elements of L. But a similar argument as above shows that elements $\alpha \in K_S$ such that for all $\sigma \in \mathrm{Hom}(K, \mathbf{Q}^a)$ the logs $\log|\sigma\alpha|_v$ are bounded satisfy only a finite number of equations with rational coefficients, and so form a finite set, as desired.

The present book deals in qualitative results, and Theorem 6.4 will suffice for our purposes. However, it is much weaker than the standard unit theorem of algebraic number theory, which states:

Theorem 6.5. *Let K be a number field, S a finite subset of M_K containing the archimedean absolute values, and s the number of elements of S. Let K_S be the S-units. Then $\log(K_S)$ is an $(s - 1)$-dimensional lattice.*

Let K be a field with a proper family M_K of absolute values. For each M_K-divisor $\mathfrak{d}$ we shall denote by $L(\mathfrak{d})$ the set of elements $x \in K$ such that for each $v \in M_K$ we have

$$|x|_v \leqq \mathfrak{d}(v).$$

If $\alpha \in K^*$, and if we denote by $\alpha\mathfrak{d}$ the product $\mathfrak{d}_\alpha \mathfrak{d}$ of the principal M_K-divisor α times $\mathfrak{d}$, then $L(\alpha\mathfrak{d})$ and $L(\mathfrak{d})$ are in canonical bijection under the mapping

$$x \mapsto \alpha x, \qquad x \in L(\mathfrak{d}).$$

We denote the number of elements of $L(\mathfrak{d})$ by $\lambda(\mathfrak{d})$. Then $\lambda(\alpha\mathfrak{d}) = \lambda(\mathfrak{d})$. If we think of $\mathfrak{d}$ as prescribing the sides of a box, all but a finite number of which are 1, then $\lambda(\mathfrak{d})$ may be interpreted as the number of field elements in the box.

We define $|\mathfrak{d}|_v = \mathfrak{d}(v)$, and when we have multiplicities N_v,

$$\|\mathfrak{d}\|_v = \mathfrak{d}(v)^{N_v}.$$

We define the K-size, or size of $\mathfrak{d}$ to be

$$\|\mathfrak{d}\|_K = \prod_{v \in M_K} \|\mathfrak{d}\|_v.$$

If M_K satisfies the product formula with multiplicities N_v, then the size of $\mathfrak{d}$ is the same as that of $\alpha\mathfrak{d}$. This size may be interpreted as giving the volume of our box. In number fields, the number of elements in the box is approximately equal to the volume.

Theorem 6.6. *Let K be a number field with its canonical proper set of absolute values M_K. There exist two numbers $c_1, c_2 > 0$ depending only on K, such that for any M_K-divisor $\mathfrak{d}$, we have*

$$c_1 \|\mathfrak{d}\|_K < \lambda(\mathfrak{d}) \leqq \sup[1, c_2 \|\mathfrak{d}\|_K].$$

The result as stated in Theorem 6.6 is taken from Artin–Whaples [A–W] and will be used in the proof of Roth's theorem in Chapter 7. It is a weak version of a more general statement giving the asymptotic behavior as follows:

Theorem 6.7. *Let γ_K be the constant*

$$\gamma_K = \frac{2^{r_1}(2\pi)^{r_2}}{d_K^{1/2}},$$

where r_1 is the number of real absolute values of K, r_2 the number of imaginary ones, and d_K the absolute value of the discriminant of K. Let $N = [K:\mathbf{Q}]$. Then

$$\lambda(\mathfrak{d}) = \gamma_K \|\mathfrak{d}\|_K + O(\|\mathfrak{d}\|_K^{1-1/N}) \quad \text{for } \|\mathfrak{d}\|_K \to \infty.$$

For proofs of both versions, see [L 5], Chapter V, §2. This asymptotic theorem corresponds to the Riemann part of the Riemann–Roch theorem for function fields in one variable. It can also be refined to give an exact formula. Indeed, K can be viewed as a lattice in the adele ring. If f is the characteristic function of a parallelotope, then one can apply to f the Poisson summation formula for the adeles modulo this lattice to get the Riemann–Roch theorem in number fields. See Tate's thesis, for instance in [L 5], Chapter XIV, or the original version reprinted in [C–F].

In this chapter, we emphasized the trivial nature of the qualitative results by giving simple proofs for the finiteness of the ideal class group and the finite generation of the units, so that we could use these in the last section §7 in the context of finitely generated rings and fields over the prime field to obtain similar qualitative results. More quantitative bounds or estimates depend on deeper considerations, and not only on making more powerful use of the geometry of convex sets. Indeed, let h be the **class number** (order of the ideal class group) and let R be the **regulator** of K (the volume of a fundamental domain for the lattice of the logs of units in euclidean $(s - 1)$-space). Siegel [Sie 2] (following a general idea of Landau) showed that

$$hR \leqq c^N w_K d_K^{1/2} (\log d_K)^{N-1},$$

with an explicit very good constant c, where $N = [K : \mathbf{Q}]$; w_K is the number of roots of unity in K; and d_K is the absolute value of the discriminant. Since h is an integer, this gives a corresponding bound for R. In light of Hermite's theorem giving an almost orthogonalized basis for a lattice (sometimes called a reduced basis for the lattice), one then obtains a corresponding bound for the absolute values of a suitably chosen basis for the units modulo roots of unity. A proof of the Hermite theorem will be recalled in Theorem 7.7 of Chapter 5. For analogous conjectures on elliptic curves, see [L 15].

On the other hand, one can use a somewhat different point of view to describe some of the above notions, more in line with the language of algebraic-differential geometry (some would say in the spirit of "giving a far greater number of definitions than is usual with other writers..."). I am indebted to L. Szpiro for the rest of this section, inspired by Arakelov [Ar 2].

First, we shall work with a single number field K, of degree $N = [K : \mathbf{Q}]$. It will then be convenient to write divisors additively. Then by the definition of §5, a divisor is a formal linear combination

$$D = \sum_{\mathfrak{p}} n_{\mathfrak{p}} \log(\mathbf{N}\mathfrak{p})\mathfrak{p} + \sum_{v \in S_\infty} \gamma(v)v,$$

where $n_{\mathfrak{p}}$ is an integer for each prime ideal $\mathfrak{p}$ of $\mathfrak{o}$ $(= \mathfrak{o}_K)$; almost all $n_{\mathfrak{p}} = 0$; and $\gamma(v)$ is an arbitrary real number. Each $v \in S_\infty$ corresponds to an embedding

$$\sigma_v : K \to \mathbf{C}$$

into the complex numbers, inducing the corresponding absolute value on K. Thus the group of divisors is isomorphic to the group of fractional ideals plus $(r_1 + r_2)$ copies of **R**. The **degree** of a divisor is defined as usual to be the sum of its coefficients, so

$$\deg D = \sum_{\mathfrak{p}} n_{\mathfrak{p}} \log(N\mathfrak{p}) + \sum_{v \in S_\infty} \gamma(v).$$

To each element $f \in K^*$ we have its associated divisor, where

$$n_{\mathfrak{p}} = \mathrm{ord}_{\mathfrak{p}} f \quad \text{and} \quad \gamma_v(f) = -\log\|f\|_v.$$

The product formula says that $\deg(f) = 0$.

Now let L be a projective module of rank 1 over $\mathfrak{o}$. As $\mathfrak{o}$-module, such L is isomorphic to some fractional ideal $\mathfrak{a}$, but we prefer to keep L separate from any embedding in K. Alternatively, if the reader does not wish to use the words projective or rank, we let L be a finitely generated torsion free module over $\mathfrak{o}$. I call such L a **line module** over $\mathfrak{o}$ (to avoid calling L a line bundle, since the bundle itself would require further terminology to define). Let v be archimedean. Let K_v be the completion, which can be identified with **C** or **R**. For each v we suppose given a norm $|\ |_v$ on $L \otimes K_v = L_v$. As usual, we then use the normalized form

$$\|x\|_v = \begin{cases} |x|_v & \text{if } v \text{ is real,} \\ |x|_v^2 & \text{if } v \text{ is complex.} \end{cases}$$

A line module with a norm at each v will be called a **metrized line module**.

Given $s \in L$, $s \neq 0$ we can **associate** a **divisor** to L, as follows. First we define an ideal $\mathfrak{a}_s$. There is a unique injection of $\mathfrak{o}$ into L as $\mathfrak{o}$-module, sending 1 to s. Then L can be identified with a fractional ideal in K, and we define $\mathfrak{a}_s = L^{-1}$. Then we define the **associated divisor**.

$$D_s = \sum_{\mathfrak{p}} v_{\mathfrak{p}}(\mathfrak{a}) \log(N\mathfrak{p})\mathfrak{p} - \sum_v \log\|s\|_v \cdot v.$$

Let L_1, L_2 be two metrized line modules over $\mathfrak{o}$. Then the notion of **isomorphism** is defined in the only possible way: it is an isomorphism of $\mathfrak{o}$-modules which preserves the v-norm for each $v \in S_\infty$. Observe that if $L_1 = L_2 = L$, then the only automorphisms of L as $\mathfrak{o}$-modules are of the form

$$x \mapsto ux$$

for some unit in $\mathfrak{o}$, and therefore the $\mathfrak{o}$-module isomorphism preserves the norms if and only if

$$|x|_v^{(1)} = |x|_v^{(2)} |u|_v \quad \text{for all } x \in L.$$

We denote by Pic(o) the group of isomorphism classes of metrized line modules, and by Cl(o) the usual group of o-module isomorphism classes of line modules. Then there is an exact sequence

$$0 \to \mathbf{R}^{r_1 + r_2}/\log \mathfrak{o}^* \to \mathrm{Pic}(\mathfrak{o}) \to \mathrm{Cl}(\mathfrak{o}) \to 0,$$

where $\log \mathfrak{o}^*$ is the usual embedding of $\mathfrak{o}^*/\mu(K)$ into $\mathbf{R}^{r_1 + r_2}$ by the logs of absolute values on each component.

The degree function is a real valued function

$$\deg : \mathrm{Pic}(\mathfrak{o}) \to \mathbf{R},$$

if we define deg L to be the degree of its associated divisor,

$$\deg L = \log \mathbf{N}\mathfrak{a}_s - \sum_v \log \|s\|_v.$$

This sum is independent of the choice of s, and is defined on the group of metrized module classes, or divisor classes by the product formula.

We define $H^0(L)$ to be the set of elements $s \in L$ such that

$$|s|_v \leqq 1 \quad \text{for all } v.$$

The product formula immediately shows that

$$H^0(\mathfrak{o}) = \mu(K) \cup \{0\}.$$

On the other hand, we have the following proposition.

Proposition 6.8

(i) *If* deg $L = 0$ *and* $H^0(L) \neq 0$, *then* $L \approx \mathfrak{o}$ *with the standard norms.*
(ii) *If* deg $L < 0$ *then* $H^0(L) = \{0\}$.

Proofs. Both proofs are immediate by using the product formula, and will be left to the reader.

Let B be the unit ball in the product

$$\prod_{v \in S_\infty} L_v.$$

By vol(L) we mean the volume of a fundamental domain for L in this product. We let vol(B) be the volume of the unit ball. The quotient vol(B)/vol(L) is in fact independent of the metrics up to scalar factors, since in this quotient, proportionality constants disappear. We define the **Euler characteristic**

$$\chi(L) = \log \frac{\mathrm{vol}(B)}{2^N \, \mathrm{vol}(L)}.$$

Theorem 6.7 can be used to show that

$$\frac{\mathrm{vol}(B)}{\mathrm{vol}(L)} = \lim_{t \to \infty} \frac{|L \cap B(t)|}{t^N},$$

where $B(t)$ is the ball of radius t. Then we have:

Proposition 6.9. $\chi(\mathfrak{o}) = -(r_2 \log(2/\pi) + \log d_K^{1/2})$.

The proof follows from the evaluation of the constant γ_K in Theorem 6.7, and will be left to the reader.

Proposition 6.10. *If* $\deg L \geq -\chi(\mathfrak{o})$, *then* $H^0(L) \neq \{0\}$.

Proof. This follows at once by using Minkowski's theorem that in a symmetric convex closed set, there exists a lattice point $\neq 0$ if the volume is sufficiently large. See [L 5], Chapter V, §3, Theorem 3. Note that the 2^N in the definition of the Euler characteristic has been inserted so that we can apply Minkowski's theorem in a suitably normalized form.

In addition one also has a weak and trivial form of the Riemann–Roch theorem:

$$\chi(L) = \deg L - \chi(\mathfrak{o}).$$

As Szpiro also remarks, one can deduce Minkowski's theorem that $d_K > 1$ if $N > 1$ from the above propositions. For instance if $r_2 \neq 0$ we use Proposition 6.8(ii) and Proposition 6.9, which imply that $-\chi(\mathfrak{o}) \geq 0$. Then

$$\log d_K^{1/2} \geq \log(\pi/2)^{r_2},$$

and we are done. If $r_2 = 0$ and $r_1 > 1$, if $-\chi(\mathfrak{o}) > 0$, then the same easy proof works. If $\chi(\mathfrak{o}) = 0$, then Proposition 6.8(i) and Proposition 6.9 give a contradiction.

Finally, Szpiro also remarks that the same techniques prove Hermite's theorem that there is only a finite number of number fields of given degree, with discriminant divisible only by a fixed set of primes (the multiplicities are irrelevant). Only a weaker statement is given in [L 5].

As already stated, the above considerations are formulated in a language which parallels that of Arakelov [Ar 2] for metrized line bundles over absolute two-dimensional schemes (curves over rings of integers of number fields). The finiteness of isomorphism classes of curves with bad reduction in a fixed finite set of primes is an unsolved problem in the case of genus $\geqq 2$. (For genus 1, the result is due to Shafarevich.) The analogous problem for curves over a one-dimensional base over the complex numbers was solved by Parsin [Par 1] and Arakelov [Ar 1].

§7. Relative Units and Divisor Classes

The results of this section will not be used anywhere else in this book, except in Chapter 8, §5. They are taken from Roquette [Ro].

Let K be a field with a proper set of discrete valuations M. We denote by U_M, D_M, C_M the associated groups of units, divisors (or ideals, same thing in this case), and divisor classes. We then have by definition, an exact sequence

$$(1) \qquad\qquad 0 \to U_M \to K^* \to D_M \to C_M \to 0.$$

Let M' be a subset of M. Then we have the obvious surjection

$$D_M \to D_{M'} \to 0$$

whose kernel is $D_{M-M'}$. Hence we have the following exact and commutative diagram:

$$(2)$$

$$
\begin{array}{c}
0 \\
\downarrow \\
D_{M-M'} \\
\downarrow \\
0 \to U_M \to K^* \to D_M \to C_M \to 0 \\
\downarrow \\
0 \to U_{M'} \to K^* \to D_{M'} \to C_{M'} \to 0 \\
\downarrow \\
0.
\end{array}
$$

From this one gets in the usual manner two exact sequences:

$$(3) \qquad\qquad 0 \to U_M \to U_{M'} \to D_{M-M'},$$

$$(4) \qquad\qquad D_{M-M'} \to C_M \to C_{M'} \to 0,$$

obtained by trivial chasing around the diagram, which we shall leave to the reader. As an application, we get:

Proposition 7.1. *Suppose that M' differs from M by a finite number of elements. Then U_M (resp. C_M) is finitely generated if and only if $U_{M'}$ (resp. $C_{M'}$) is finitely generated.*

Theorem 7.2. *Let k be a field, V a normal variety over k, and M the set of discrete valuations of $k(V)$ corresponding to the prime divisors of V rational over k. Let U_M be its set of units in $k(V)^*$. Then U_M/k^* is finitely generated.*

Proof. If V is complete, then $U_M = k^*$ and our assertion is trivial. In general, by considering an affine open subset of V whose set of prime rational divisors differs from that of V by a finite set, we may assume that V is affine. Then one may embed V in a projective k-normal variety $\bar{V}$ whose set of prime divisors over k again differs from that of V by a finite set. We can now use the exact sequence (3), modified as follows:

$$0 \to U_M/k^* \to U_{M'}/k^* \to D_{M-M'}$$

to conclude the proof.

Corollary 7.3. *Let k be a field, K a finitely generated regular extension of k, and S a subalgebra of K finitely generated as an algebra over k. Let S^* be the invertible elements of S. Then S^*/k^* is a finitely generated group.*

Proof. Our algebra determines a variety, and by going to an affine open subset of this variety corresponding to a larger ring S_1, we may assume that this variety is non-singular in codimension 1. (Indeed, S^*/k^* is canonically contained in S_1^*/k^*.) The invertible elements of S_1 coincide with the units for its set of prime divisors, and we can apply the theorem.

Let $\mathfrak{o}$ be a Noetherian ring, integrally closed, let V be a variety scheme over $\mathfrak{o}$ and assume that V is normal. Using the notation of §4, let $M'_{\mathfrak{o}} = M'_{\mathfrak{o}, V}$ be the subset of $M_{\mathfrak{o}, V}$ consisting of those elements of $M_{\mathfrak{o}}$ having a unique extension to M_V, and let $M'_V = M'_{\mathfrak{o}} \cup M_{k, V}$ where k is the quotient field of $\mathfrak{o}$. Then we have an exact and commutative diagram:

$$
\begin{array}{ccccccccc}
 & & 0 & & 0 & & 0 & & \\
 & & \downarrow & & \downarrow & & \downarrow & & \\
 & & U_{M'_{\mathfrak{o}}} & & U_{M'_V} & & U_{M_{k,V}}/k^* & & \\
 & & \downarrow & & \downarrow & & \downarrow & & \\
0 & \to & k & \to & K^* & \to & K^*/k^* & \to & 0 \\
 & & \downarrow & & \downarrow & & \downarrow & & \\
0 & \to & D_{M'_{\mathfrak{o}}} & \to & D_{M'_V} & \to & D_{M_{k,V}} & \to & 0 \\
 & & \downarrow & & \downarrow & & \downarrow & & \\
 & & C_{M'_{\mathfrak{o}}} & & C_{M'_V} & & C_{M_{k,V}} & & \\
 & & \downarrow & & \downarrow & & \downarrow & & \\
 & & 0 & & 0 & & 0 & &
\end{array}
$$

(5)

and consequently an exact sequence

$$(6) \qquad U_{M'_0} \to U_{M'_V} \to U_{M_{k,V}}/k^* \to C_{M'_0} \to C_{M'_V} \to C_{M_{k,V}},$$

where the middle map is the usual coboundary operator (snake lemma).

If we now apply Theorem 7.2 to the first part of this sequence, together with the finite generation of the unit group in number fields, we get:

Theorem 7.4. *Let V be an irreducible normal scheme of finite type over $\mathbf{Z}$. Then the group of units U_{M_V} in its field of rational functions is finitely generated.*

Corollary 7.5. *Let R be a finitely generated ring over $\mathbf{Z}$. Then the invertible elements R^* of R form a finitely generated group.*

Proof. Using Proposition 4.1, we may embed R in an integrally closed ring of finite type over $\mathbf{Z}$, and it clearly suffices to prove our assertion in this case. Our ring then determines a variety scheme as in the theorem, and its invertible elements are equal to the M_V-units, so that the theorem applies.

It will be a comparatively deep theorem that the group $C_{M_{k,V}}$ is finitely generated whenever k is a field of finite type over the prime field. Once this will be proved, however, (Chapter 6) we shall see, from the end of the exact sequence (6) together with the finiteness of class number theorem in number fields (Theorem 6.3), that we can derive a similar statement for the divisor classes:

Theorem 7.6. *Let V be an irreducible normal scheme of finite type over $\mathbf{Z}$. Then $\mathrm{Pic}(V)$ is finitely generated.*

Proof. We let M_V be the set of discrete valuations corresponding to the prime divisors on V as in §4. Then

$$\mathrm{Pic}(V) = C_{M_V}.$$

We apply sequence (6) using the set M'_V which differs from M_V by a finite subset. This of course makes no difference to the finite generation in view of sequence (4).

Corollary 7.7. *Let R be a finitely generated normal ring over $\mathbf{Z}$ and let C be its group of divisor classes. Then C is finitely generated. There exists an element $a \in R$ such that $\mathrm{Pic}\, R[1/a] = 0$, in other words every divisor of $R[1/a]$ is principal.*

Proof. The first statement is a repetition of Theorem 7.6 for the affine case. As to the second, let $W_1, \ldots, W_r$ be divisors, generating the divisor class group. Replacing these by their prime components, we may assume without loss of generality that they are prime divisors. For each i let t_i be a prime element in the local ring of W_i. We can find an element $a \in R$ such that the minimal prime ideal of the restriction of W_i to $R[1/a]$ is principal, generated by t_i. Such an element a satisfies our requirements.

Remark. In connection with such finitely generated groups of divisor classes and units, the case of a regular ring R is especially interesting (recall that **regular** means that every local ring is regular, and therefore every divisor is locally principal). First one observes that the difference

$$\text{rank } R^* - \text{rank } \text{Pic}(R)$$

is the same for all regular rings R finitely generated over $\mathbf{Z}$, having the same quotient field K. Let P be a closed point in $V = \text{spec}(R)$, or equivalently, a maximal ideal in R. Then R/P is a finite field, and we let $\mathbf{N}P$ be the number of elements in R/P. One defines the **Hasse–Weil zeta function**

$$\zeta(R, s) = \zeta(V, s) = \prod_P (1 - \mathbf{N}P^{-s})^{-1}, \quad \text{for } \text{Re}(s) > s_0(R).$$

Such a function reflects many arithmetic and geometric properties of V, and in particular, we have the following conjecture of Tate [Ta 4], assuming the analytic continuation:

Let R be a regular ring of finite type over $\mathbf{Z}$. Then the order of $\zeta(R, s)$ at $s = \dim R - 1$ is equal to

$$\text{rank } R^* - \text{rank } \text{Pic}(R).$$

See also [Ta 2] for the connection with the Birch–Swinnerton-Dyer conjecture.

§8. The Chevalley–Weil Theorem

In this section, we are concerned with a morphism of normal varieties

$$\varphi: V \to W$$

defined over a field k with a proper set of discrete absolute values M_K.

We shall assume some foundational (and essentially simple) material as follows. Suppose first that φ is a **covering**. Given our assumptions on

V, W this means that if U is an affine open subset of W and $k[U]$ is the affine coordinate ring, then the inverse image $U' = \varphi^{-1}(U)$ is affine, and its coordinate ring is the integral closure of $k[U]$ in the function field $k(V)$. We let

$$n = [k(V):k(W)] = [V:W]$$

be the **degree** of φ. For all points $x \in W(k^a)$ outside some open set, the inverse image $\varphi^{-1}(x)$ consists of precisely n distinct points on V. If this is true for all points $x \in W(k^a)$, then we say that φ is **unramified**. Given our assumptions of normality, this has a number of implications as follows.

Let $\{g_1, \ldots, g_n\}$ be a basis of $k(V)$ over $k(W)$ such that $g_i \in k[U']$ for all i. Then the discriminant

$$\mathbf{D}_{V/W}(g_1, \ldots, g_n)$$

is defined in the usual way as $\det(\sigma_i g_j)^2$ where $\sigma_1, \ldots, \sigma_n$ are the distinct conjugate embeddings of $k(V)$ over $k(W)$ in an algebraic closure of $k(W)$. The ideal in $k[U]$ generated by all such discriminants is called the **discriminant ideal**. That φ is unramified implies that this discriminant ideal is the unit ideal. In particular, locally in the Zariski topology the discriminant ideal is the unit ideal in some neighborhood of a given point. This implies V is locally free over W, or in other words, given any point there is an affine open set U containing that point such that $k[\varphi^{-1}(U)]$ is free of dimension n over $k[U]$, and the discriminant of a basis is a unit in the ring $k[U]$.

On the other hand, let K be a finite extension of k, and let v be one of the discrete absolute values on k. Let $\mathfrak{o}_v = \mathfrak{o}_v(k)$ be the discrete valuation ring in k corresponding to v. Let $\mathfrak{o}_v(K)$ be the integral closure of $\mathfrak{o}_v$ in K. Then $\mathfrak{o}_v(K)$ is a Dedekind ring with only a finite number of maximal ideals, whose local rings are the valuation rings of K lying above $\mathfrak{o}_v(k)$. Let K' be a finite extension of K and let $\mathfrak{o}_v(K')$ be the integral closure of $\mathfrak{o}_v$ in K'. Then $\mathfrak{o}_v(K')$ is a free $\mathfrak{o}_v(K)$-module. The discriminant of a basis of $\mathfrak{o}_v(K')$ over $\mathfrak{o}_v(K)$ generates a principal ideal, which is called the **discriminant** of K' over K at v. If K' is separable, then for all but a finite number of v in M_k this ideal is the unit ideal. The discriminant ideal having a unique factorization into prime ideals ($\mathfrak{o}_v(K)$ is a Dedekind ring) we may also view it as a divisor. The direct sum (or product) over all v is also called the **discriminant ideal** or **divisor** of K' over K.

The following theorem is due to Chevalley-Weil, see [Ch-W] and [We 6]. We shall write divisors multiplicatively. We deal only with discrete absolute values.

Theorem 8.1. *Let $\varphi: V \to W$ be an unramified covering of projective normal varieties defined over a field k with a proper set of discrete absolute* *values M_K. There exists an element $d \in k$ such that for any point $x \in W(K)$ (where K is a finite extension of k), the discriminant of $K(\varphi^{-1}(x))$ over K divides d.*

The proof will result from a sequence of lemmas. First observe that we may assume that the covering is Galois. Indeed, let L be the smallest Galois extension of $k(W)$ containing $k(V)$, and let k' be the algebraic closure of k in L. Let V' be the normalization of V in L. Then $\psi: V' \to W$ is an unramified covering defined over k'. If we have shown that the discriminant of $Kk'(\psi^{-1}(x))$ over Kk' is bounded, then the same result holds for $K(\varphi^{-1}(x))$ over K.

Lemma 8.2. *Let U be an affine variety over k and let its coordinate ring be*

$$R = k[U] = k[f_1, \ldots, f_m].$$

Let δ be a unit in R. There exist M_k-divisors $\mathfrak{d}_1$, $\mathfrak{d}_2$ such that for any point $P \in U(k^a)$ and $v \in M(k^a)$ where $f_i(P)$ is v-integral for $i = 1, \ldots, m$ we have

$$\mathfrak{d}_1(v) \leqq |\delta(P)|_v \leqq \mathfrak{d}_2(v)$$

Proof. Any $f \in R$ is a polynomial in $f_1, \ldots, f_m$ with coefficients in k. By the non-archimedean nature of v it follows that $|\delta(P)|_v$ is bounded from above, and in fact by $\mathfrak{d}_2(v)$ for some M_k-divisor $\mathfrak{d}_2$. The opposite inequality follows by considering δ^{-1} instead of δ. This concludes the proof.

Lemma 8.3. *Let $\varphi: U' \to U$ be an unramified Galois covering of affine normal varieties, defined over k. Assume that $k[U']$ is free over $k[U]$, and that the discriminant of a basis*

$$\mathbf{D}(g_1, \ldots, g_n)$$

is a unit in $k[U]$. Let $k[U] = k[f_1, \ldots, f_m]$. There exist M_k-divisors $\mathfrak{d}_1$, $\mathfrak{d}_2$ such that for any finite extension K of k, any point $P \in U(K)$, and any absolute value $v \in M(K)$ where $f_1(P), \ldots, f_m(P)$ are v-integral we have

$$\mathfrak{d}_1(v) \leqq |\mathbf{D}(K(\varphi^{-1}(P))/K)|_v \leqq \mathfrak{d}_2(v).$$

Proof. Let $B = k[U']$ and $A = k[U]$. Let $\mathfrak{m}$ be the maximal ideal in A which is the kernel of the map $f \mapsto f(P)$ for $f \in k[U]$. Then $B/\mathfrak{m}B = B(\mathfrak{m})$ is a finite algebra of dimension n over $A(\mathfrak{m}) = A/\mathfrak{m}A = K$. The Galois group G operates on $B(\mathfrak{m})$ over K, and

$$\det(\sigma_i g_j(\mathfrak{m})) = \delta(P)$$

is the discriminant of a basis of $B(\mathfrak{m})$ over K, and is non-zero. Thus $B(\mathfrak{m})$ is a separable semisimple algebra over K, and we have

$$B(\mathfrak{m}) = \bigoplus_{\lambda=1}^{r} K_{\lambda},$$

where K_{λ} is a finite separable extension of K.

Let $v \in M_k$, and let $\mathfrak{o}_v = \mathfrak{o}_v(k)$ be the corresponding discrete valuation ring in k. Let $\mathfrak{o}_v^{(\lambda)} = \mathfrak{o}_v(K_{\lambda})$ be its integral closure in K_{λ}. Then the integral closure of $\mathfrak{o}_v$ in $B(\mathfrak{m})$, namely

$$\mathfrak{o}_v(B(\mathfrak{m})) = \prod_{\lambda=1}^{r} \mathfrak{o}_v^{(\lambda)}$$

is a free module of dimension n over $\mathfrak{o}_v(K)$.

There exists a finite set $S \subset M_k$ such that for all $v \notin S$ the functions $g_1, \ldots, g_n$ are integral over $\mathfrak{o}_v[f_1, \ldots, f_m]$ and $\mathbf{D}(g_1, \ldots, g_n)$ is a unit in $\mathfrak{o}_v[f_1, \ldots, f_m]$. For $v \notin S$ and such that $f_\alpha(\mathfrak{m})$ (or equivalently $f_\alpha(P)$) is v-integral for all $\alpha = 1, \ldots, m$, it follows that

$$\mathbf{D}(g_1(\mathfrak{m}), \ldots, g_n(\mathfrak{m})) = \det(\sigma_i g_j(\mathfrak{m}))$$

is a unit in $\mathfrak{o}_v(K)$, and hence that $\{g_1(\mathfrak{m}), \ldots, g_n(\mathfrak{m})\}$ is a basis of $\mathfrak{o}_v(B(\mathfrak{m}))$ over $\mathfrak{o}_v(K)$. For such v, we see that $K(\varphi^{-1}(P))$ is unramified at v over K.

Next let $v \in S$ but still $f_\alpha(\mathfrak{m})$ is v-integral for all α. There exists an element $a_v \in \mathfrak{o}_v$ (so a_v is v-integral in k) such that $a_v g_j$ is integral over

$$\mathfrak{o}_v[f_1, \ldots, f_m].$$

Then

$$a_v g_1(\mathfrak{m}), \ldots, a_v g_n(\mathfrak{m}) \in \mathfrak{o}_v(B(\mathfrak{m})).$$

and

$$\mathbf{D}(a_v g_1(\mathfrak{m}), \ldots, a_v g_n(\mathfrak{m})) = a_v^{2n} \mathbf{D}(g_1(\mathfrak{m}), \ldots, g_n(\mathfrak{m})).$$

By Lemma 8.2, $\mathbf{D}(g_1(\mathfrak{m}), \ldots, g_n(\mathfrak{m}))$ lies between two fixed bounds determined by an M_k-divisor. The v-integral transformation which sends an $\mathfrak{o}_v(K)$-basis of $\mathfrak{o}_v(B(\mathfrak{m}))$ on the elements

$$a_v g_1(\mathfrak{m}), \ldots, a_v g_n(\mathfrak{m})$$

shows that the discriminant of $\mathfrak{o}_v(B(\mathfrak{m}))$ over $\mathfrak{o}_G(K)$ divides

$$\mathbf{D}(a_v g_1(\mathfrak{m}), \ldots, a_v g_n(\mathfrak{m})).$$

But the discriminant of $\mathfrak{o}_v(B(\mathfrak{m}))$ over $\mathfrak{o}_v(K)$ is the product of the discriminants of $\mathfrak{o}_v^{(\lambda)}$ over $\mathfrak{o}_v(K)$. Hence these discriminants are bounded by an M_k-divisor, as was to be shown.

Lemma 8.3 proves the theorem "locally". We conclude the full proof by showing that there exists a covering of W by a finite number of affine open sets U with affine coordinates $f_1, \ldots, f_m$ (depending on U) having the following properties:

- (i)   For each U, $k[\varphi^{-1}(U)]$ is free over $k[U]$ and its discriminant is the unit ideal in $k[U]$.
- (ii) Given a point $P \in V$ and $v \in M_k$ there exists some U such that $f_1(P), \ldots, f_m(P)$ are v-integral.

These are the two conditions forming the hypotheses of Lemma 8.2, and if we can show that they are satisfied, then we have taken care of all points P and all v, thus proving Theorem 8.1.

The first condition is certainly true locally in the neighborhood of any given point, and since a variety is compact (satisfies the property that any open covering has a finite subcovering), we can find a finite affine open covering satisfying the first property.

To satisfy the second, we have to refine this covering, using the property that W is projective. One can find homogeneous polynomials $h_0, \ldots, h_r$ of sufficiently high degree (the same degree) such that if W_i is the hypersurface complement of $h_i = 0$ in W, then each W_i is contained in one of the open sets of the covering satisfying condition (i), and such that $W_0, \ldots, W_r$ have no point in common. Let $x_0, \ldots, x_s$ be the standard coordinates in the given projective embedding, and let U_{ij} be the complement of

$$h_i x_j = 0.$$

Then U_{ij} is an affine open set, equal to the complement of one of the coordinate hypersurfaces in the projective embedding determined by the $h_i x_j$ ($i = 0, \ldots, r; j = 0, \ldots, s$). Such U_{ij} give a refinement of the given open covering, and we have found projective coordinates $(z_0, \ldots, z_m)$ with $m = (r + 1)(s + 1) - 1$ such that the complement of the hyperplane $z_i = 0$ is contained in one of the open sets of the covering satisfying (i).

Now given any point P and absolute value v on k^a, there is some index α such that $|z_\alpha(P)|_v$ is maximal. Then the affine coordinates

$$(z_0/z_\alpha, \ldots, z_m/z_\alpha)$$

evaluated at P are v-integral, thus showing that (ii) is satisfied and concluding the proof.

Remark. In this last part of the proof, we have simply done ad hoc what will be a general principle used in a more general context in the theory of Néron divisors, Chapter 10, Lemma 1.1 and Proposition 1.2. I thought it might be more useful to put the Chevalley–Weil theorem rather at the beginning, to illustrate the connection with M_k-divisors, independently of the more extensive language used to deal with general Néron divisors and functions. Historically, however, the Chevalley–Weil theorem arose as an application of Weil's decomposition theorem.

CHAPTER 3

Heights

The possibility of defining the height of a point on a variety lies at the base of all possibilities of counting such points. In this book, this allows us to get qualitative results, to the effect that certain sets of points are finite, or, if they form a group, a finitely generated one.

One can give the definition of the height just after the axiom that we have a field with a proper set of absolute values satisfying the product formula. We can also prove important geometric properties of the height (how it varies under certain mappings) using only this axiom. The formalism we shall develop will then be used in the infinite descent of Chapter 6, and in the counting of Siegel's theorem in Chapter 8.

In number fields, the set of points in a given projective space of bounded degree and bounded height is finite. We shall give in §3 the analogue in function fields, where we prove that certain varieties lie in a finite number of algebraic families.

§1. Definitions

Let $\mathbf{F}$ be a field with a proper set of absolute values $M_{\mathbf{F}}$, satisfying the product formula. Then for each finite extension K of $\mathbf{F}$, the set of absolute values on K extending those of $M_{\mathbf{F}}$ is a proper set M_K, satisfying the product formula with multiplicities $N_v = [K_v : \mathbf{F}_v]$ for $v \in M_K$. As before, we set

$$\|x\|_v = |x|_v^{N_v}, \qquad x \in K.$$

We now consider projective n-space $\mathbf{P}^n$, and let $P \in \mathbf{P}^n(K)$ be a point in projective space, rational over K. Let $(x_0, \dots, x_n)$ be a set of coordinates for P, with each $x_i \in K$. Then we define the (**multiplicative**) **height of** P **relative to** M_K by the formula

$$H_K(P) = \prod_{v \in M_k} \sup_i \|x_i\|_v.$$

Here and elsewhere we write H_K instead of H_{M_K}. In view of the product formula, one sees immediately that our product is independent of the choice of coordinates chosen: If $y \in K^*$ then

$$\sup_i \|yx_i\|_v = \|y\|_v \sup_i \|x_i\|_v$$

and taking the product, the term involving y will be equal to 1.

In particular, we define the height of an element $x \in K$ to be the height of the point $(1, x)$ in $\mathbf{P}^1$, so that we have

$$H_K(x) = \prod_{v \in M_K} \sup(1, \|x\|_v),$$

and we see that if $x \neq 0$, then

$$H_K(x) = H_K(x^{-1}).$$

Furthermore, we have trivially

$$H_K(x_1 \cdots x_n) \leq H_K(x_1) \cdots H_K(x_n)$$

and for $x \in K$,

$$H_K(x^n) = H_K(x)^n.$$

Let us suppose that our set M_F is fixed. Let $E \supset K$ be two finite extensions of F, and let $(x_0, \ldots, x_n)$ be coordinates of a point P in $\mathbf{P}^n$, rational over K. Then one has

$$H_E(P) = H_K(P)^{[E:K]}.$$

Indeed, letting $N_w = [E_w : F_w]$ for $w \in M_E$, we get

$$H_E(P) = \prod_{w \in M_E} \sup_i |x_i|_w^{N_w} = \prod_{v \in M_K} \prod_{w|v} \sup_i |x_i|_w^{N_w}.$$

On the other hand,

$$[E_w : F_w] = [E_w : K_w][K_w : F_w].$$

For all absolute values w inducing the same absolute value v on K, the term $[K_w : F_w]$ is constant, and we have

$$\sum_{w|v} [E_w : K_w] = [E : K].$$

From this our assertion is obvious.

The formula we have just proved shows how the height changes as we extend the field. It is therefore natural to define an **absolute height** of a point, rational over the algebraic closure of **F**, by

$$H(P) = H_M(P) = H_K(P)^{1/[K:F]}$$

for any finite extension K of **F** over which P is rational. The argument we have just given shows that $H(P)$ is independent of the choice of this field.

Let σ be an isomorphism of K over **F** (i.e. leaving **F** fixed). Let P be a point as above, with coordinates $(x_0, \dots, x_n)$ rational over K. Then we can define the point P^σ, rational over K^σ, and having coordinates $(x_0^\sigma, \dots, x_n^\sigma)$. By transport of structure, we get immediately

$$H_K(P) = H_{K^\sigma}(P^\sigma)$$

whence in particular,

$$H(P) = H(P^\sigma).$$

We define the **logarithmic height** (or simply **height** if it is the only one occurring throughout a discussion) to be the logarithm of the other height:

$$h_K(P) = \log H_K(P) \quad \text{and} \quad h(P) = \log H(P).$$

Let us now consider an example. We take $\mathbf{F} = \mathbf{Q}$, the rational numbers. We can write any non-zero rational number in the form a/b where a, b are relatively prime integers. If p is a prime number then $|a|_p$ and $|b|_p$ are ≤ 1, and one of them $= 1$. Taking the canonical set of absolute values on $\mathbf{Q}$ to define the height, we see that

$$H_{\mathbf{Q}}(a/b) = \sup(|a|, |b|).$$

More generally, any point in projective n-space rational over $\mathbf{Q}$ has a set of coordinates $(x_0, \dots, x_n)$ which are relatively prime integers, and we then see that

$$H_{\mathbf{Q}}(P) = \sup_i |x_i|$$

the absolute value being the ordinary one. In particular, the set of points in $\mathbf{P}^n$ rational over $\mathbf{Q}$, and of height $\leq$ a fixed number is finite. We shall generalize this later to a number field. *We agree from now on that the heights in number fields will always be those relative to the canonical set of absolute values on* $\mathbf{Q}$.

Next let K be a number field. Let S_∞ be the set of archimedean absolute values in M_K. Let

$$(x_0, \ldots, x_n) \in \mathbf{P}^n(K)$$

be a point in projective space, and assume that the coordinates x_i are *relatively prime algebraic integers in K*. Then directly from the definition of the height, we get

$$H_K(x) = \prod_{v \in S_\infty} \sup_i \|x_i\|_v.$$

Thus again the height is described entirely in terms of the absolute values at infinity.

On the other hand, let $\alpha \in K$, $\alpha \neq 0$ and let

$$(\alpha) = \mathfrak{b}/\mathfrak{d}$$

be an ideal factorization for α where $\mathfrak{b}$, $\mathfrak{d}$ are relatively prime ideals in $\mathfrak{o}_K$. Then

$$\boxed{H_K(\alpha) = \mathbf{N}\mathfrak{d} \prod_{v \in S_\infty} \sup(1, \|\alpha\|_v)}$$

where $\mathbf{N}$ denotes the absolute norm (cf. my *Algebraic Number Theory*, Chapter II). Indeed, we have $\max(1, |\alpha|_v) > 1$ for a $\mathfrak{p}$-adic absolute value v if and only if $\mathfrak{p}$ divides the denominator $\mathfrak{d}$. In this case, let π be a prime element at $\mathfrak{p}$, and let $e = e(\mathfrak{p})$ and $f = f(\mathfrak{p})$ be the ramification index and residue class degree as usual. Then

$$|\pi^e|_{\mathfrak{p}} = |p|_{\mathfrak{p}} \qquad \text{so} \qquad |1/\pi|_{\mathfrak{p}} = p^{1/e}.$$

We have $N_{\mathfrak{p}} = e_{\mathfrak{p}} f_{\mathfrak{p}}$ so that

$$|1/\pi|_{\mathfrak{p}}^{N_{\mathfrak{p}}} = p^{f_{\mathfrak{p}}} = N\mathfrak{p}.$$

This proves that $\mathbf{N}\mathfrak{d}$ is the contribution to the height from the non-archimedean absolute values.

Remark. Suppose that α is algebraic of degree d over the rational numbers, and let

$$f(X) = a_d X^d + \cdots + a_0 = 0, \qquad a_d > 0,$$

be its irreducible equation, with coefficients $a_i \in \mathbf{Z}$, and $a_0, \ldots, a_d$ relatively prime. Let $K = \mathbf{Q}(\alpha)$. Then one also has the formula

$$H_K(\alpha) = a_d \prod_{i=1}^{d} \max(1, |\alpha_i|),$$

where $|\ |$ is the complex absolute value, and $\alpha_1, \ldots, \alpha_d$ are the distinct conjugates of α in $\mathbf{C}$. Indeed, the product is just the same as the product over $v \in S_\infty$ in the previous formula. One sees that $a_d = \mathbf{N}\mathfrak{d}$ by using the standard Gauss lemma, which will be recalled in the next section. Namely,

$$1 = |f(X)|_p = |a_d|_p \prod_p \max(1, \|\alpha\|_p) = |a_d|_p \mathbf{N}\mathfrak{d}_p^{-1},$$

so $a_d = \mathbf{N}\mathfrak{d}_p$, where $\mathfrak{d}_p$ is the p-part of $\mathfrak{d}$.

Similarly, suppose $x_0, \ldots, x_n$ are elements of K not all 0, and let $\mathfrak{a} = (x_0, \ldots, x_n)$ be the fractional ideal generated by them. Then we get

$$H_K(x) = \mathbf{N}\mathfrak{a}^{-1} \prod_{v \in S_\infty} \sup \|x_i\|_v.$$

This is proved in the same way as above for one element.

Proposition 1.1. *Let K be a number field, and $x \in K$, $x \neq 0$. Then $H_K(x) = 1$ (or $h(x) = 0$) if and only if x is a root of unity.*

Proof. The definition of the height shows first that x is an algebraic integer, and second that the archimedean absolute values of x and all its conjugates are bounded (by 1). The coefficients of the irreducible equation of x over $\mathbf{Z}$ are elementary symmetric functions of the conjugates of x and are therefore bounded. The set of elements $x \neq 0$ in K with $H_K(x) = 1$ is a multiplicative group, and the bound on the equations for such elements shows that this group is finite, so x is a root of unity, as was to be shown.

Remark. Later we shall deal with canonical heights on abelian varieties, and we shall meet the analogous property in Theorem 6.1 of Chapter 5, that the canonical height of a point is 0 if and only if the point is a torsion point.

§2. Gauss' Lemma

Let K be a field with a non-trivial absolute value v. If $f(X) = \sum a_v \mathbf{M}_v(X)$ is a polynomial in several variables with coefficients $a_v \in K$, and

$$\mathbf{M}_v(X) = \mathbf{M}_v(X_1, \ldots, X_n) = X_1^{v_1} \cdots X_n^{v_n}$$

a monomial, then we define

$$|f|_v = |f| = \sup_v |a_v|.$$

Gauss' lemma for valuations then asserts that this is a valuation if v is a valuation.

Proposition 2.1. *Let K be a field with a non-trivial valuation. If $f, g \in K[X_1, \ldots, X_n]$ then*

$$|fg| = |f||g|.$$

Proof. Let us first assume that f, g are polynomials in one variable, so we can write

$$f(X) = a_d X^d + \cdots + a_0,$$
$$g(X) = b_e X^e + \cdots + b_0.$$

Say a_r is a coefficient of f such that $|a_r| \geq |a_i|$ for all i, and is the one which lies furthest to the left that has this property. Similarly for b_s with respect to the b_j. Then

$$(1/a_r)f = (a_d/a_r)X^d + \cdots + X^r + \cdots + a_0/a_r,$$
$$(1/b_s)g = (b_e/b_s)X^e + \cdots + X^s + \cdots + b_0/b_s,$$

and the coefficients of these polynomials each have a value ≤ 1. Those to the left of X^r (resp. X^s) have a value < 1. If we take the product

$$(1/a_r b_s)fg$$

then we see that $X^r X^s$ has a coefficient of type $1 + c$ where $|c| < 1$. The coefficient of every power X^i with $i > r + s$ has a value < 1, while if $i < r + s$, then its value is ≤ 1. From this one sees that

$$|(1/a_r b_s)fg| = 1$$

and hence that $|fg| = |a_r b_s|$, which proves our proposition for polynomials in one variable.

Now let f be a polynomial in n variables $X_1, \ldots, X_n$, of degree $< d$. Then the polynomial in one variable

$$(S_d f)(Y) = f(Y, Y^d, \ldots, Y^{d^{n-1}})$$

has the same set of non-zero coefficients as f. Thus, if f and g are two polynomials in n variables, such that the sum of their degrees is $< d$, then

$$S_d(fg) = S_d(f)S_d(g)$$

has the same non-zero coefficients as fg. From this our reduction of the n-variable case to the 1-variable case is clear.

For an archimedean absolute value, we do not get the full multiplicativity, but we shall see that we get a modified version of it. Let us first consider the case of one variable.

Lemma 2.2. *Let K be a field of characteristic 0, with an absolute value which coincides with the ordinary one on $\mathbf{Q}$. Let $f(X) \in K[X]$ be a polynomial of degree d, and let*

$$f(X) = \prod_{i=1}^{d} (X - \alpha_i)$$

be a factorization in K^a. We assume that our absolute value is extended to K^a. Then

$$\frac{1}{2^d} \prod_{i=1}^{d} \sup(1, |\alpha_i|) \leq |f| \leq 2^d \prod_{i=1}^{d} \sup(1, |\alpha_i|).$$

Proof. The right inequality is trivially proved by induction, estimating the coefficients in a product of a polynomial $g(X)$ by $(X - \alpha)$. We prove the other by induction on the number of indices i such $|\alpha_i| > 2$. If $|\alpha_i| \leq 2$ for all i, our assertion is obvious. Suppose now that

$$f(X) = g(X)(X - \alpha)$$

with $|\alpha| > 2$ and suppose that our assertion is true for

$$g(X) = X^d + b_{d-1}X^{d-1} + \cdots + b_0.$$

We have

$$f(X) = X^{d+1} + (b_{d-1} - \alpha)X^d + (b_{d-2} - \alpha b_{d-1})X^{d-1} + \cdots + (-\alpha)b_0.$$

We can assume $|g| = |b_i| \geq |b_{i-1}|$ for some $i, 0 \leq i \leq d$ (with the convention $b_d = 1, b_{-1} = 0$). Then

$$|f| \geq |\alpha b_i - b_{i-1}| \geq |\alpha||b_i| - |b_{i-1}|$$

$$\geq |\alpha||b_i| - |b_i| = (|\alpha| - 1)|b_i|$$

$$\geq |\alpha|/2 \cdot |b_i| = |\alpha|/2 \cdot |g|$$

and our lemma is now obvious, since $|\alpha| > 2$.

Proposition 2.3. *Let K be a field of characteristic 0, with an absolute value which coincides with the ordinary one on $\mathbf{Q}$. Let d be an integer ≥ 0. If f and g are two polynomials in $K[X_1, \ldots, X_n]$ such that $\deg f + \deg g < d$, then*

$$\frac{1}{4^{d^n}} |fg| \leq |f||g| \leq 4^{d^n} |fg|.$$

Proof. Using the substitution S_d as before, we are reduced to the case of one variable with $n = 1$. Say

$$f(X) = a \prod (X - \alpha_i),$$
$$g(X) = b \prod (X - \beta_j).$$

Without loss of generality, we may assume $a = b = 1$, and that we have extended our absolute value to K^a. Using the lemma, we get

$$|f||g| \leq 2^d \prod \sup(1, |\alpha_i|) \prod \sup(1, |\beta_j|)$$
$$\leq 2^d 2^d |fg|.$$

The other inequality is proved in the same way.

Let $\mathbf{F}$ again be a field with a proper set of absolute values $M_\mathbf{F}$ satisfying the product formula. We consider a polynomial

$$f(X) = \sum a_j \mathbf{M}_j(X)$$

in several variables, with coefficients in $\mathbf{F}^a$, and thus in a finite extension K of $\mathbf{F}$. We define its **absolute height** $H(f)$ to be the height $H(P)$ of the point P having the a_j (in any order) as coordinates, and we define its **relative height** $H_K(f)$ in a similar way.

From Proposition 2.3 we can now deduce analogous results for heights.

Proposition 2.4. *Let $\mathbf{F}$ be a field with a proper set of absolute values $M_\mathbf{F}$ satisfying the product formula, and let s be the number of archimedean absolute values in $M_\mathbf{F}$. Let f and g be two polynomials in n variables with coefficients in $\mathbf{F}^a$, with $\deg f + \deg g < d$. Then*

$$\frac{1}{4^{sd^n}} H(fg) \leq H(f)H(g) \leq 4^{sd^n} H(fg).$$

Proof. Let $c_1 = 1/4^{d^n}$ and $c_2 = 4^{d^n}$. Let K be a finite extension of F in which f and g have their coefficients. For $v \in M_K$, put $\|f\|_v = |f|_v^{N_v}$.

We have

$$H_K(fg) = \prod_{v \in M_K} \|fg\|_v \geqq \prod_{v \in M_K} \|f\|_v \|g\|_v c_v,$$

where $c_v = 1$ if v is non-archimedean, and otherwise, $c_v = c_1^{N_v}$. Let S_K be the set of archimedean absolute values in M_K. Then

$$\sum_{v \in S_K} N_v = [K : F]s,$$

whence

$$\prod_{v \in M_K} c_v = c_1^{[K:F]s}$$

and the inequality on the left follows immediately. The one on the right follows in a similar way.

Let $f(K)$ be as above, with coefficients in a finite extension K of $\mathbf{F}$. If σ is an isomorphism of K over $\mathbf{F}$ then we get the polynomial

$$f^\sigma(X) = \sum a_i^\sigma \mathbf{M}_j(X),$$

and thus, as for points, we have

$$H(f) = H(f^\sigma).$$

Let α be algebraic over $\mathbf{F}$, and let $f(X)$ be its irreducible polynomial over $\mathbf{F}$. Then

$$f(X) = \prod_{j=1}^d (X - \alpha_j),$$

where d is the degree of α over $\mathbf{F}$ (i.e. $[\mathbf{F}(\alpha) : \mathbf{F}]$) and the α_j are the conjugates of α, each repeated with a suitable multiplicity in characteristic p. In view of the above results we get:

Proposition 2.5. *Let $\mathbf{F}$ be a field with a proper set of absolute values satisfying the product formula. Let d be an integer > 0. Then there exist two numbers $c_1, c_2 > 0$ depending on d, such that if α is algebraic over $\mathbf{F}$ of degree d, and $f(X)$ is its irreducible polynomial over $\mathbf{F}$, then*

$$c_1 H(\alpha)^d \leqq H(f) \leqq c_2 H(\alpha)^d.$$

Let us consider the special case where $\mathbf{F} = \mathbf{Q}$ is the rational numbers. Then $H(f)$ is the maximum of the absolute values of the coefficients of the irreducible polynomial of α over $\mathbf{Z}$. More generally, consider a point $(\alpha_0, \ldots, \alpha_n)$ in projective space, rational over a finite extension K of $\mathbf{Q}$, of degree d. Consider the polynomial

$$f(X) = \alpha_0 X_0 + \cdots + \alpha_n X_n.$$

Then $H(f)$ is the height of our point. Let

$$g(X) = \prod_\sigma f^\sigma$$

the product being taken over all distinct isomorphisms σ of K over $\mathbf{Q}$. Then g has coefficients in $\mathbf{Q}$, and $H(g)$, $H(f)^d$ have the same order of magnitude. We have already seen in §1 that the number of points with height less than a fixed number, in a projective space, and rational over $\mathbf{Q}$ is bounded. Consequently we now get:

Theorem 2.6. *Let d_0, H_0 be two fixed numbers > 0. Then the set of points P in $\mathbf{P}^n$ algebraic over $\mathbf{Q}$, and such that*

$$[\mathbf{Q}(P):\mathbf{Q}] < d_0 \qquad and \qquad H(P) < H_0$$

is finite.

The above theorem is due to Northcott [No].

We note the strong uniformity in the statement of the above theorem. Not only is the set of points in $\mathbf{P}^n$ rational over a *given* number field and of bounded height finite, but so is the set of points of bounded degree and bounded height. Actually, in the final applications, this stronger uniformity will not be used here, because we shall consider only points of varieties which are rational over a fixed number field (for instance in the Mordell–Weil theorem, or the Siegel theorem).

The proof of the Gauss lemma for archimedean absolute values was carried out by a simple induction. By using slightly more elaborate means, one can give better inequalities which are useful in various number-theoretic applications, see for instance Gelfond's book on transcendental numbers. Although we shall not need them in this book, Michel Waldschmidt recommended their inclusion, and I am indebted to him for the following exposition. Let

$$f(X) = a_d X^d + \cdots + a_0 = a_d \prod_{j=1}^d (X - \alpha_j)$$

be a polynomial with complex coefficients, of degree d. We define:

$$L_2(f) = \left(\sum_{j=0}^d |a_j|^2 \right)^{1/2}; \qquad M(f) = |a_d| \prod_{j=1}^d \max(1, |\alpha_j|).$$

We have the trivial inequalities:

$$|f| \leq L_2(f) \leq (d+1)^{1/2}|f|.$$

Proposition 2.7 (Landau's inequality). $M(f) \leq L_2(f)$.

Proof. By Jensen's formula, we have

$$M(f) = \exp \int_0^1 \log|f(e^{2\pi i t})| \, dt.$$

On the other hand,

$$L_2(f) = \left(\int_0^1 |f(e^{2\pi i t})|^2 \, dt \right)^{1/2}.$$

Finally, Jensen's formula and the convexity of the exponential function give

$$\exp \int_0^1 u(t) \, dt \leq \int_0^1 \exp u(t) \, dt,$$

where $u(t) = 2 \log|f(e^{2\pi i t})|$. This proves Landau's inequality.

Theorem 2.8. $(d+1)^{-1/2}M(f) \leq |f| \leq 2^d M(f)$.

Proof. For $0 \leq j \leq d$ we have

$$|a_j| \leq \binom{d}{j} M(f).$$

Indeed, we write a_j as a function of the roots, and make a trivial upper estimate. The inequality of the theorem follows from this and Landau's inequality.

Corollary 2.9. *Let $f_1, f_2 \in \mathbb{C}[X]$ be polynomials of degrees d_1, d_2 respectively. Then*

$$(1 + d_1)^{-1}|f_1 f_2| \leq |f_1||f_2| \leq e^{d_1 + d_2}|f_1 f_2|.$$

Proof. The inequality $|f_1 f_2| \leq (1 + d_1)|f_1||f_2|$ is trivial. For the other inequality, note that

$$M(f_1 f_2) = M(f_1)M(f_2),$$

and $M(f) = |f|$ if f has degree 1. Furthermore,

$$2^d(d + 1)^{1/2} \leqq e^d \quad \text{if } d \geqq 2.$$

The corollary follows at once.

Similar arguments also apply to polynomials in several variables. Indeed, let $f(X_1, \ldots, X_n) \in \mathbf{C}[X_1, \ldots, X_n]$ be a non-zero polynomial of total degree d, and degree d_j in X_j, say

$$f(X) = \sum_{j_1 = 0}^{d_1} \cdots \sum_{j_n = 0}^{d_n} a_{j_1 \ldots j_n} X_1^{j_1} \cdots X_n^{j_n}.$$

We define $|f| = \max |a_{(j)}|$ as usual, and:

$$L_2(f) = \left(\sum_{(j)} |a_{(j)}|^2 \right)^{1/2};$$

$$M(f) = \exp\left(\int_0^1 \cdots \int_0^1 \log |f(e^{2\pi i t_1}, \ldots, e^{2\pi i t_n})| \, dt_1 \cdots dt_n \right).$$

Again we have the trivial inequalities:

$$|f| \leqq L_2(f) \leqq (d_1 + 1)^{1/2} \cdots (d_n + 1)^{1/2} |f|.$$

The same proof as for one variable gives the inequality for several variables, namely:

Proposition 2.10. $M(f) \leqq L_2(f)$.

By induction on n, we find

$$|a_{j_1 \ldots j_n}| \leqq \binom{d_1}{j_1} \cdots \binom{d_n}{j_n} M(f).$$

Hence we get:

Proposition 2.11. $(d + 1)^{-n/2} M(f) \leqq |f| \leqq 2^{nd} M(f)$.

Here we majorized each d_j by d. We then obtain the same application to the product of polynomials:

Proposition 2.12. Let $f_1, f_2 \in \mathbf{C}[X_1, \ldots, X_n]$ with $f_1 f_2$ of total degree $\leqq d$. Then

$$(1 + d)^{-n} |f_1 f_2| \leqq |f_1||f_2| \leqq e^{nd} |f_1 f_2|.$$

Besides Gelfond's book for such inequalities, see (among many other references) K. Mahler, On some inequalities for polynomials in several variables, *J. London Math. Soc.* **37** (1962), pp. 341–344.

§3. Heights in Function Fields

We shall consider the height function from a geometric point of view. This point of view was already used by Néron in his proof of the theorem of the base [Ne 1], [Ne 2], and was taken up again in Lang–Néron [L–N].

Let W be a projective variety in $\mathbf{P}^r$, non-singular in codimension 1, and defined over a field k. Let c be a number, $0 < c < 1$. Let $M_K = M_{k(W)}$ be the set of discrete absolute values of the function field $k(W)$ obtained from the prime rational divisors of W over k. We then have by definition

$$|x|_\mathfrak{p} = c^{(\text{ord}_\mathfrak{p} x)\,\deg(\mathfrak{p})}$$

for each such prime divisor $\mathfrak{p}$, and our set M_K satisfies the product formula. If P is a point in $\mathbf{P}^n$ rational over $k(W)$ with coordinates $(y_0, \ldots, y_n)$ in $k(W)$, then

$$H_{k(W)}(P) = H_W(P) = \prod_\mathfrak{p} \sup_i |y_i|_\mathfrak{p}.$$

Proposition 3.1. *Let* $d = \deg \sup_i(y_i)_\infty$ *be the projective degree in* $\mathbf{P}^r$ *of the sup of the polar divisors of the* y_j. *Then*

$$H_W(P) = (1/c)^d.$$

Proof. Without loss of generality, we may assume that one of the y_i is equal to 1. In our product defining the height, we will have a contribution $\neq 1$ only if $\mathfrak{p}$ is a pole of one of the y_i. For such a pole $\mathfrak{p}$ we have

$$\sup_i c^{(\text{ord}_\mathfrak{p} y_i)\deg(\mathfrak{p})} = \sup_i (1/c)^{(-\text{ord}_\mathfrak{p} y_i)\deg(\mathfrak{p})}$$
$$= (1/c)^{\sup_i(-\text{ord}_\mathfrak{p} y_i)\deg(\mathfrak{p})}.$$

Our assertion is now obvious.

We define the **logarithmic height**, or simply **height** if we work with it throughout a discussion to be

$$\boxed{h_W(P) = \deg \sup_i(y_i)_\infty.}$$

It can also be interpreted in the following manner.

Proposition 3.2. *Let W be a projective variety in $\mathbf{P}^m$, non-singular in codimension 1, defined over the field k, and let P be a point in projective space $\mathbf{P}^n$, rational over $k(W)$. Let $f: W \to \mathbf{P}^n$ be the rational map defined over k, determined by P. Then*

$$h_W(P) = \deg f^{-1}(L)$$

for any hyperplane L of $\mathbf{P}^n$, such that $f^{-1}(L)$ is defined, the degree being that in the given projective embedding of W in $\mathbf{P}^m$.

Proof. Without loss of generality, we may assume none of the coordinates $(y_0, \ldots, y_n)$ of P is 0. Let $(Y_0, \ldots, Y_n)$ be the variables of $\mathbf{P}^n$. Each $Y_i/L(Y)$ determines a function φ_i on $\mathbf{P}^n$, i.e. a rational map $\varphi_i: \mathbf{P}^n \to \mathbf{P}^1$. We have two successive maps:

$$W \xrightarrow{f} \mathbf{P}^n \xrightarrow{\varphi_i} \mathbf{P}^1$$

and if $L(y) \neq 0$ (i.e. $f^{-1}(L)$ is defined) then $y_i/L(y)$ determines the composite function $\varphi_i \circ f = \psi_i$ of W into $\mathbf{P}^1$. By AV, Corollary 1 of Proposition 2, Appendix, §1, we get

$$(\psi_i) = f^{-1}(L_i) - f^{-1}(L),$$

where L_i is the hyperplane $Y_i = 0$ of $\mathbf{P}^n$. Since the L_i are without common point, the divisors $f^{-1}(L_i)$ are without common component, and hence

$$\sup_i(\psi_i)_\infty = f^{-1}(L).$$

This proves our proposition. .

We observe that our logarithmic height, and thus the height, are geometric, i.e. essentially do not depend on the field of definition of W. In particular, our point P being also rational over $k^a(W)$, we have

$$h_{k^a(W)}(P) = h_{k(W)}(P).$$

Let W be as before, and let P be a point in $\mathbf{P}^n$ rational over $k(W)$. Then P defines a generically surjective rational map

$$g: W \to T$$

of W, such that if $j: T \to \mathbf{P}^n$ is the inclusion, then $j \circ g = f$. If x is a generic point of W over k, we have $f(x) = g(x)$. This is a generic point of T over k.

Since T is contained in $\mathbf{P}^n$, it has a degree in $\mathbf{P}^n$ which we shall also write deg T or $\deg_{\mathbf{P}^n} T$ if there is any chance of confusion with respect to the embedding. We wish to give a rough comparison of $h_W(P)$ and deg T. To do this, we need some lemmas.

We start therefore with a generically surjective rational map $g: W \to T$ defined over k. Let dim $T = s$, and let $L_{u_1}, \ldots, L_{u_s}$ be generic independent hyperplanes of $\mathbf{P}^n$ over k. This means that $u_1, \ldots, u_s$ are independent generic over k (cf. IAG, Chapter VIII, §6). We also write L_i instead of L_{u_i}. Then $T \cdot L_1 \cdots L_s$ is a cycle on $\mathbf{P}^n$ consisting of deg T points, which are all generic points of T over k, by IAG, Proposition 10 of Chapter VIII, §6. Furthermore, each one of these points is separable over $k(u_1, \ldots, u_s)$, say by the criterion for multiplicity 1.

If y is a generic point of T over k and $y = f(x) = g(x)$, then $g^{-1}(y)$ is the prime rational cycle on W equal to the locus of x over $k(y)$, taken with multiplicity 1 by F^1-VII_6 Th. 12, i.e. it is the "generic fiber". (It is not necessarily a variety, i.e. absolutely irreducible.)

Lemma 3.3. *Let W be a projective variety in $\mathbf{P}^m$. Let $g: W \to T$ be a generically surjective rational map defined over k. Let y be a generic point of T over k, and let $U = g^{-1}(y)$. Let $j: T \to \mathbf{P}^n$ be an inclusion, and let $L_1 \cdots L_s$ be generic independent hyperplanes of $\mathbf{P}^n$ over k. Then*

$$(\deg T)(\deg U) = \deg g^{-1}(T \cdot L_1 \cdots L_s).$$

Proof. Let $T \cdot L_1 \cdots L_s = \sum (y_i)$. Then deg T is the number of these points (y_i), and the cycle on the right is precisely $\sum U_i = \sum g^{-1}(y_i)$. The U_i are all conjugate over k (i.e. each one can be transformed into the other by an automorphism of the universal domain) and hence have the same degree in $\mathbf{P}^m$, namely the degree of U. Our assertion is therefore clear.

Lemma 3.4. *Let the notation be as in the previous lemma, and $f = j \circ g$. Suppose that y is one of the points of $T \cdot L_1 \cdots L_s$. Then each component of $U = g^{-1}(y)$ is a proper component of*

$$f^{-1}(L_1) \cap \cdots \cap f^{-1}(L_s)$$

and we have

$$g^{-1}(T \cdot L_1 \cdots L_s) \leqq f^{-1}(L_1) \cdots f^{-1}(L_s).$$

Proof. Let U_0 be a component of U and let U_0' be a component of $f^{-1}(L_1) \cap \cdots \cap f^{-1}(L_s)$ containing U_0 (which is itself contained in their intersection). Observe that f and g are both defined along U_0 (since a generic point for U_0 is also one for U over $k(y)$ and is a generic point for W over k). Hence f and g are defined along U_0'. Let x' be a generic point of U_0' over the algebraic closure of $k(u_1, \ldots, u_s)$ (notation as in the beginning of the discussion). Then $f(x')$ lies in $T \cap L_1 \cap \cdots \cap L_s$ and hence is one of the points y of $T \cdot L_1 \cdots L_s$. Since $f(x') = g(x')$, it follows that x' lies in supp (U), and thus is one of the components of U. Hence U_0' itself is contained in this component, which must therefore be U_0, and we get $U_0' = U_0$.

To prove the second assertion, note that

$$f^{-1}(L_1) \cdots f^{-1}(L_s)$$

is rational over $k(u_1, \ldots, u_s)$ and as we mentioned before, each point y in $T \cdot L_1 \cdots L_s$ is separable over this field. Since $g^{-1}(y)$ is prime rational over $k(u_1, \ldots, u_s)(y)$ (which is the composite of $k(y)$ and a purely transcendental extension of k, free from $k(y)$ over k), it follows that

$$g^{-1}(y) \leq f^{-1}(L_1) \cdots f^{-1}(L_s).$$

If y_1, y_2 are two distinct points y as above, then $g^{-1}(y_1)$ has no component in common with $g^{-1}(y_2)$. Hence we get

$$g^{-1}(T \cdot L_1 \cdots L_s) \leq f^{-1}(L_1) \cdots f^{-1}(L_s)$$

as desired.

Lemma 3.5. *Let W be a projective variety in a projective space $\mathbf{P}^m$. Let $X_1, \ldots, X_s$ be positive divisors on W. Then*

$$\deg(X_1 \cdots X_s) \leq (\deg X_1) \cdots (\deg X_s).$$

Proof. Let W and the X_i be rational over an algebraically closed field k. We take a generic linear direction over k as in $\text{F}^2\text{-IX}_3$ such that the join of this linear variety with a divisor on V is a divisor on $\mathbf{P}^m$. We call this join also the projecting cone over the divisor, as usual. We recall the following facts. If A is a subvariety of W defined over k, and $\tilde{A}$ the projecting cone over A, then $\deg A = \deg \tilde{A}$. If A and B are subvarieties of W defined over k, and $A \neq B$, then $\tilde{A} \neq \tilde{B}$ by $\text{F}^2\text{-IX}_3$, Proposition 6, Corollary 3. If A is a proper component of $X_1 \cdots X_s$ of multiplicity m, then $\tilde{A}$ is also a proper component of $\tilde{X}_1 \cdots \tilde{X}_s$ of multiplicity m, by the corollary to Proposition 7 of $\text{F}^2\text{-IX}_3$, combined with $\text{F}^2\text{-VIII}_4$, Theorems 10 and 11. The intersection of the X_i is of course taken on $\mathbf{P}^m$. From these facts, it follows that

$$\deg(\tilde{X}_1 \cdots \tilde{X}_s) \leq \deg(\tilde{X}_1 \cdots \tilde{X}_s)$$

and this reduces the proof to the case where W is equal to all of projective space.

We then proceed by induction on s. By definition, we have

$$(X_1 \cdots X_s) = (X_1 \cdots X_{s-1}) \cdot X_s.$$

Thus it will suffice to prove our assertion for two factors, say X and Y, one of which is a divisor in $\mathbf{P}^m$. If every component of the intersection $X \cap Y$ is proper, we can apply Bezout's theorem. Otherwise, we proceed by induction on the sum of the multiplicities of the components of X and Y. If both X and Y are varieties with X of codimension 1 in $\mathbf{P}^m$, then either $X \supset Y$ in which case $X \cdot Y = 0$, or $X \not\supset Y$ and Bezout's theorem applies. In any case, our assertion is obvious. Say now $X = X_1 + A$ where A is a variety, and $X_1 > 0$. Then

$$X \cdot T = (X_1 + A) \cdot Y = X_1 \cdot Y + A \cdot Y.$$

By induction, we get

$$\deg(X_1 \cdot Y) + \deg(A \cdot Y) \leqq (\deg X_1)(\deg Y) + (\deg A)(\deg Y),$$

which proves the lemma.

Theorem 3.6 (Néron). *Let W^r be a projective variety, non-singular in codimension 1 and defined over a field k. Let P be a point in $\mathbf{P}^n$ rational over $k(W)$, and let T be the locus of P over k, so that we have a generically surjective rational map $g: W \rightarrow T$. Then*

$$\deg_{\mathbf{P}^n} T \leqq h_W(P)^r.$$

Proof. This follows immediately from the preceding lemmas, using $X_i = f^{-1}(L_i)$ and applying Proposition 3.2.

In particular, we see that if P ranges over a set of points of bounded heights, then T has bounded degree in $\mathbf{P}^n$. Cf. Néron [Ne 2].

§4. Heights on Abelian Groups

This section is entirely different from the preceding three. In some applications, e.g. to abelian varieties, the height function becomes related to the group law in certain ways, and it is convenient to make an abstraction of this situation. The results proved below are purely combinatorial. We shall axiomatize certain properties of a function which are weaker than those satisfied by the height.

Let Γ be an abelian group. Let d be a real number > 0. Let $h: \Gamma \to \mathbf{R}$ be a real-valued function on Γ. We shall say that h is of **quasi-degree** d if $h(P) \geq 0$ for all $P \in \Gamma$ and if it satisfies the following condition:

Given an integer $m > 0$, and ε real > 0, and an element $a \in \Gamma$, there exists a number $c > 0$ such that for all $P \in \Gamma$ we have

$$-c + (m^d - \varepsilon)h(P) \leq h(mP + a) \leq c + (m^d + \varepsilon)h(P).$$

We shall say that h is **non-degenerate** if given $B > 0$ there exists only a finite number of elements $P \in \Gamma$ such that $h(P) \leq B$.

The arguments used to prove the next proposition are known as the **infinite descent**.

Proposition 4.1. *Let Γ be an abelian group and h a real valued function on Γ of quasi-degree $d > 0$. Let m be an integer > 1 such that $\Gamma/m\Gamma$ is finite, and let $a_1, \ldots, a_s$ be representatives in Γ of $\Gamma/m\Gamma$. There exists a number c_1 and a subset $\mathfrak{S}$ of Γ such that:*

(i)   *$h(P) \leq c_1$ for all $P \in \mathfrak{S}$.*
(ii) *For any $P_0 \in \Gamma$, there exist integers $n_0, n_1, \ldots, n_s$ and a point P in $\mathfrak{S}$ such that*

$$P_0 = n_0 P + n_1 a_1 + \cdots + n_s a_s.$$

Proof. In the definition of quasi-degree, the number c depends on m and a, so let us write $c(m, a)$ for it. In the present application, we let

$$c = \sup_i c(m, a_i).$$

Consider a sequence $(P_0, P_1, \ldots)$ of points of Γ constructed by starting with our point P_0, and such that

$$mP_{\nu+1} = P_\nu - a_{i_\nu}.$$

By definition, there is a number $\lambda > 1$ (in fact, $\lambda = m^d - \varepsilon$) such that

$$h(P_{\nu+1}) \leq h(P_\nu)/\lambda + c/\lambda.$$

Proceeding inductively, we get

$$h(P_\nu) \leq 1/\lambda^\nu \cdot h(P_0) + c(1 + 1/\lambda + 1/\lambda^2 + \cdots)$$

and from this our assertion is obvious.

Corollary 4.2. *Let Γ be an abelian group and h a real-valued function on Γ of quasi-degree $d > 0$, and non-degenerate. Let m be an integer > 1 such that $\Gamma/m\Gamma$ is finite. Then Γ is finitely generated.*

Proof. Immediate consequence of the definition of non-degeneracy.

The procedure in the proof of Proposition 4.1 can be reversed and leads to an infinite ascent, which we use to estimate the number of elements $P \in \Gamma$ such that $h(P) \leqq B$ as B becomes large, following Néron [Ne 1].

Theorem 4.3. *Let Γ be a finitely generated abelian group, of rank r (maximum number of linearly independent elements over $\mathbf{Z}$). Let h be a real-valued function on Γ, which is of quasi-degree $d > 1$ and non-degenerate. Given $\varepsilon > 0$ there exist two numbers $c_1(\varepsilon), c_2(\varepsilon) > 0$ and a number δ with $0 < \delta < \varepsilon$ such that for all $B > B_0(\varepsilon)$, the number $N(B)$ of elements of Γ such that $h(P) \leqq B$ satisfies the inequalities*

$$c_1(\varepsilon)B^{r/d-\delta} \leqq N(B) \leqq c_2(\varepsilon)B^{r/d+\delta}.$$

Proof. We can write Γ as a direct sum of a free abelian group and a finite torsion group. From the definitions, and the approximate nature of our estimates, we may assume without loss of generality that Γ is free over $\mathbf{Z}$.

Let m be an integer $\geqq 2$ and $x_1, \ldots, x_r$ a basis for Γ over $\mathbf{Z}$. Let $\{a_j\}$ range over those elements of Γ which can be written in the form

$$n_1 x_1 + \cdots + n_r x_r$$

with $-(m - 1) \leqq n_i \leqq (m - 1)$. By σ_v we denote a sequence

$$(P_0, P_1, \ldots, P_v)$$

of *distinct* points of Γ such that

$$P_0 = a_{j_0},$$
$$P_1 = mP_0 + a_{j_1},$$
$$\ldots$$
$$P_v = mP_{v-1} + a_{j_v}.$$

Taking into account the m-adic expansion of an integer, we see that the number of elements of Γ which can occur in some sequence σ_v is precisely $(2m^v - 1)^r$ because their coefficients in $\mathbf{Z}$, in terms of our basis, range from $-(m^v - 1)$ to $+(m^v - 1)$.

Since h is non-degenerate, given $\varepsilon > 0$ there exists an integer $v_0 = v_0(m, \varepsilon)$ such that for all $v \geq v_0$ and any point P_v in a sequence σ_v we have

$$(m^d - \varepsilon)h(P_{v-1}) \leq h(P_v) \leq (m^d + \varepsilon)h(P_{v-1}).$$

Inductively, we get

(1) $$(m^d - \varepsilon)^{v - v_0}h(P_{v_0}) \leq h(P_v) \leq (m^d + \varepsilon)^{v - v_0}h(P_{v_0}).$$

Put $c_0(m, \varepsilon) = \log h(P_{v_0})/\log(m^d + \varepsilon)$.
  If $v \geq v_0$ and if

(2) $$v \leq \frac{\log B}{\log(m^d + \varepsilon)} + v_0 + c_0(m, \varepsilon),$$

then by an obvious logarithm, we see from the right inequality of (1) that

$$h(P_v) \leq B.$$

Consequently, for such v, we get a lower bound for $N(B)$, namely $(2m^v - 1)^r$ which we replace for convenience by $m^{vr}/2^r$. We see that for all $B \geq B_0(m, \varepsilon)$ we get

$$c_1(m, \varepsilon)B^{r\lambda} \leq N(B),$$

where $\lambda = \log m/\log(m^d + \varepsilon)$.
  On the other hand, if $v \geq v_0$ and

(3) $$v \geq \frac{\log B}{\log(m^d - \varepsilon)} - c_0'(m, \varepsilon) + v_0$$

with an obvious number $c_0'(m, \varepsilon)$, then by the left inequality of (1) we see that

$$h(P_v) > B.$$

Consequently, all the P with $h(P) \leq \beta$ must lie in sequences σ_v for v satisfying

(4) $$v \leq \frac{\log B}{\log(m^d - \varepsilon)} - c_0'(m, \varepsilon) + v_0.$$

This gives us an upper bound for $N(B)$ using $2^r m^{vr}$ as an upper bound for the number of elements in such sequences, namely,

$$N(B) \leq c_2(m, \varepsilon)B^{r\lambda'}$$

where $\lambda' = \log m/\log(m^d - \varepsilon)$. Since m, ε were fixed at the beginning of our proof, we can now take m large and ε small, so that the assertion in our theorem is clear.

In fact, in the application to abelian varieties, one gets a more precise result which will be stated in Chapter 5, §6.

§5. Counting Points of Bounded Height

It is interesting to give an estimate for the number of elements of a number field K of height $\leq B$ for $B \to \infty$. This question can be asked of elements of K, of integers in $\mathfrak{o}_K$, and of units. In each case, the method of proof consists of determining the number of lattice points in a homogeneously expanding domain which has a sufficiently smooth boundary, and using the basic fact:

Theorem 5.1. *Let D be a subset of $\mathbf{R}^n$ and L a lattice in $\mathbf{R}^n$, with fundamental domain F. Assume that the boundary of D is $(n-1)$-Lipschitz parametrizable. Let $N(t) = N(t, D, L)$ be the number of lattice points in tD for t real > 0. Then*

$$N(t) = \frac{\mathrm{Vol}(D)}{\mathrm{Vol}(F)} t^n + O(t^{n-1}),$$

where the constant in O depends on L, n and the Lipschitz constants.

For the proof, see [La 5], Chapter VI, §2. The expression "**$(n-1)$-Lipschitz parametrizable**" means that there exists a finite number of mappings

$$\rho: [0, 1]^{n-1} \to \text{Boundary of } D,$$

whose images cover the boundary of D, and such that each mapping satisfies a Lipschitz condition. In practice, such mappings exist which are even of class C^1 (that is, with continuous derivatives).

Theorem 5.2. *Let K be a number field. Let $r = r_1 + r_2 - 1$ where r_1 is the number of real absolute values and r_2 the number of complex ones.*

(i)   *The number of algebraic integers $x \in \mathfrak{o}_K$ with height $H_K(x) \leq B$ is*

$$\gamma_0 B(\log B)^r + O(B(\log B)^{r-1})$$

for some constant γ_0.

(ii)   *The number of units $u \in \mathfrak{o}_K^*$ with $H_K(u) \leq B$ is*

$$\gamma_0^*(\log B)^r + O(\log B)^{r-1}.$$

for some constant γ_0^*.

Proofs of both parts of Theorem 5.2 come from a straightforward application of Theorem 5.1. On the other hand, Schanuel [Sch] has determined
the somewhat harder asymptotic behavior of field elements of bounded
height in projective space as follows.

Theorem 5.3. *Let $N_K(B)$ be the number of elements in $\mathbf{P}^{n-1}(K)$ with
$H_K(x) \leqq B$. Let $[K : \mathbf{Q}] = d$. Then:*

$$N_K(B) = \frac{hR/w}{\zeta_K(n)}\, \gamma_{K,n} B^n + \begin{cases} O(B \log B) & \text{if } d = 1, n = 2, \\ O(B^{n-1/d}) & \text{otherwise.} \end{cases}$$

*As usual, h is the class number; R is the regulator; w is the number of roots
of unity in K; and*

$$\gamma_{K,n} = \left(\frac{2^{r_1}(2\pi)^{r_2}}{d_K^{1/2}}\right)^n n^r,$$

where d_K is the absolute value of the discriminant.

Schanuel's proof is more complicated because one has to determine a
fundamental domain modulo the action of the units, and because of the higher
dimensionality. However, we observe that the result over the rational numbers with $d = 1$ and $n = 2$ is classical. It amounts to determining the number
of relatively prime pairs of integers of absolute value $\leqq B$, which is easily
determined to be

$$\frac{6}{\pi^2} B^2 + O(B \log B).$$

For the convenience of the reader, I sketch Schanuel's proof. Let
$x = (x_0, \ldots, x_{n-1}) \in \mathbf{P}^{n-1}(K)$ and let $\mathfrak{a}_x$ be the fractional ideal generated by
the coordinates, assumed to be in K. Multiplying x by a non-zero element
of K changes the ideal by this element, but does not change the ideal class
$\mathrm{Cl}(\mathfrak{a}_x)$. Then Theorem 5.3 is reduced to the corresponding theorem for each
class.

Theorem 5.4. *Let c be an ideal class of K, and N(c, B) the number of
points $x \in \mathbf{P}^{n-1}(K)$ such that $\mathrm{Cl}(\mathfrak{a}_x) = c$ and $H_K(x) \leqq B$. Then*

$$N(c, B) = \frac{R/w}{\zeta_K(n)}\, \gamma_n B^n + \textit{same error terms as in Theorem 5.3,}$$

where $\gamma_n = \gamma_{K,n}$.

To handle each class, let $\mathfrak{a}$ be an ideal of $\mathfrak{o}(K)$. Define two points x and y in K^n to be **equivalent** if there is a unit $u \in \mathfrak{o}^*$ such that

$$x = uy = (uy_0, \dots, uy_{n-1}).$$

Recall from §1 that if $\mathfrak{a}_x = \mathfrak{a}$ then

$$H_K(x) = \mathbf{N}\mathfrak{a}^{-1}H_{K,\infty}(x) = \mathbf{N}\mathfrak{a}^{-1}\prod_{v \in M_K^\infty}\sup\|x_i\|_v.$$

Let $N(\mathfrak{a}, B) =$ number of elements $(x_1, \dots, x_n)$ with $x_i \in \mathfrak{a}$ for all i, modulo the above equivalence relation, such that

$$H_{K,\infty}(x) \leqq B\mathbf{N}\mathfrak{a}.$$

One first proves:

Theorem 5.5.

$$N(\mathfrak{a}, B) = \frac{R}{w}\gamma_n B^n + O(B^{n-1/d}).$$

Proof. In this formulation, the theorem becomes a direct application of the counting principle for a lattice in a homogeneously expanding domain, as follows. Let

$$K_\infty = \mathbf{R} \otimes K = \prod_{v \in M_\infty} K_v.$$

Let Δ be a fundamental domain for K_∞^* modulo the units. Consider the mapping

$$\eta : K_\infty^n \to \prod_{v \in M_\infty} K_v$$

given by

$$\eta(x) = \left(\dots, \sup_i \|x_i\|_v^{1/d}, \dots\right).$$

Let $D(B)$ be the region in K_∞^n consisting of all points $x = (x_1, \dots, x_n)$ such that

$$\eta(x) \in \Delta \quad \text{and} \quad H_\infty(x) \leqq B\mathbf{N}\mathfrak{a}.$$

Then the evaluation of $N(\mathfrak{a}, B)$ amounts to counting the number of points of $\mathfrak{a}^n$ (which is a lattice in K_∞^n) in the region $D(B)$. But for real positive t, we have

$$H_{K, \infty}(tx) = t^d H_{K, \infty}(x),$$

and hence we get at once that

$$D(B) = BD(1).$$

Thus $D(B)$ is a homogeneously expanding domain. The volume of the fundamental domain Δ is evaluated classically, for instance by Hecke (cf. also [L 5], Chapter V, §3 and Chapter VI, §3.) One must prove that the boundary of $D(1)$ is C^1-parametrizable by a finite number of cubes of the appropriate dimension. The details are in Schanuel, as well as the computation of the volume which generalizes Hecke's classical computation. This concludes the proof.

Note that the above computations are generalizations of the computations giving the number of ideals with bounded norm, and also of the theorem giving the number of field elements in a parallelotope recalled as Theorem 6.7 of Chapter 2.

Now we have $N(c, B) = N^*(\mathfrak{a}, B)$ where

$N^*(\mathfrak{a}, B)$ = number of $x \in \mathfrak{a}^n$ such that $\mathfrak{a}_x = \mathfrak{a}$ and $H_{K, \infty}(x) \leqq B N \mathfrak{a}$.

But for $x \in \mathfrak{a}^n$ we have $\mathfrak{a}_x = \mathfrak{a}\mathfrak{b}$ for some ideal $\mathfrak{b}$. Therefore

$$N(\mathfrak{a}, B) = \sum_{\mathfrak{b}} N^*(\mathfrak{a}\mathfrak{b}, B N \mathfrak{b}^{-1}).$$

By the Moebius inversion formula, we get formally

$$N^*(\mathfrak{a}, B) = \sum_{\mathfrak{b}} \mu(\mathfrak{b}) N(\mathfrak{a}\mathfrak{b}, B N \mathfrak{b}^{-1})$$

$$\sim \sum_{\mathfrak{b}} \mu(\mathfrak{b}) \frac{R}{w} \gamma_n B^n N \mathfrak{b}^{-n}$$

$$\sim \frac{R}{w} \gamma_n B^n \sum_{\mathfrak{b}} \mu(\mathfrak{b}) N \mathfrak{b}^{-n},$$

whence Theorem 5.4 follows since

$$\zeta_K(s)^{-1} = \sum_{\mathfrak{b}} \mu(\mathfrak{b}) N \mathfrak{b}^{-s}.$$

We now redo the formalism taking into account the error terms. We have

$$N^*(\mathfrak{a}, B) = \sum_{\mathrm{N}\mathfrak{b} \leq B} + \sum_{\mathrm{N}\mathfrak{b} > B} \mu(\mathfrak{b}) N(\mathfrak{a}\mathfrak{b}, B\mathrm{N}\mathfrak{b}^{-1})$$

$$= \sum_{\mathrm{N}\mathfrak{b} \leq B} \mu(\mathfrak{b}) \frac{R}{w} \gamma_n B^n \mathrm{N}\mathfrak{b}^{-n} + \sum_{\mathrm{N}\mathfrak{b} \leq B} O(B\mathrm{N}\mathfrak{b}^{-1})^{n-1/v}$$

$$+ \sum_{\mathrm{N}\mathfrak{b} > B} N(\mathfrak{a}\mathfrak{b}, B\mathrm{N}\mathfrak{b}^{-1}).$$

There is only a finite number of algebraic integers in K of bounded height, and so the third sum is actually finite. It is bounded by $N(\mathfrak{o}, 1)$ times the number of ideals $\mathfrak{b}$ with $B < \mathrm{N}\mathfrak{b} \leq CB$ with some constant C depending only on K. The number of such ideals is easily seen to be $O(B)$, which is within the desired error term.

The first sum differs from the main term by a tail end, which is bounded by

$$B^n \sum_{\mathrm{N}\mathfrak{b} \leq B} \mathrm{N}\mathfrak{b}^{-n} = B^n \sum_{k=B}^{\infty} k^{-n} g(k),$$

where $g(k)$ is the number of ideals $\mathfrak{b}$ such that $\mathrm{N}\mathfrak{b} = k$. Note that

$$g(k) = G(k) - G(k-1),$$

where $G(k)$ is the number of ideals $\mathfrak{b}$ with $\mathrm{N}\mathfrak{b} \leq k$. But a trivial estimate reducing to the rational integers shows that $G(k) = O(k)$. Summing by parts, one sees that

$$\sum_{k=B}^{\infty} k^{-n} g(k) = O(B^{1-n}),$$

so the missing tail end of the first sum is $O(B)$.

Finally the middle term is bounded by a constant times

$$B^{n-1/d} \sum_{\mathrm{N}\mathfrak{b} \leq B} \mathrm{N}\mathfrak{b}^{-(n-1/d)}.$$

In the case $n = 2, d = 1$ we have

$$\sum_{\mathrm{N}\mathfrak{b} \leq B} \mathrm{N}\mathfrak{b}^{-1} = O(\log B).$$

In every other case, $n - 1/d \geq \frac{3}{2}$ and the sum $\sum \mathrm{N}\mathfrak{b}^{-(n-1/d)}$ is uniformly bounded from above by $\zeta_K(n - 1/d)$. This concludes the proof.

Schanuel obtains more general results in two ways. First he carries out the computations with S-units; and second by imposing bounds at the archimedean absolute values more general than those of the height, he can give a theorem which includes both the counting of points in projective space and the number of points in a parallelotope. Cf. the last lines of his paper, leaving an open problem in this direction.

CHAPTER 4

Geometric Properties of Heights

*Throughout this chapter, **F** is a field with a proper set of absolute values $M_\mathbf{F}$ satisfying the product formula. We denote by K some finite extension of **F**, so that the set M_K satisfies the product formula with multiplicities $N_v, v \in M_K$.*

If V is a variety defined over K, then for each morphism of V into a projective space also defined over K, we can define the absolute height of a point of V algebraic over K, and its relative height if it is rational over K.

We shall assume that V is complete and normal. Maps into projective spaces are then given by linear systems, and it will be our main purpose to give conditions under which the corresponding height functions are equivalent or quasi-equivalent (for the definition, see §1). We shall formulate our theorems for the absolute height H, and it will then be obvious that they remain valid for the relative height H_K which is a fixed power of the absolute one (namely the $[K:\mathbf{F}]$-th power). In the applications, we need only the equivalence criteria for the relative height.

§1. Functorial Properties

Let V be a variety. Two real-valued functions λ and λ' on a set of points of V will be said to be **equivalent** if there exist two numbers $c_1, c_2 > 0$ such that for all points P in our set, we have

$$c_1 \lambda(P) \leqq \lambda'(P) \leqq c_2 \lambda(P).$$

This is a notion of "multiplicative" equivalence, and we use the sign

$$\lambda \gg \ll \lambda'$$

to denote this equivalence.

The above notion of equivalence is useful when we deal with the multiplicative height H. To deal with the logarithmic height h, we define two functions to be (additively) **equivalent** if their difference is a bounded function.

We use the sign $\sim$ to denote this (additive) equivalence, that is we write $h_1 \sim h_2$ to mean that $h_1 - h_2$ is a bounded function.

Let $\mathbf{F}$ be a field with a proper set of absolute values $M_\mathbf{F}$ satisfying the product formula. Then for each finite extension K of $\mathbf{F}$, M_K satisfies the product formula with multiplicities N_v. Let V be a variety defined over K. Let $\varphi: V \to \mathbf{P}^m$ be a morphism of V into projective space, defined over K. Then for each point P of V_K, $\varphi(P)$ is a point of $\mathbf{P}^m$, rational over K, and we can thus define its height which we shall denote by $H_{K,\varphi}(P)$. If P is algebraic over $\mathbf{F}$, then there exists a finite extension K of $\mathbf{F}$ over which it is rational, and we can then define its **absolute height**

$$H_\varphi(P) = H(\varphi(P)).$$

Thus $H_{K,\varphi}$ is a function on V_K while H_φ is a function on

$$V(K^a) = V(\mathbf{F}^a),$$

i.e. on the set of points on V rational over the algebraic closure of K.

Let $f: U \to V$ and $g: V \to \mathbf{P}^n$ be morphisms. Then trivially

$$H_{g \circ f} = H_g \circ f \qquad \text{and} \qquad h_{g \circ f} = h_g \circ f.$$

We shall now give some elementary inequalities for the height under linear projections and simple morphisms.

Linear Projections

Let V be projective, defined over K. Suppose its points are represented by homogeneous coordinates $(x_0, \ldots, x_n)$. Let (a_{ij}) ($i = 0, \ldots, m$ and $j = 0, \ldots, n$) be a matrix with coefficients in K, and put

$$y_i = a_{i0} x_0 + \cdots + a_{in} x_n.$$

Then the map $(x) \mapsto (y)$ defines a rational map $\varphi: V \to \mathbf{P}^m$. If P is a point with coordinates (x) such that not all y_i are equal to 0 in the above formula, then φ is defined at P. A map φ obtained in the manner just described is called a **linear projection**, defined over K.

Proposition 1.1. *Let V be a projective variety defined over K. Let $\varphi: V \to \mathbf{P}^m$ be a linear projection also defined over K. There exists a number $c > 0$ depending only on φ, such that if P is a point of V, algebraic over K, such that not all the coordinates y_i above are 0, then*

$$H_\varphi(P) \leqq cH(P)$$

(H being the height in the given projective embedding of V). Or, for the logarithmic height,

$$h_\varphi \leq h + c_1$$

with some constant c_1.

Proof. Let S_K be the subset of M_K containing all those absolute values v for which some $|a_{ij}|_v$ is not 1, and all archimedean absolute values. Then S_K is a finite set. If our coordinates are in a finite extension E of K, then for any $w \in M_E$ extending some $v \in M_K$, we have

$$\sup_i |y_i|_w \leq c_v \sup_j |x_j|_w,$$

where $c_v = (n + 1) \sup_{ij}[|a_{ij}|_v, 1]$ for all $v \in S_K$, and $c_v = 1$ for all $v \notin S_K$. From this we get

$$H_\varphi(P) \leq cH(P),$$

where

$$c = \left[\prod_{v \in M_K} \prod_{w|v} c_v^{N_w} \right]^{1/[E:F]} \leq \prod c_v.$$

From our proposition, we see in particular that if V is again a complete variety, and if φ, φ' are two morphisms into $\mathbf{P}^m$ which differ by a projective transformation defined over a finite extension K of $\mathbf{F}$, then H_φ is equivalent to $H_{\varphi'}$. In Proposition 2.1, we shall prove a much stronger criterion for equivalence.

In the next propositions we assume that all varieties and morphisms are defined over K.

Proposition 1.2. *If $f: V \to \mathbf{P}^n$ and $g: V \to \mathbf{P}^m$ are morphisms such that f is obtained from g by a linear projective transformation followed by a linear projection, and g is similarly obtained from f, then*

$$H_f \gg \ll H_g \qquad or \qquad h_f \sim h_g.$$

Proof. This follows by symmetry from Proposition 1.1.

Proposition 1.3. *Let L be a linear variety in $\mathbf{P}^n$ and let V be a subvariety of $\mathbf{P}^n$ whose intersection with L is empty. Let $\pi: \mathbf{P}^n \to \mathbf{P}^r$ be a linear projection from L (depending on some choice of linear coordinates in $\mathbf{P}^n$). Then for $x \in V(K^a)$ we have*

$$h(\pi(x)) = h(x) + O(1).$$

Proof. By induction we may suppose that $r = n - 1$ because the linear projection of a variety not meeting L is again a variety since projective space is complete. We may then choose the coordinates $(x_0, \ldots, x_n)$ such that L is defined by $x_n = 0$. If $x \in V(K^a)$ then by assumption $x_n \neq 0$ and hence we may assume without loss of generality that $x_n = 1$. In this case it follows from the definition that $h(\pi(x)) = h(x)$, thus proving the proposition.

Morphisms

Let

$$f: V \to \mathbf{P}^n \qquad \text{and} \qquad g: V \to \mathbf{P}^m$$

be two morphisms of a variety into projective spaces, say given by projective coordinates

$$f = (f_0, \ldots, f_n) \qquad \text{and} \qquad g = (g_0, \ldots, g_m).$$

Then we may form the **join**

$$f \otimes g: V \to \mathbf{P}^{(n+1)(m+1)-1}$$

whose projective coordinates are $f_i g_j$ $(i = 0, \ldots, n; j = 0, \ldots, m)$. It is also a morphism.

Proposition 1.4. *We have the relations*

$$H_{f \otimes g} = H_f H_g \qquad \text{and} \qquad h_{f \otimes g} = h_f + h_g.$$

Proof. This is immediate, because for any absolute value v, and coordinates $(x_0, \ldots, x_n)$, $(y_0, \ldots, y_m)$ in K^a we have

$$\sup_{i, j} \|x_i y_j\|_v = \sup_i \|x_i\|_v \sup_j \|y_j\|_v.$$

An important special case is given by the embedding of $\mathbf{P}^n$ into a higher dimensional projective space by the monomials of a given degree, say d. Indeed, if $f_0, \ldots, f_M$ are these monomials in some selected ordering, then the map

$$x \mapsto (f_0(x), \ldots, f_M(x))$$

is an embedding of $\mathbf{P}^n$ into $\mathbf{P}^M$, called the **embedding of degree** d.

Proposition 1.5. *If* $\varphi: \mathbf{P}^n \to \mathbf{P}^M$ *is the embedding of degree d, then*

$$h_\varphi = d h_{\mathbf{P}^n}$$

Proof. Special case of Proposition 1.4, applied d times.

Lemma 1.6. *Let* $f_0, \ldots, f_m$ *be homogeneous polynomials of degree d with coefficients in K, and in $n + 1$ variables $X_0, \ldots, X_n$. Let S be the set of points $x = (x_0, \ldots, x_n)$ in projective space $\mathbf{P}^n(K^a)$ such that not all polynomials $f_i(x)$ vanish, $i = 0, \ldots, m$. Let*

$$f: S \to \mathbf{P}^m(K^a)$$

be the morphism defined by $x \mapsto (f_0(x), \ldots, f_m(x))$. Then

$$h_f(x) \leqq dh(x) + c_1$$

for some constant c_1 independent of $x \in S$.

Proof. Trivial estimates using the triangle inequality show that for any point $P \in \mathbf{P}^n(K)$, $P \in S$ we have

$$H_K(f(P)) \leqq C_1^{[K:F]} H_K(P)^d.$$

Taking the $[K : F]$ root and the logarithm yield the lemma.

Proposition 1.7. *Let V be a locally Zariski closed algebraic subvariety of $\mathbf{P}^n$, let h be the height on $\mathbf{P}^n$, and let*

$$f: V \to \mathbf{P}^m$$

be a morphism, all defined over K. Then there exist constants $c_1 > 0$ and $c_2 > 0$ such that on $V(K^a)$ we have

$$h_f \leqq c_1 h + c_2.$$

Proof. By the compactness of the Zariski topology, it suffices to prove the proposition locally in a neighborhood of a point. The rational map f is represented by coordinate functions which are rational functions, and since f is a morphism, near a given point we may write these functions in the form

$$(\varphi_0, \ldots, \varphi_m) \quad \text{with } \varphi_0 = 1,$$

such that each φ_j lies in the local ring at that point. But we can write

$$\varphi_j = f_j/f_0, \qquad j = 1, \ldots, m,$$

such that $f_0, \ldots, f_m$ are homogeneous polynomials of the same degree, and $f_0(x) \neq 0$. We can then apply Lemma 1.6 to conclude the proof.

The next result gives a less trivial inequality on the other side in the situation of Lemma 1.6 when the polynomials have only the trivial common zero.

Theorem 1.8. *Let $f : \mathbf{P}^n \to \mathbf{P}^N$ be a morphism of degree d, defined over K. Then*

$$|h \circ f - dh|$$

is bounded on $\mathbf{P}^n(K^a)$, that is $h \circ f \sim dh$.

Proof. One inequality was proved in Lemma 1.6.

The main point of the theorem is to prove an inequality on the other side. Since the polynomials $f_0, \ldots, f_N$ have no common zero except the origin, by the Hilbert Nullstellensatz, there exist polynomials

$$g_{ij} \in K[X_0, \ldots, X_n]$$

and a positive integer m such that

$$X_i^{m+d} = \sum_j g_{ij} f_j.$$

Disregarding the monomials in g_{ij} of degree $\neq m$, we can assume without loss of generality that g_{ij} is homogeneous of degree m. It is also convenient to clear denominators, so we pick an element c in K, integral at all valuations of M_K, such that c is a denominator for all the coefficients of the polynomials g_{ij}. Multiplying by c, we may assume without loss of generality that we have the equation

$$cX_i^{m+d} = \sum_j g_{ij} f_j,$$

where the coefficients of g_{ij} are integral in K. Let

$$x \in \mathbf{P}^n(K), \quad \text{and} \quad x = (x_0, \ldots, x_n) \quad \text{with } x_i \text{ integral in } K.$$

If $v \in M_K$ is non-archimedean, then

$$|c|_v |x_i|_v^{m+d} \leq \max_j |f_j(x)|_v |x_i|_v^m,$$

whence

$$|c|_v^{N_v} \max_i |x_i|_v^{N_v(m+d)} \leq \max_j |f_j(x)|_v^{N_v} \max_i |x_i|_v^{N_v m}.$$

If $v \in M_K$ is archimedean, then

$$|c|_v |x_i|_v^{m+d} \leq (m+1)^n \max_{i,j} C_{ij} |x_i|_v^m \max_j |f_j(x)|_v,$$

where C_{ij} are constants giving bounds for the absolute values at infinity for the coefficients of the polynomials involved. We then obtain

$$|c|_v^{N_v} \max_i |x_i|_v^{N_v(m+d)} \leqq C_2^{n_v} \max_i |x_i|_v^{mN_v} \max_j |f_j(x)|_v^{N_v}.$$

Taking the product yields

$$H_K(x)^{m+d} \leqq C_2^{[K:F]} H_K(x)^m H_K(f(x)).$$

This proves the theorem.

Observe that in multiplicative notation, which is the way the theorem has been proved, it expresses the inequalities in the form

$$\boxed{C_2^{-[K:F]} H_K(x)^d \leqq H_K(f(x)) \leqq C_1^{[K:F]} H_K(x)^d.}$$

The constants C_1, C_2 depend only on f.

One can perform an averaging procedure on any function satisfying the conclusion of the preceding theorem, a result due to Tate.

Proposition 1.9. *Let S be a set and $f: S \to S$ a map of S into itself. Let $h: S \to \mathbf{R}$ be a function. Assume that there exists a positive integer d such that*

$$h \circ f - dh \text{ is bounded.}$$

Then there exists a unique function h_f such that:

(i) $h_f - h$ *is bounded.*
(ii) $h_f \circ f = dh_f$.

Proof. If h'_f and h''_f are two such functions, then $h'_f - h''_f$ is bounded, and it is clear from (ii) that $h'_f - h''_f = 0$, so uniqueness is obvious. As to existence, we let

$$h_f(x) = \lim_{n \to \infty} \frac{1}{d^n} h(f^n(x)).$$

If $|h \circ f - dh| \leqq C$, then

$$\left| \frac{1}{d^{n+1}} h \circ f^{n+1}(x) - \frac{1}{d^n} h \circ f^n(x) \right| \leqq C/d^{n+1}.$$

Hence

$$\left| \frac{1}{d^{n+k}} h \circ f^{n+k}(x) - \frac{1}{d^n} h \circ f^n(x) \right| \leq \frac{C}{d^n} \frac{1}{1 - 1/d}.$$

Hence the sequence $\left\{ \frac{1}{d^n} h \circ f^n \right\}$ is Cauchy, its limit exists and trivially satisfies (i) and (ii), as was to be shown.

The proposition may be applied to a morphism of projective space into itself, as in Theorem 1.8 with $n' = n$.

§2. Heights and Linear Systems

Let V be a complete normal variety. Mappings

$$f: V \to \mathbf{P}^n$$

are obtained by means of linear systems (cf. IAG, Chapter VI or Weil's *Foundations*, Chapter IX). We recall briefly how this is done. We use the sign $X \sim Y$ to denote that a divisor X is linearly equivalent to Y.

Let X_0 be a divisor on V (in the sense of Severi, that is a formal linear combination of subvarieties of codimension 1). By $L(X_0)$ we mean the vector space of functions f on V such that

$$(f) \geqq -X_0,$$

where (f) is the divisor associated with f. If $f \neq 0$, then

$$(f) = X - X_0,$$

where X is a positive divisor. The set of all positive divisors X obtained in this fashion, that is linearly equivalent to X_0, is called a **linear system**, and is denoted by $\mathscr{L}$, or $\mathscr{L}(X_0)$.

Suppose X_0 is a positive divisor. Let

$$(f_0, \ldots, f_m)$$

be a family of generators of $L(X_0)$ over the constant field (it is easy to prove that $L(X_0)$ is finite dimensional). Then

$$\varphi = (f_0, \ldots, f_m)$$

defines a rational map of V into projective space $\mathbf{P}^m$. This map is defined at any point P which does not lie in all the elements of $\mathcal{L}$, or in other words, which is not a **fixed point** (or **base point**) of $\mathcal{L}$. Thus by definition, if $\mathcal{L}$ is without fixed point, then the associated rational map φ is a morphism. Conversely, a linear system may have a fixed component, that is a positive divisor Y such that $X \geq Y$ for all X in $\mathcal{L}$. Then the set of divisors $\{X - Y\}$ for $X \in \mathcal{L}$ is also a linear system, without fixed component.

Let

$$\varphi: V \to \mathbf{P}^n$$

be a morphism. The hyperplane sections form a linear system on $\mathbf{P}^n$. If $\{Y\}$ is the family of hyperplane sections which do not contain $\varphi(V)$, then the family

$$\{\varphi^{-1}(Y)\}$$

is a linear system on V, without fixed point.

Morphisms of V into projective space and linear systems without fixed points thus essentially correspond to each other. If $\mathcal{L}$ is a linear system on V without fixed point, and $f_0, \ldots, f_n$ is a set of generators (not necessarily a basis) for the space of functions $L(X_0)$ $(X_0 \in \mathcal{L})$, then we call

$$\varphi = (f_0, \ldots, f_n)$$

a **morphism associated with the linear system.** Any two such maps differ by a projective linear transformation and a linear projection, to which we can apply Proposition 1.2. Thus we may reformulate that proposition by saying that if φ, ψ are morphisms associated with the same linear system, then the corresponding heights are equivalent, that is,

$$h_\varphi \sim h_\psi.$$

Let $\mathcal{M}$ be another linear system on V without fixed point. Then we can define the sum $\mathcal{L} + \mathcal{M}$, which is a linear system. If $g_0, \ldots, g_m$ is a set of generators for $L(Y_0)$ where Y_0 is a divisor in $\mathcal{M}$, then the functions

$$\{f_i g_j\} \qquad (i = 1, \ldots, n; \ j = 1, \ldots, m)$$

form a family of generators for the vector space $L(X_0 + Y_0)$. Thus

$$\varphi \otimes \psi: V \to \mathbf{P}^{(n+1)(m+1)-1}$$

is a morphism associated with $\mathcal{L} + \mathcal{M}$. If

$$(f_i) = X_i - X_0 \qquad \text{and} \qquad (g_j) = Y_j - Y_0,$$

then

$$(f_i g_j) = X_i + Y_j - X_0 - Y_0.$$

Proposition 1.4 may now be interpreted in terms of linear systems to state that if φ is a morphism associated with $\mathscr{L}$ and ψ is a morphism associated with $\mathscr{M}$, then $\varphi \otimes \psi$ is associated with $\mathscr{L} + \mathscr{M}$ and

$$h_{\varphi \otimes \psi} = h_\varphi + h_\psi.$$

The next property due to Weil is the main property of heights. We have taken it and its proof from [We 1].

Proposition 2.1. *Let V be a complete, abstract, normal variety defined over K. Let $\mathscr{L}$, $\mathscr{M}$ be two linear systems on V, also defined over K, and without fixed points. Let φ, ψ be two rational maps of V into $\mathbf{P}^m$, $\mathbf{P}^n$ respectively, derived from these systems. If the divisors of $\mathscr{L}$ and $\mathscr{M}$ are linearly equivalent to each other, then h_φ and h_ψ are equivalent.*

Before giving the proof, we make a notational remark. In light of the proposition, if X is a divisor in $\mathscr{L}$, then we denote by h_X any one of the height functions h_φ, where φ is associated with $\mathscr{L}$. This avoids introducing extra symbols when a divisor or its class are prescribed in advance. Furthermore, one often prefers to add divisors or divisor classes rather than take the join of mappings.

Proof of Proposition 2.1. We shall first need a lemma. In the course of its proof, we shall use the fact that if a function $f \in K^{\mathrm{a}}(V)$ is not defined at a point Q of V, then it has a pole passing through Q (cf. Proposition 4.3 of Chapter 2).

As usual, we let ∞ be greater than any real number. Also, the product formula is irrelevant for the lemma, so we restate the hypotheses completely.

Lemma 2.2. *Let K be a field with a proper set of absolute values M_K. Let V be a complete normal variety defined over K. Let $f_1, \ldots, f_m$ be rational functions in $K^{\mathrm{a}}(V)$ whose divisors of zeros have no point in common. Then there exists an M_K-divisor c such that for all points P where the functions are defined and all $w \in M(K^{\mathrm{a}})$ we have*

$$\sup_i |f_i(P)|_w \geq c(w).$$

Proof. In the ring $K[f_1, \ldots, f_m]$ consider the ideal $(f_1, \ldots, f_m)$. It must be the unit ideal, for otherwise there would exist a homomorphism of our ring over K mapping all the f_i on 0. This homomorphism can be taken to be K^{a}-valued, corresponding to a maximal ideal in $K[f_1, \ldots, f_m]$. Such a homomorphism extends to a K^a-valued place of the function field $K(V)$. Since V is complete the place induces a point P on some affine open set U of V, and this point P is then a common zero of $f_1, \ldots, f_m$ contrary to assumption. Hence there exists a relation

$$1 = \sum z_q M_q(f_1, \ldots, f_m)$$

where M_q are monomials with coefficient 1, degree ≥ 1; and the coefficients z_q lie in K. Let S_K be the finite set consisting of all archimedean absolute values in M_K and all those v where some $|z_q|_v \neq 1$. We can take $c(v) = 1$ for $v \notin S_K$. The above relation shows that for $v \in S_K$ and $w | v$, $w \in M(K^a)$, the values

$$|f_i(P)|_w$$

cannot all lie below an obvious lower bound, determined by the absolute values $|z_q|_v$. This proves the lemma.

Let X_0 be a divisor of $\mathscr{L}$, rational over K. Then the corresponding space of functions L_0 has a basis $(f_0, \ldots, f_m)$ defined over K. For each f_i we can write

$$(f_i) = X_i - X_0, \qquad X_i \in \mathscr{L}.$$

We denote by φ_i the rational map of V into the affine K-open subset of $\mathbf{P}^m$ determined by the functions $(f_0/f_i, \ldots, f_m/f_i)$. We let V_i be the K-open subset $V - \text{Supp}(X_i)$ of V. Then the V_i cover V, and φ_i is defined at every point of V_i.

We proceed similarly with $\mathscr{M}$. In view of our assumption, we may take the divisor $Y_0 = X_0$ to get the derived space of functions M_0 with a basis $(g_0, \ldots, g_n)$ defined over K. Then

$$(g_j) = Y_j - X_0, \qquad Y_j \in \mathscr{M},$$

and the $(g_0, \ldots, g_n)$ determine our map $\psi : V \to \mathbf{P}^n$.

We apply our lemma to the functions $f_0/g, \ldots, f_m/g$ where g is any one of the g_j. Then

$$(f_i/g) = X_i - Y,$$

with $Y > 0$ and the hypothesis of the lemma is satisfied. For any point $P \notin \text{supp}(X_0)$, such that $g(P) \neq 0$, we get

$$\sup_i |f_i(P)/g(P)|_w \geq c(v)$$

for any $w \in M(K^a)$ extending $v \in M_K$, and hence

$$\sup_i |f_i(P)|_w \geq c(v)|g(P)|_w.$$

This last relation remains valid whether $g(P)$ is 0 or not (always provided $P \notin \text{supp}(X_0)$). Since g was selected arbitrarily among the g_j, we get for any $P \notin \text{supp}(X_0)$

$$\sup_i |f_i(P)|_w \geqq c(v) \sup_j |g_j(P)|_w.$$

If P is rational over the finite extension E of K, then w may be viewed as being in M_E. We raise this inequality to the power $N_w = [E_w : F_w]$, take the product over all $w \in M_E$, and take the $1/[E : F]$-th root. In view of the facts that $c(v) = 1$ for v not lying in some fixed finite subset S_K of M_K, and that our absolute values are proper, we get

$$c_1 H_\psi(P) \leqq H_\varphi(P),$$

where c_1 is the fixed number

$$\prod_{v \in S_K} c(v)$$

which does not depend on the field E, and thus holds for all points P algebraic over K, $P \notin \text{supp}(X_0)$.

Our constant c_1 of course depends on V_0, and we may write $c_1 = c_1(V_0)$. Since V is covered by $V_0, \ldots, V_m$, we can now take for c_1 the smallest of the $c_1(V_i)$, so that our inequality above holds for all P in V algebraic over K.

The inequality on the right follows by symmetry.

§3. Ample Linear Systems

A divisor X is said to be **ample** if there exists a positive integer m such that the linear system $\mathscr{L}(mX)$ gives a projective embedding of V. [*Note*: this represents a shift of terminology due to Grothendieck, and now more or less generally adopted.] If the linear system $\mathscr{L}(X)$ itself gives a projective embedding, then X is called **very ample**.

The next lemmas will be applied several times. The first use we make of them is to show that if the divisors of two linear systems are algebraically equivalent, then the associated height functions are quasi-equivalent. The next application, in a subsequent section, will be to define a height function associated with any linear equivalence class.

Lemma 3.1. *Let V be a non-singular projective variety, and X a hyperplane section. Given an integer d, there exists a positive integer e depending only on V and d such that for any positive divisor Y on V of degree d, the divisors $Y + eX$ and $-Y + eX$ are very ample.*

Proof. This lemma belongs to the foundations of algebraic geometry. It is even true on singular varieties, provided one works with the notion of Cartier divisors, cf. Chapter 10, §2. We shall recall a proof for this more general case when we deal with Cartier divisors later. A proof in the geometric style of Weil's foundations for the assertion concerning $-Y + eX$ is given in F^2-IX$_6$, Theorem 13. The other assertion is an immediate consequence of this one: namely, we first find $e_1 > 0$ such that $-Y + e_1 X$ is very ample. It is then linearly equivalent to a positive divisor T, and by cutting with generic linear varieties of suitable dimension, we see that the degree of T depends only on that of Y, and e_1. Repeating our application of Theorem 13, we find an integer $e_2 > 0$ such that $-T + e_2 X$ is very ample, and e_2 then depends only on d. This proves our lemma.

Lemma 3.2. *Let V be a complete non-singular variety. Let X be a divisor on V such that some positive multiple of X is very ample. Then there exists an integer $e > 0$ such that for any divisor Z on V algebraically equivalent to 0, the divisor $Z + eX$ is very ample.*

Proof. Let A' be the Picard variety of V. We can always find a Poincaré divisor D on $V \times A'$ which is positive. If V is an abelian variety, this is AV, Theorem 10 of Chapter IV, §4, and otherwise, making a generic translation on D, the pull-back method gives a positive Poincaré divisor on $V \times A'$ (AV, Theorem 1 of Chapter VI, §1).

By hypothesis, $Z \sim {}'D(a) - {}'D(0)$ for some point $a \in A'$. The intersections are defined after making a generic translation on D. By Lemma 3.1, there exists an integer $e_1 > 0$ such that $-{}'D(0) + e_1 X$ is very ample. Furthermore, the divisors ${}'D(a)$ as a ranges over A' are all algebraically equivalent to each other, are positive divisors, and have the same projective degree. Hence there exists an integer $e_2 > 0$ such that ${}'D(a) + e_2 X$ is very ample, again by Lemma. 3.1. From this our lemma is immediate.

The following notation will be convenient. If Y is a divisor on V defined over K, such that its complete linear system is without fixed points, then we shall denote by h_Y the height function h_φ associated with any one of the maps φ derived from this linear system. Thus h_Y is well defined, up to equivalence.

Let λ, λ' be two positive real valued functions on a set of points. We define them to be **quasi-equivalent** (multiplicatively) if given $\varepsilon > 0$ there exist two numbers $c_1, c_2 > 0$ depending on ε, such that for all points P in our set we have

$$c_1 \lambda(P)^{1-\varepsilon} \leqq \lambda'(P) \leqq c_2 \lambda(P)^{1+\varepsilon}.$$

Similarly, we define **additive quasi-equivalence** for two positive functions h_1, h_2 by

$$(1 - \varepsilon)h_1 - c_1 \leqq h_2 \leqq (1 + \varepsilon)h_1 + c_2.$$

When we compare heights H_1, H_2 we use the notion of multiplicative quasi-equivalence. When we compare the logarithmic heights h_1, h_2 then we use the notion of additive quasi-equivalence.

Proposition 3.3. *Let V be a complete non-singular variety defined over K. Let $X, Y > 0$ be two positive divisors on V, rational over K. Assume that a positive multiple of each is very ample, and that the linear systems $\mathscr{L}(X)$ and $\mathscr{L}(Y)$ are without fixed points. If X and Y are algebraically equivalent, then h_X and h_Y are quasi-equivalent.*

Proof. Using Proposition 2.1 we shall reduce our assertion to a statement concerning linear equivalence of divisors on V. By Lemma 3.2, there exists an integer $e > 0$ such that for all $n > 0$ we have

$$n(X - Y) + eX \sim Z_n,$$

where Z_n is a positive divisor on V, and $\mathscr{L}(Z_n)$ is without fixed points. Since $n(X - Y) + eX$ is rational over K, and one may take Z_n rational over K. We get $nX + eX \sim nY + Z_n$ and taking heights, this yields in view of Proposition 1.2,

$$(n + e)h_X \sim nh_Y + h_{Z_n}.$$

Since the logarithmic height corresponding to a morphism into projective space is always $\geqq 0$, after dividing by n we see that given $\varepsilon > 0$ there exists a number $c > 0$ such that

$$c + (1 + \varepsilon)h_X(P) \geqq h_Y(P)$$

for all points P on V, algebraic over K. The other inequality is obtained in a similar way, or by symmetry.

Corollary 3.4. *Let X, Y be ample and algebraically equivalent. Then*

$$\lim_{h_X(P) \to \infty} h_Y(P)/h_X(P) = 1.$$

Proof. Immediate from the definition of quasi-equivalence.

An important special case of Proposition 3.3 is when V is a curve. In that case, algebraic equivalence is determined by the degree, and hence we get:

Corollary 3.5. *Let V be a complete non-singular curve, and X, Y divisors on V. If $\deg(X) = d$ and $\deg(Y) = d'$, then h_X is quasi-equivalent to $(d/d')h_Y$.*

The case of curves is due to Siegel [Si], who used it in the theory of integral points. He could use the Riemann–Roch theorem for the proof instead of the Picard variety. Thus when V has dimension 1, very little machinery is needed to establish Proposition 3.3. I gave the generalization to higher dimension in [L 4].

§4. Projections on Curves

In the application of the theory of heights in Chapter 8 we shall deal with curves, and it is therefore convenient to investigate more closely how the height behaves under projection. The techniques depend on the same type of considerations as above, *but this section will be used only in Chapter 8, and may therefore be omitted without impairing the understanding of the rest of the book.*

Let V be a projective non-singular curve defined over K. The height h is taken with respect to this embedding.

Proposition 4.1. *Let $(y_0, \ldots, y_n)$ with $y_0 = 1$, be functions in $K(V)$ determining our embedding in $\mathbf{P}^n$. Let Φ be a finite set of points of V in K^a. Then there is a polynomial equation such that if $(a_0, \ldots, a_n, b_0, \ldots, b_n)$ are elements of K not satisfying this equation, then the function*

$$y = \frac{a_0 y_0 + \cdots + a_n y_n}{b_0 y_0 + \cdots + b_n y_n}$$

has the following properties:

(i) *The function y is not constant, and has no zero or pole among the points of Φ.*

(ii) *If h_y is the height determined by the mapping of V into $\mathbf{P}^1$ arising from the function y, then $h_y \sim h$.*

(iii) *The mapping $Q \mapsto y(Q)$ gives an injection of Φ into K^a.*

Proof. Let Q be a point of V in K^a. If t is a function of order 1 at Q, then each y_i has an expansion as a power series in t, say $y_i = \xi_i t^{e_i} + \cdots$ with an integer e_i, which may be negative, and $\xi_i \in K^a$. We see that there is a polynomial G_Q (linear, in fact) such that for any set of elements $(a_0, \ldots, a_n)$ in K for which $G_Q(a) \neq 0$, then $a_0 y_0 + \cdots + a_n y_n$ has order e at Q, where $e = \inf_i e_i$. Taking Q from a finite set Φ, we then take the product of the G_Q for $Q \in \Phi$ and achieve the same thing for all $Q \in \Phi$.

If $\mathfrak{a}$ denotes the sup of the polar divisors of our given y_i, then we see that almost all linear combinations $a_0 y_0 + \cdots + a_n y_n$ have precisely $\mathfrak{a}$ as polar divisor. Furthermore, applying the above remarks to zeros instead of poles, and taking into account that the linear system determined by $(1, y_1, \ldots, y_n)$ is without a fixed point, we see that we can make a sufficiently general choice of a_i and b_i such that the function y has no zero or pole in Φ, and its divisor

$$(y) = (y)_0 - (y)_\infty$$

is such that $(y)_\infty$ lies in the above linear system. According to Proposition 2.1 it follows that h_y is equivalent to h.

To insure that the map $Q \mapsto y(Q)$ is injective on Φ, we select among the y_i (for each Q) one of the functions having the highest order pole at Q, and denote it by y_Q. All quotients y_i/y_Q are defined at Q, and we have

$$y(Q) = \frac{a_0 w_0(Q) + \cdots + a_n w_n(Q)}{b_0 w_0(Q) + \cdots + b_n w_n(Q)}$$

where $w_i = y_i/y_Q$. (Strictly speaking, each w_i should carry Q also as an index.) We can choose the b_i so that the denominator does not vanish, and the condition that $y(Q) \neq y(Q')$ when $Q \neq Q'$ are two distinct points of Φ is immediately seen to be implied by the nonvanishing of a polynomial in the a_i and b_i. This concludes the proof of our three statements.

§5. Heights Associated with Divisor Classes

Let V be a projective non-singular variety, defined over K. Let X be a divisor rational over K. Given a finite set of simple points S there exists a function f in the function field $K(V)$ such that the support of $X + (f)$ does not meet S, by Proposition 4.2 of Chapter 2. Let:

$\text{Div}(V)$ = group of divisors on V, rational over K^a;

$\text{Div}_a(V)$ = subgroup of divisors algebraically equivalent to 0;

$\text{Div}_l(V)$ = subgroup of divisors linearly equivalent to 0.

To save space sometimes we write $D(V)$, $D_a(V)$ and $D_l(V)$ respectively. We let the **Picard group** be

$$\text{Pic}(V) = \text{Div}(V)/\text{Div}_l(V) \qquad \text{and} \qquad \text{Pic}_0(V) = \text{Div}_a(V)/\text{Div}_l(V).$$

If X is a divisor on V we let $\text{Cl}(X)$ be its class in $\text{Pic}(V)$. We let the **Néron–Severi group** be

$$NS(V) = \text{Div}(V)/\text{Div}_a(V).$$

Let $f: V \to W$ be a morphism of projective non-singular varieties defined over K. Then f induces an inverse mapping

$$f^*: \text{Pic}(W) \to \text{Pic}(V).$$

Indeed, let Y be a divisor on W. By Proposition 4.2 of Chapter 2, there exists a function g on W such that $Y + (g)$ does not meet some point in $f(V)$, and therefore the inverse image

$$f^{-1}(Y + (g))$$

is defined, and is a divisor on V. All such divisors (keeping Y fixed but changing g) are linearly equivalent on V, and therefore the linear equivalence class of $f^{-1}(Y + (g))$ is determined by $\text{Cl}(Y)$.

Let $c \in \text{Pic}(V)$ be a divisor class, and suppose at first that c contains a divisor X such that $\mathscr{L}(X)$ is very ample. If $\mathscr{L}$ is any linear system without fixed point containing a divisor linearly equivalent to X, then Proposition 2.1 implies that the heights are equivalent, that is

$$h_{\mathscr{L}} \sim h_{\mathscr{L}(X)}.$$

In particular, the equivalence class of the height is determined by the divisor class c, and a height in the equivalence class will be denoted by h_c. It is determined only up to a bounded function denoted $O(1)$.

Next, let Y be any divisor. By Lemma 3.1 there exists a very ample divisor X such that $Y + X$ is very ample. Let $c = \text{Cl}(Y)$. Then we can write

$$c = c' - c'',$$

where c' and c'' contain very ample divisors. The equivalence class of functions

$$h_{Y+X} - h_X$$

or $h_{c'} - h_{c''}$ is independent of the choice of X, again by Proposition 2.1. Namely, if X_1 is very ample such that $Y + X_1$ is very ample, then we get an equivalence of functions

$$h_{Y+X} - h_X \sim h_{Y+X_1} - h_{X_1},$$

because we have a linear equivalence of divisors

$$(Y + X) + X_1 \sim (Y + X_1) + X,$$

and therefore

$$h_{Y+X} + h_{X_1} \sim h_{Y+X_1} + h_X.$$

Thus we may summarize our construction as follows.

Theorem 5.1. *There exists a unique homomorphism*

$$\text{Pic}(V) \to \text{real valued functions on } V(\mathbf{F}^a)$$
$$\text{modulo bounded functions,}$$

denoted by $c \mapsto h_c + O(1)$, *such that: if* c *contains a very ample divisor, then* h_c *is equivalent to the height associated with a projective embedding obtained from the linear system of that divisor.*

In addition, if $f: V \to W$ *is a morphism of projective non-singular varieties defined over* K, *then for* $c \in \text{Pic}(W)$ *we have*

$$h_{f^*(c)} = h_c \circ f + O(1).$$

Proof. We already proved existence and uniqueness. As to the last assertion, it follows from the trivial functorial formula of §1.

In the next chapter, we shall use standard criteria for linear equivalence on abelian varieties to get corresponding equivalences between height functions on abelian varieties.

Proposition 5.2. *Let* $c \in \text{Pic}(V)$ *and suppose that* c *contains a positive divisor* Z. *Then one can choose* h_c *in its class modulo bounded functions such that*

$$h_c(P) \geq 0$$

for all $P \in V$, $P \notin \text{supp}(Z)$.

Proof. There exist very ample positive divisors X and Y such that $Z + X \sim Y$. We have to show that

$$h_Y - h_X$$

is bounded from below on the complement of Z. But let $\{f_0, \ldots, f_n\}$ be a basis for $L(X)$. Then we can extend this choice of functions to a basis

$$\{f_0, \ldots, f_n, g_1, \ldots, g_m\}$$

of $L(Y)$ because Z is positive. The functions f_i $(i = 0, \ldots, n)$ have divisors

$$(f_i) = X_i - X$$

and the divisors X_i have no point in common. Let the morphisms associated with the linear systems $\mathscr{L}(X)$ and $\mathscr{L}(Y)$ be

$$f = (f_0, \ldots, f_n) \quad \text{and} \quad g = (f_0, \ldots, f_n, g_1, \ldots, g_m).$$

For any point P not in the support of Z it then follows from the definition of the height that

$$h_g(P) \geqq h_f(P).$$

This proves the proposition. Essentially we have reproved Proposition 1.1 in the present context.

We may also reformulate Proposition 3.3 in the present context, under less restrictive hypotheses.

Proposition 5.3. *Let c, c' be algebraically equivalent classes, and ample. Then h_c is quasi-equivalent to $h_{c'}$.*

Proof. After multiplying divisors in c, c' by some positive integer, the classes contain very ample divisors which are algebraically equivalent. One can then apply Proposition 3.3 to complete the proof.

Note. In the statement of Proposition 5.3, it is understood that we have selected h_c and $h_{c'}$ to be positive functions in their equivalence classes.

Proposition 5.4. *Let c be an ample class and c' any class. Choose h_c in its equivalence class so that $h_c \geqq 0$. There exist numbers γ_1, γ_2 such that*

$$|h_{c'}| \leqq \gamma_1 h_c + \gamma_2.$$

Proof. This is an immediate consequence of Proposition 1.7. We can write c' as a difference $c' = c_1 - c_2$ where c_1, c_2 are very ample, and we apply Proposition 1.7 to each one of h_{c_1} and h_{c_2}.

CHAPTER 5

Heights on Abelian Varieties

Néron at the Edinburgh International Congress had conjectured that the (logarithmic) height on an abelian variety differed from a quadratic function by a bounded function. He proved this in [Ne 3], as well as proving an analogous statement for local components for the height. Tate showed that a direct argument applied to the global height could be used, by-passing the local considerations. We shall give Tate's argument in this chapter, as well as a few consequences.

We shall apply systematically the result at the end of Chapter 4: any relation between divisor classes yields a relation between corresponding heights. The theory of abelian varieties provides a number of interesting relations in this context, first for divisor classes, and then automatically for the corresponding height functions.

The content of this chapter is not needed in the subsequent chapters if one is willing to work with quasi-equivalence in the proof of the Mordell–Weil theorem. Using the canonical quadratic form instead of a quasi-quadratic function would not make the infinite descent argument essentially shorter, and so the reader may skip this chapter if interested in seeing a proof of the Mordell–Weil theorem first.

§1. Some Linear and Quasi-Linear Algebra

Let A be an abelian group. Let h be a function from A into some abelian group on which multiplication by 2 is invertible. We define Δh on $A \times A$ by

$$\Delta h(x, y) = h(x + y) - h(x) - h(y).$$

We say that h is **quadratic** if Δh is bilinear. If in addition h is even, that is h satisfies $h(-x) = h(x)$, then we say that h is a **quadratic form**. If h is quadratic and if we put

$$B(x, y) = \Delta h(x, y),$$

then one verifies at once that the function of x given by

$$h(x) - \tfrac{1}{2}B(x, x)$$

is linear, so any quadratic functions is expressible as a quadratic form plus a linear function. If in addition h is even, and $h(0) = 0$, then it is immediate that the linear function is 0, so h is a quadratic form. We call Δh the **derived bilinear form of** h.

Next suppose we are dealing with real valued functions. Let L be a real valued function on $A \times A$. We call L **quasi-bilinear** if the functions on $A \times A \times A$ given by

$$\Delta_1 L(x, y, z) = L(x + y, z) - L(x, z) - L(y, z),$$
$$\Delta_2 L(x, y, z) = L(x, y + z) - L(x, z) - L(x, y)$$

are bounded. Similarly, a real valued function L on A is called **quasi-linear** if $L(x + y) - L(x) - L(y)$ is bounded.

A function h on A is called **quasi-quadratic** if the function

$$\Delta h(x, y) = h(x + y) - h(x) - h(y)$$

is quasi-bilinear.

Formal relations of linear algebra relating to quadratic and bilinear functions apply in the quasi-quadratic and quasi-bilinear case. In addition, we have the following lemmas, applying to real valued functions, giving bilinear or quadratic representatives in a quasi-bilinear or quasi-quadratic equivalence class.

Lemma 1.1. *If L is quasi-bilinear, then there exists a unique bilinear function L' such that $L - L'$ is bounded, that is L is equivalent to L'. In fact,*

$$L'(x, y) = \lim_{n \to \infty} \frac{L(2^n x, 2^n y)}{4^n}.$$

Proof. Uniqueness is obvious. For existence, write

$$L_n(x, y) = \frac{L(2^n x, 2^n y)}{4^n}.$$

Suppose that C is a positive number such that

$$|\Delta_1 L(x, y, z)| \qquad \text{and} \qquad |\Delta_2 L(x, y, w)| \leq C$$

for all x, y, z, w. From quasi-bilinearity we find

$$|L_{n+1}(x, y) - L_n(x, y)| = |\Delta_1 L(2^n x, 2^n x, 2^{n+1} y) + 2\Delta_2 L(2^n x, 2^n y, 2^n y)|/4^{n+1}$$
$$\leq 3C/4^{n+1}.$$

Hence $\{L_n(x, y)\}$ is a Cauchy sequence, and it is immediately verified that the limit is bilinear.

We note that we get an explicit bound on $L' - L$, namely C, if we know a bound for the $O(1)$ expressing the quasi-bilinearity of L. The same holds for the quadratic case, with a similar low integer multiplying $O(1)$.

If h is a function on an abelian group, we let h^- be the function such that

$$h^-(x) = h(-x).$$

A function is said to be **quasi-even** if h is equivalent to h^-, that is $h - h^-$ is bounded. Similarly to Lemma 1.1 we get:

Lemma 1.2. *Let h be quasi-quadratic. Then*:

(i)   *There exists a unique quadratic form q and a unique linear function l such that*

$$h = q + l + O(1);$$

   or in other words, h is equivalent to $q + l$.

(ii) *If h is quasi-even, then the linear function l in (i) is 0, and h is equivalent to a quadratic form.*

Proof. Let

$$L(x, y) = h(x + y) - h(x) - h(y).$$

Then L is quasi-bilinear, so by Lemma 1.1 is equivalent to a bilinear function, say $B(x, y)$. It follows that

$$h(x) - \tfrac{1}{2}B(x, x)$$

as function of x is quasi-linear, whence again taking a limit as before, this quasi-linear function is equivalent to a unique linear function, thus proving existence. The uniqueness is immediate. It is also immediate that if h is quasi-even then the linear function is 0. This proves the lemma.

If h is quasi-quadratic, then the quadratic form q can be obtained directly from the limit

$$q(x) = \lim \frac{h(2^n x)}{4^n}.$$

Note. The averaging process was applied by Néron to the local component of the height function [Ne 3], whence he obtained the height as a quadratic function; and it was shown by Tate how to apply it directly to the global height to get a simpler proof in the global case, cf. [L 6].

The next result gives a criterion for a function to be a quasi-quadratic form. We say that h **quasi-satisfies the parallelogram law** if

$$h(x + y) + h(x - y) = 2h(x) + 2h(y) + O(1).$$

Lemma 1.3. *If h quasi-satisfies the parallelogram law, then h is a quasi-quadratic form, and if $L = \Delta h$, then*

$$h(x) - \tfrac{1}{2}L(x, x)$$

is bounded, that is h is equivalent to a quadratic form.

Proof. The usual arguments of linear algebra work with the extra term $O(1)$. We reproduce these arguments. First we note that h is quasi-even, as one sees by putting $x = 0$ in the quasi-parallelogram law, which yields

$$h(-y) = h(y) + O(1).$$

From the definition

$$L(x, y) = \Delta h(x, y) = h(x + y) - h(x) - h(y),$$

replacing y by $-y$ and using the quasi-parallelogram law, we conclude that L is quasi-odd in the variable y, that is

$$L(x, -y) = -L(x, y) + O(1),$$

whence L is also quasi-odd in the variable x by symmetry.
Let

$$
\begin{aligned}
M(x, y, z) = \Delta_1 L(x, y, z) &= L(x + y, z) - L(x, z) - L(y, z) \\
&= h(x + y + z) - h(x + y) - h(y + z) - h(x + z) \\
&\quad + h(x) + h(y) + h(z).
\end{aligned}
$$

Then $M(x, y, z)$ is symmetric in x, y, z and is quasi-odd in z since L is quasi-odd in each variable by what we have just shown. By symmetry, M is quasi-odd in each variable, whence

$$M(-x, -y, -z) = (-1)^3 M(x, y, z) + O(1).$$

However, since h is quasi-even it follows that $M = O(1)$, so M is bounded, whence L is quasi-bilinear. The lemma follows, since the final conclusion is a repetition of Lemma 1.2(ii) in the present context.

Note. If we omit all references to $O(1)$ in the preceding lemma, the arguments apply to any function with values in an abelian group on which multiplication by 2 is invertible, to give exact relations, as already mentioned in the case of a quadratic form. In other words, if a function satisfies the parallelogram law and is even, then it is a quadratic form. The converse, that a quadratic form satisfies the parallelogram law, is of course obvious.

§2. Quadraticity of Endomorphisms on Divisor Classes

Let now A be an abelian variety defined over a field k. As before we denote by $\approx$, $\sim$ the equivalences corresponding to algebraic and linear equivalence respectively. We put

$$\text{Pic}(A) = \text{Div}(A)/\text{Div}_l(A) \qquad \text{and} \qquad \text{Pic}_0(A) = \text{Div}_a(A)/\text{Div}_l(A).$$

If $\alpha: A \to B$ is a homomorphism defined over k, then for any element $c \in \text{Pic}(B)$ the inverse image $\alpha^*(c)$ is defined as an element of $\text{Pic}(A)$.

From the fundamental theory of abelian varieties, we now accept the following result.

Theorem 2.0. *Let $c \in \text{Pic}(A)$. Let c_u be the translation of c by a point u, and let*

$$\varphi_c: A \to \text{Pic}_0(A)$$

be the homomorphism $\varphi_c(u) = c_u - c$. Then the kernel of the homomorphism

$$\text{Pic}(A) \to \text{Hom}(A, \text{Pic}_0(A)) \quad \text{given by } c \mapsto \varphi_c$$

is precisely $\text{Pic}_0(A)$. Furthermore, if c is ample, then φ_c maps A onto $\text{Pic}_0(A)$ and its kernel is finite.

The property that φ_c is a homomorphism is sometimes called the **theorem of the square** on abelian varieties.

The above theorem is proved in AV, Chapter IV, §1 and §2, except for possible p-torsion in characteristic p: if $\varphi_c = 0$ then some p-power multiple of c is algebraically equivalent to 0. That actually c is algebraically equivalent to 0 follows from the fact that A is the Picard variety of its Picard variety (biduality), cf. AV, Chapter VIII, §4, second Theorem 10 (sorry about the numbering error). The biduality is due to Cartier and Nishi, and is done in Mumford's book, Corollary at the end of §13. Mumford's definitions are such that his biduality theorem includes the fact that the Néron–Severi group $NS(A)$ does not have torsion.

Given a class $c \in \mathrm{Pic}(B)$, we obtain a map of $\mathrm{Hom}(A, B)$ into $\mathrm{Pic}(A)$ by

$$\alpha \mapsto \alpha^*(c).$$

The fundamental fact about this map is that it is quadratic.

Theorem 2.1. *For α, $\beta\colon A \to B$ homomorphisms of abelian varieties, and $c \in \mathrm{Pic}(B)$, define*

$$D_c(\alpha, \beta) = (\alpha + \beta)^*(c) - \alpha^*(c) - \beta^*(c).$$

Then the map

$$(\alpha, \beta) \mapsto D_c(\alpha, \beta)$$

is bilinear.

Proof. In fact, the bilinearity comes from an even more universal relation as follows. We have various mappings from the product of A with itself into A.

The sum

$$s_2\colon A \times A \to A \qquad \text{and} \qquad s_3\colon A \times A \times A \to A$$

given by

$$s_2(x, y) = x + y \qquad \text{and} \qquad s_3(x, y, z) = x + y + z.$$

The partial sums

$$s_{12}(x, y, z) = x + y, \qquad s_{13}(x, y, z) = x + z, \qquad s_{23}(x, y, z) = y + z.$$

The projections

$$p_1\colon A \times A \to A \qquad \text{and} \qquad p_2\colon A \times A \to A$$

on the first and second factor, as well as the partial projections

$$p_1(x, y, z) = x, \qquad p_2(x, y, z) = y, \qquad p_3(x, y, z) = z.$$

The context will always make clear whether p_i applies to projections from a double or triple product.

For the convenience of the reader, we recall here two formulations of the **seesaw principle**. The first is from AV, Appendix 2, Theorem 6.

Let U, V be varieties, and assume that V is complete, non-singular in codimension 1. Let D be a divisor on $U \times V$, and k a field of definition for U, V over which D is rational. Let u be a generic point of U over k, and assume that there exists a divisor Y on V, rational over k, such that

$$D(u) \sim Y$$

(where $D(u) = \mathrm{pr}_2[D.(u \times V)]$. Then there exists a rational function φ on $U \times V$, defined over k, and a divisor X on U, rational over k, such that

$$D = X \times V + U \times Y + (\varphi).$$

The first form is essentially birational in nature. The reader may prefer the second form.

Let U, V be complete non-singular varieties, and let $c \in \mathrm{Pic}(U \times V)$. Let $u_0 \in U$ and $v_0 \in V$. If the restriction of c to each vertical subvariety

$$u_0 \times V \qquad and \qquad U \times v_0$$

is 0 then $c = 0$.

If one replaces c by a line bundle, then it is not even necessary to assume non-singularity. For the proof, see Mumford [Mum 2], Chapter II, §6.

Lemma 2.2. *For any divisor X on A, the following divisor on $A \times A \times A$ is linearly equivalent to 0:*

$$s_3^{-1}(X) - s_{12}^{-1}(X) - s_{13}^{-1}(X) - s_{23}^{-1}(X) + p_1^{-1}(X) + p_2^{-1}(X) + p_3^{-1}(X).$$

Proof. Let D be the divisor as shown. By the seesaw principle, it suffices to prove that the intersections (for u, v generic independent)

$$D.(u \times v \times A), \qquad D.(A \times u \times v), \qquad D.(u \times A \times v)$$

are linearly equivalent to 0. But we have for instance

$$D.(u \times v \times A)$$
$$= u \times v \times X_{-u-v} - u \times v \times X_{-u} - u \times v \times X_{-v} + u \times v \times X$$
$$\sim 0$$

by the theorem of the square. Here, as usual, X_a is the translation of X by a point a of A. Note that we have used the fact that since u, v are generic, $s_{12}^{-1}(X)$, $p_1^{-1}(X)$, and $p_2^{-1}(X)$ all have trivial intersection with $u \times v \times A$. This proves the lemma.

We can then apply Lemma 2.2 to the proof of Theorem 2.1 by considering the homomorphism

$$\alpha: A \to B \times B \times B$$

of A into the triple product, given by

$$\alpha(u) = (\alpha_1 u, \alpha_2 u, \alpha_3 u)$$

for three homomorphisms $\alpha_i: A \to B$. We then get for instance

$$s_3 \circ \alpha = \alpha_1 + \alpha_2 + \alpha_3,$$

and similarly with other partial sums. Then

$$(s_3 \circ \alpha)^*(c) = (\alpha_1 + \alpha_2 + \alpha_3)^*(c)$$

for any class $c \in \mathrm{Pic}(B)$. Since

$$(s_3 \circ \alpha)^*(c) = \alpha^* \circ s_3^*(c),$$

and similarly for the other terms, we see that Theorem 2.1 is equivalent with the statement that

$$\alpha^*(d) = 0,$$

where d is the linear equivalence class of the divisor in Lemma 2.2. But this linear equivalence class is the trivial class by Lemma 2.2, thus proving Theorem 2.1.

We may reformulate Theorem 2.1 by saying that

for c fixed, the map

$$\alpha \mapsto \alpha^*(c) \quad \text{of} \quad \mathrm{Hom}(A, B) \to \mathrm{Pic}(A)$$

is a quadratic map.

Let $R = \mathbf{Z}[\frac{1}{2}]$. Tensoring $\mathrm{Pic}(A)$ with R yields a group of values in which 2 is invertible. We can then express the map

$$\alpha \mapsto \alpha^*(c)$$

as a quadratic form plus a linear map according to the general results of linear algebra of the preceding section.

We have already seen formally the importance of even maps, to get rid of the linear term. In practice it is useful to decompose a divisor class into an even one and an odd one, and to deal with these separately. We say that an element $c \in \text{Pic}(A)$ is **even** if

$$[-1]^*(c) = c,$$

where $[-1]$ is the endomorphism of A given by

$$[-1](x) = -x.$$

If X is a divisor we let $X^- = [-1]^*(X)$. To say that $\text{Cl}(X)$ is even is equivalent to saying that $X^- \sim X$.

We say that c is **odd** if

$$[-1]^*(c) = -c.$$

This oddness condition is equivalent to saying that for any divisor X in the class c, we have

$$X^- \sim -X.$$

Proposition 2.3. *For any* $c \in \text{Pic}(A)$ *we have* $c \equiv c^-$ *mod* $\text{Pic}_0(A)$, *and the following conditions are equivalent.*

O 1. c *is odd.*

O 2. $c \in \text{Pic}_0(A)$.

O 3. $s_2^*(c) - p_1^*(c) - p_2^*(c) = 0$.

O 4. *The map* $\alpha \mapsto \alpha^*(c)$ *is linear, that is for any homomorphisms* $\alpha, \beta: B \to A$

$$(\alpha + \beta)^*(c) = \alpha^*(c) + \beta^*(c).$$

Proof. For any class c we have the intersection

$$[s_2^*(c) - p_1^*(c) - p_2^*(c)].(u \times A) = u \times (c_{-u} - c).$$

Suppose $c \in \text{Pic}_0(A)$. Write $c = c_b' - c'$ for some c' (which can be done by Theorem 2.0). The theorem of the square shows that the above intersection is 0, and similarly taking the intersection with $(A \times u)$. By the seesaw principle, it follows that **O 3** is true.

Assume that c satisfies **O 3.** Let $\alpha, \beta: B \to A$ be homomorphisms. Taking the composite of the maps s_2, p_1, p_2 with the map of $A \to A \times A$ given by

$x \mapsto (\alpha x, \beta x)$, and taking pull backs of c to A by these composites we see that

$$(\alpha + \beta)^*(c) - \alpha^*(c) - \beta^*(c) = 0.$$

In particular, the map $\alpha \mapsto \alpha^*(c)$ is linear, and therefore $[-1]^*(c) = -c$. This proves that c is odd.

Next let c be any class. To show that $c^- \equiv c \bmod \mathrm{Pic}_0(A)$, it suffices to prove that $\varphi_{c^-} = \varphi_c$. But we have

$$c_u^- - c^- = (c_{-u} - c)^- = -(c_{-u} - c),$$

because $c_{-u} - c \in \mathrm{Pic}_0(A)$ and we can use the linearity of the representation which we have already proved to get this last equality. The theorem of the square then shows that $\varphi_{c^-} = \varphi_c$, whence $c^- \equiv c \bmod \mathrm{Pic}_0(A)$ by Theorem 2.0.

Finally, assume that c is odd. Then $c^- = -c$, and we also know that $c^- \equiv c \bmod \mathrm{Pic}_0(A)$. Hence $2c \in \mathrm{Pic}_0(A)$ whence $c \in \mathrm{Pic}_0(A)$, thereby proving the proposition.

Proposition 2.4. *Let* $c \in \mathrm{Pic}(A)$. *The following properties are equivalent.*

E 1.   *c is even.*

E 2.   *Let d_2 be the difference homomorphism $d_2(x, y) = x - y$. Then*

$$s_2^*(c) + d_2^*(c) = 2p_1^*(c) + 2p_2^*(c).$$

E 3.   **(Parallelogram Law)** *For any homomorphisms $\alpha, \beta : B \to A$,*

$$(\alpha + \beta)^*(c) + (\alpha - \beta)^*(c) = 2\alpha^*(c) + 2\beta^*(c).$$

Proof. One can either deduce these properties along the formal arguments of linear algebra over $\mathbf{Z}$ by using the bilinearity of $D_c(\alpha, \beta)$ proved in Theorem 2.1; or one can reprove them ab ovo. A direct proof from Theorem 2.0 is similar to the proof of Theorem 2.1 and Lemma 2.2, except that the proof occurs on the double product $A \times A$ instead of the triple product, and so is somewhat easier. To get the statement concerning α, β we compose the various homomorphisms s_2, d_2, p_1, p_2 with the homomorphism

$$(\alpha, \beta) : A \to A \times A$$

given by $x \mapsto (\alpha x, \beta x)$. The technique is entirely similar to that already used for the preceding results, including the odd case, and consequently the details will be left to the reader.

Remark. We have deduced certain formal properties from Theorem 2.0. In the actual development of the theory, some of these properties may be proved earlier. We have selected here an ordering best suited to summarize standard properties, in a more or less axiomatized way minimizing the requirements on the reader.

Proposition 2.5. *Any divisor class c can be expressed as a sum*

$$c = c' + c'',$$

where c' is even and c'' is odd.

Proof. We want to write the usual expression

$$c = \frac{c + c^-}{2} + \frac{c - c^-}{2}.$$

The problem is with the division by 2. But $c - c^-$ is algebraically equivalent to 0, and $\text{Pic}_0(A)$ is isomorphic to the Picard variety which is divisible by 2. Hence there exists an element $c'' \in \text{Pic}_0(A)$ such that

$$2c'' = c - c^-.$$

We then define

$$c' = c - c''.$$

Then

$$
\begin{aligned}
(c')^- &= c^- - (c'')^- \\
&= c + 2(c'')^- - (c'')^- \quad \text{by using } 2(c'')^- = c^- - c \\
&= c + (c'')^- \\
&= c'.
\end{aligned}
$$

Hence c' is even, thus proving the proposition.

The only non-unique part of the proposition comes from elements of order 2 in the Picard variety.

Finally we have a lemma of Mumford which gives an explicit expression for $h(n)$ if h is a quadratic function.

Lemma 2.6. *Let h be a quadratic function from the integers $\mathbf{Z}$ into an abelian group on which multiplication by 2 is invertible. Then*

$$h(n) = \frac{n(n + 1)}{2} h(1) + \frac{n(n - 1)}{2} h(-1).$$

Proof. This is immediate from the representation of a quadratic function as a sum of a quadratic form and a linear function.

In particular, we may apply this to the quadratic representation

$$n \mapsto [n]^*(c)$$

with a given divisor class c on A. We find **Mumford's formula**

$$[n]^*(c) = \frac{n(n+1)}{2} c + \frac{n(n-1)}{2} c^-.$$

If c is even, then of course

$$[n]^*(c) = n^2 c.$$

§3. Quadraticity of the Height

Again let A be an abelian variety defined over K, where K is a finite extension of our field $\mathbf{F}$ having a set of absolute values satisfying the product formula as before. By §5 of Chapter 5 we know that to each element $c \in \mathrm{Pic}(A)$ we can associate a height function h_c up to equivalence. Furthermore, if $\alpha \in \mathrm{End}(A)$, then

$$h_c \circ \alpha = h_{\alpha^*(c)} + O(1).$$

We can also apply directly the functorial formulas to Lemmas 2.2 and 2.5. In particular, also using Lemma 1.2, we find:

Theorem 3.1. *For any divisor class $c \in \mathrm{Pic}(A)$, the height h_c is quasi-quadratic. There exists a unique quadratic form q_c and linear form l_c such that*

$$h_c = q_c + l_c + O(1).$$

If c is even, that is $c = c^-$, then $l_c = 0$.

The sum $q_c + l_c$ which is uniquely determined by c will be denoted by $\hat{h}_c$ and will be called the **Néron–Tate height**, or **canonical height, associated with the class c. Fos simplicity of notation, we shall omit the roof and denote by h_c this unique quadratic function in the equivalence class unless otherwise specified.**

The existence and uniqueness of the canonical height may then be summarized as follows.

Theorem 3.2. *There exists a unique homomorphism*

$$\mathrm{Pic}(A) \to \text{quadratic real valued functions on } A(\mathbf{F}^a)$$

noted by $c \mapsto h_c$, such that: if c contains a very ample divisor X, then h_c is equivalent to the height associated with a projective embedding obtained from the linear system $\mathscr{L}(X)$.

The proof merely puts together §5 of Chapter 4 and the quasi-linear algebra of §1, together with the divisor class relations of §2.

Proposition 3.3. *Let $\alpha: A \to B$ be a homomorphism, and let $c \in \mathrm{Pic}(B)$. Then*

$$h_{\alpha^*(c)} = h_c \circ \alpha.$$

Proof. By the general functoriality of the height, the above relation holds up to a bounded function, that is

$$h_{\alpha^*(c)}(x) = h_c(\alpha(x)) + O(1).$$

But we have trivially:

Lemma 3.4. *If a bounded function is equal to the sum of a quadratic form and a linear form then the function is 0.*

Proof. Let $g(x)$ be the bounded function. The assertion is immediate by looking at $g(nx)$ with n a positive integer tending to infinity. Thus Proposition 3.3 is proved.

It is useful to consider separately the cases when the class c is odd or even.

Suppose that c is odd. Equivalently, c is algebraically equivalent to 0. Then the quadratic form q_c is 0, so h_c is a linear function.

Suppose that c is even. Then the linear form is 0, so h_c is a quadratic form.

The next theorem gives the behavior of h_c under a translation

$$T_a: A \to A$$

defined by $T_a(x) = x + a$. If $c \in \mathrm{Pic}(A)$, and if X is a divisor in c, then $T_a c$ is the class containing the divisor X_a. The functorial formula for heights with respect to morphisms is contravariant, and

$$T_a c = T^*_{-a}(c),$$

so that in particular,

$$h_{c_a}(x) = h_c(x - a) + O(1).$$

This formula applies both when c is even or odd, so for general c.

Proposition 3.5. *Let* $f: A \to A$ *be a homomorphism followed by a translation (such are all morphisms of* A *into itself). Then*

3.5.1. $$h_{f^*(c)} = h_c \cdot f - h_c(f(0)).$$

More specifically, for translations, let

$$L_c(x, y) = h_c(x + y) - h_c(x) - h_c(y)$$

be the bilinear form derived from h_c. *Then*

3.5.2. $$h_{c_a}(x) = h_c(x) - L_c(x, a).$$

3.5.3. $$h_{c_a}(x) = h_c(x - a) - h_c(a).$$

If $\varphi_c: A \to \operatorname{Pic}_0(A)$ *is the usual map* $\varphi_c(a) = c_a - c$, *then*

3.5.4. $$h_{\varphi_c(a)}(x) = -L_c(x, a).$$

If c *is odd, then* $h_{c_a} = h_c$.

Proof. We already know 3.5.1 when f is a homomorphism, by Proposition 3.3, so it suffices to consider translations. We have

$$h_{c_a}(x) = h_c(x - a) + O(1)$$
$$= h_c(x) + h_c(-a) + L_c(x, -a) + O(1)$$
$$= h_c(x) + L_c(x, -a) + g(x),$$

where $g(x)$ is a bounded function, sum of a quadratic form and a linear form, so equal to 0. This proves 3.5.2. The other statements are trivially proved in a similar manner.

Proposition 3.6. *Let* c *be even and suppose* c *contains an ample divisor. Then* h_c *is a positive quadratic form (not necessarily definite).*

Proof. Multiplying c by a large positive integer if necessary, we may assume that c contains a very ample divisor X, so

$$h_c = h_\varphi + O(1),$$

where h_φ is the height associated with any one of the projective embeddings φ of A by means of the complete linear system determined by X. Then we have

$$h_\varphi \geqq 0.$$

For any positive integer n and any point $P \in A(\mathbf{F}^a)$ we have

$$0 \leqq h_\varphi(nP) = h_c(nP) + O(1) = n^2 h_c(P) + O(1).$$

Consequently the quadratic form h_c is positive, not necessarily definite. As usual we define the **associated seminorm** by the square root,

$$|x| = \sqrt{\langle x, x \rangle_c} \qquad \text{where} \qquad \langle x, y \rangle_c = h_c(x + y) - h_c(x) - h_c(y).$$

If x is a torsion point, then we obviously have $|x| = 0$ since by the property of a seminorm, for any positive integer n we have

$$|nx| = n|x|.$$

The converse is false in general (cf. the relative Mordell–Weil in Chapter 6) but true in number fields which we consider in a subsequent section.

In another context, a quadratic bound for the "generic" addition formula on abelian varieties was given by Altman [Al], and is applicable for instance to points with coordinates in a finitely generated ring over $\mathbf{Z}$, where the size of the coefficients is taken into account. This quadratic bound could also be obtained for commutative group varieties, as a consequence of Serre's analysis of the height in this case [Ser 2]. Altman's quadratic bound for the *size* of the coefficients occurring in the addition formula on an abelian variety is an example of the general problem posed in the introduction. It would also be interesting to see if one can give a lower bound on the size. We are not dealing here with heights, since there is no product formula, but with sizes, quite similar to absolute values. For example, consider the polynomial ring $\mathbf{Z}[X]$ in one variable. One defines the size of a polynomial

$$P(X) = \sum_{i=1}^{d} c_i X^i$$

to be $\max(\log |c_i|, \deg P)$, and similarly in several variables. Gauss' lemma already gives a bound for the size under multiplication of polynomials, even a lower bound. The theory of Weil functions as in Chapter 10 is a good candidate for generalization to the size function.

§4. Heights and Poincaré Divisors

Let A be an abelian variety, defined over a field k. Let A' be another abelian variety also defined over k, and let

$$\delta \in \operatorname{Pic}(A \times A')$$

be a divisor class on the product. A divisor class is said to be **defined over** k if it contains a divisor rational over k. For any point $a \in A$ the intersection

$$\delta.(a \times A')$$

is defined as a divisor class on $a \times A'$, and we denote its projection on A' by

$$\delta(a) = \operatorname{pr}_2[\delta.(a \times A')].$$

If D is a divisor in δ, and $D.(a \times A')$ is defined, then $\delta(a)$ is the class in $\operatorname{Pic}(A')$ of the divisor

$$\operatorname{pr}_2[D.(a \times A')].$$

The intersection $D.(a \times A')$ is defined only for a in some Zariski open set of A. On the other hand, changing D by the divisor of a function on A we always get a divisor in the class such that the intersection is defined for a given a.

Inversely, for any point $a' \in A'$ we define the **transpose**

$${}^t\delta(a') = \operatorname{pr}_1[\delta.(A \times a')].$$

By a **Picard variety**, or **dual variety** of A we mean a pair (A', δ) defined over k (that is both A' and δ defined over k) such that, for any extension K of k the map

$$a' \mapsto {}^t\delta(a')$$

gives an isomorphism of $A'(K)$ onto $\operatorname{Pic}_0(A)_K$ (the group of classes algebraically equivalent to 0 defined over K).

Theorem 4.1. *Let A be defined over k. Then a dual variety (A', δ) exists, defined over k. The abelian variety A' is uniquely determined up to k-isomorphism. Given A', the class δ is uniquely determined in $\mathrm{Pic}(A \times A')$. Finally, $(A, {}'\delta)$ is a dual variety of A' (so briefly, we have $A'' = A$).*

In light of the uniqueness, δ is called the **Poincaré class**. Any divisor in that class is called a **Poincaré divisor**.

A proof of this theorem is contained in AV, except that the self-duality is proved only up to a purely inseparable homomorphism. It was later proved by Cartier and Nishi, and is included in [Mu 2]. In AV, to overcome the technical problem of an intersection $D.(a \times A')$ not being defined, one considers the class of

$$D(u + a) - D(u), \quad \text{where } D(u) = \mathrm{pr}_2[D.(u \times A')],$$

for u generic. In that case, taking differences in this manner, the divisor D need be determined only up to trivial divisors of the form

$$X \times A' \quad \text{or} \quad A \times X',$$

where X is a divisor on A and X' a divisor on A'. It is usually better practice to consider the linear equivalence class δ, so that an intersection

$$\delta.(a \times A')$$

is always defined. Then one can add or subtract divisor classes of the form $c \times A'$ or $A \times c'$ to achieve the normalization as in Theorem 3.1, so that in particular

$$\delta(0) = 0 \quad \text{and} \quad {}'\delta(0) = 0.$$

Of course, in the expression $\delta(0)$, the 0 refers to the zero element in A, and $\delta(0)$ is the zero class in $\mathrm{Pic}_0(A')$. A similar remark applies for the transpose relation.

Corollary 4.2. *The Poincaré class is even, that is*

$$[-1]^*(\delta) = \delta.$$

Proof. Immediate from the uniqueness statement.

Suppose now that k is a finite extension of our field $\mathbf{F}$ having a proper set of absolute values satisfying the product formula. Then we have the height theory, and in particular the canonical height h_δ which is a quadratic form since δ is even. All points in A or A' will be supposed to be algebraic over k without this being made explicit any more.

Proposition 4.3. *We have $h_\delta(x, 0) = h_\delta(0, y) = 0$, and $h_\delta(x, y)$ is bilinear in x, y.*

Proof. The first assertion comes from the functoriality of the height with respect to homomorphisms, applied to the embeddings

$$A' \to 0 \times A' \quad \text{and} \quad A \to A \times 0$$

of A, A' on the factors in the product, together with the fact that the Poincaré class induces the trivial class on each factor.

Let u, v be points on $A \times A'$. Let L_δ be the bilinear form associated with h_δ, that is

$$L_\delta(u, v) = h_\delta(u + v) - h_\delta(u) - h_\delta(v).$$

For $x \in A$ and $y \in A'$ let

$$L(x, y) = L_\delta((x, 0), (0, y)).$$

Since $(x, y) = (x, 0) + (0, y)$ we get

$$h_\delta(x, y) = L(x, y)$$

by the first part of the proposition. Thus $h_\delta(x, y)$ is bilinear in x, y.

Theorem 4.4. *For $x \in A$ and $y \in A'$ we have*

$$h_{t_{\delta(y)}}(x) = h_\delta(x, y).$$

Proof. Fix $y \in A'$ and consider the embedding

$$A \to A \times y \to A \times A'$$

given by $x \mapsto (x, y)$. Then as functions of x, by the functoriality of the height, we get

$$h_{t_{\delta(y)}}(x) = h_\delta(x, y) + O(1).$$

But the function $O(1)$ is a bounded function of x, which is equal to a quadratic form plus a linear form by the remarks preceding the theorem. Hence $O(1) = 0$, thus proving the theorem.

Theorem 4.5. *Let $c \in \text{Pic}(A)$ and let $\varphi_c : A \to A'$ be the homomorphism such that $\varphi_c(a) = $ element $a' \in A'$ such that $^t\delta(a') = c_a - c$. Let L_c be the bilinear form derived from h_c. Then for x, $y \in A$ we have*

$$L_c(x, y) = -h_\delta(x, \varphi_c y).$$

If c is even, then

$$h_c(x) = -\tfrac{1}{2}h_\delta(x, \varphi_c x).$$

Proof. Immediate from Theorem 4.4 and Proposition 3.5.

§5. Jacobian Varieties and Curves

Let C be a complete non-singular curve of genus $g \geq 2$, with Jacobian variety J, defined over a field k. Then the duality on J can be described in terms of C as follows. We assume given one of the possible natural embeddings

$$\psi: C \to J.$$

For simplicity of notation, we assume that ψ is an inclusion, so $C \subset J$. This section more or less follows Mumford [Mu 1]. Let Θ be the divisor

$$\Theta = C + \cdots + C,$$

where the sum is taken $g - 1$ times on the Jacobian. We call Θ the **theta divisor**. We let θ be the class of Θ in $\mathrm{Pic}(J)$. We let $\delta_J = \delta$ be the class

$$\delta = -s_2^*(\theta) + p_1^*(\theta) + p_2^*(\theta)$$

in $\mathrm{Pic}(J \times J)$. Note the minus sign attached to $s_2^*(\theta)$! As usual, let

$$\varphi_\theta: J \to \mathrm{Pic}(J)$$

be the homomorphism $a \mapsto \theta_a - \theta$. The minus sign has been selected because of the relations

$$s_2^*(\theta).(u \times J) = u \times \theta_{-u} \quad \text{and} \quad p_2^*(\theta).(u \times J) = u \times \theta.$$

Then $\theta_{-u} - \theta = -(\theta_u - \theta)$ since φ_θ is a homomorphism.

Theorem 5.1. *The pair (J, δ) is a dual variety for J, which is therefore self dual. The homomorphism φ_θ is an isomorphism of J with $\mathrm{Pic}_0(J)$, and we have*

$$'\delta(a) = \varphi_\theta(a).$$

This allows the identification of J with J', compatible with the homomorphism φ_θ.

The above theorem, due to Weil, is also contained in AV, Theorem 3 of Chapter VI, §3. We apply these special relations on J to the theory of heights, following Mumford [Mu 1].

Theorem 5.2. *The form $h_\delta(x, y)$ is symmetric. We have*

$$-h_\delta(x, x) = h_\theta(x) + h_\theta(-x).$$

In other words, if $d: J \to J \times J$ is the diagonal map, then

$$-h_\delta \circ d = h_\theta + h_{\theta^-} = h_{\theta + \theta^-}.$$

Therefore the quadratic form $-h_\delta(x, x)$ is symmetric positive, coming from an ample divisor class on J.

Proof. The class δ is symmetric by definition, so h_δ is symmetric because of the functorial behavior of the height under the isomorphism

$$(x, y) \longmapsto (y, x).$$

Again by functoriality, we have

$$h_\delta(x, x) = h_\delta(dx) = h_{d^*(\delta)}(x).$$

But

$$
\begin{aligned}
-d^*(\delta) &= d^*(s_2^*(\theta) - p_1^*\theta - p_2^*\theta) \\
&= [2]^*(\theta) - 2\theta \\
&= 3\theta + \theta^- - 2\theta \quad \text{(by Mumford's Lemma 2.6)} \\
&= \theta + \theta^-.
\end{aligned}
$$

This proves the theorem, because it is a standard fact that θ is ample.

We apply this to the curve C. We have the embedding

$$C \times J \to J \times J,$$

which is the identity on the second factor, and we view C as contained in its Jacobian to avoid the use of one more symbol ψ to represent the embedding. The existence of the Picard variety for C can be summarized as follows.

Theorem 5.3. *Let $\delta_{C \times J}$ be the restriction of δ to $C \times J$. Suppose C, J are defined over k. For any field K containing k the map*

$$y \longmapsto {}^t\delta_{C \times J}(y)$$

gives an isomorphism of $J(K)$ with $\mathrm{Pic}_0(C)_K$.

Put $c(y) = {}^t\delta_{C \times J}(y)$. The functoriality of heights then gives the formula

5.3.1.
$$h_\delta(x, y) = h_{c(y)}(x) + O(1)$$

for $x \in C$ and $y \in J$. (As always, points are taken now in the algebraic closure of k.) In another notation, let $c \in \mathrm{Pic}_0(C)$ and let $S(c)$ be the corresponding point in J. Then the formula reads

$$h_\delta(x, S(c)) = h_c(x) + O(1).$$

For $c \in \mathrm{Pic}(C)$ the height h_c is determined only up to $O(1)$, which therefore had to be added to the relation between heights. Similarly, we also have for $x, y \in C$:

5.3.2.
$$h_\delta(x, y) = h_{\delta_{C \times C}}(x, y) + O(1),$$

where $\delta_{C \times C}$ is the restriction of δ to $C \times C$.

Proposition 5.4. *Let C be a curve of genus ≥ 1. Let c be a divisor class of degree 1 and b a divisor class of degree 0. Then*

$$h_b = O_{c, b}(h_c^{1/2}) + O_{c, b}(1).$$

Proof. Note that c is ample, so h_c can be taken in its equivalence class to be positive, whence the square root can be taken as in the statement of the lemma. By formula 5.3.2 and the Schwarz inequality, we get

$$|h_b(x)| \leq |h_\delta(x, S(b))| + O(1)$$
$$\leq |h_\delta(x, x)|^{1/2} + O(1).$$

But $-h_\delta(x, x)$ is the height of the ample class $\theta + \theta^-$, which restricts to a class of positive degree on C. We can then apply the inequality of Corollary 3.5 of Chapter 4 to conclude the proof.

For what follows, we need to deal with divisor classes of non-zero degree on C, and hence we have to fix some base class. Thus we let c_0 be an element of $\mathrm{Pic}(C)$ of degree 1. We suppose that the embedding

$$\psi: C \to J \quad \text{is given by} \quad P \mapsto \mathrm{Cl}(P) - c_0.$$

Then ψ extends by linearity to a map

$$S = S_\psi : \mathrm{Pic}(C) \to J.$$

We shall need a standard result due to Weil from the theory of Jacobians, namely AV, Chapter VI, §3, Theorem 3, which we restate here:

Theorem 5.5. *Let $x \in J$ and for any divisor class c on C let $S(c)$ be its image in the Jacobian J. Then*

$$S((\theta_x - \theta).C) = x.$$

This expresses the compatibility of the map φ_θ on J and the pull-back to the curve, with respect to the other identifications we have made.

Proposition 5.6. *The class $\delta_{C \times C}$ contains the divisor class*

$$\mathrm{Cl}(\Delta) - C \times c_0 - c_0 \times C,$$

where Δ is the diagonal on $C \times C$.

Proof. Let $\gamma = \mathrm{Cl}(\Delta) - c_0 \times C - C \times c_0$. It suffices to show that $\delta_{C \times C} - \gamma = 0$. We apply the seesaw principle. Let $P \in C$. We then have a commutative diagram

$$
\begin{array}{ccccc}
 & & C \times C & & \\
 & \nearrow & & \searrow & \\
P \times C & & & & J \times J \xrightarrow{\;s_2\;} J \\
 & \searrow & & \nearrow & \\
 & & P \times J & &
\end{array}
$$

To compute $\delta_{P \times C}$ it suffices to go through the bottom part of the diagram, and we then obtain

$$-[s_2^*(\theta) - p_1^*(\theta) - p_2^*(\theta)].(P \times C) = -[P \times (\theta_{-P} - \theta).C]$$

$$= P \times (\theta_P - \theta).C.$$

By Theorem 5.5, and the definition of the embedding ψ, this gives the same value as $\gamma.(P \times C)$. By symmetry, it follows that $\delta_{C \times C}$ and γ induce the same class on $C \times P$, whence the seesaw principle implies that $\gamma = \delta_{C \times C}$, thus proving Proposition 5.6.

As always, a relation between divisor classes provides a relation between heights, so we obtain:

Proposition 5.7. *For $x, y \in C$ we have*

$$-h_\delta(x, y) + h_\Delta(x, y) = h_{c_0}(x) + h_{c_0}(y) + O(1).$$

If x ≠ y then

$$-h_\delta(x, y) \leq h_{c_0}(x) + h_{c_0}(y) + O(1).$$

Proof. By functoriality, since

$$p_1^*(c_0) = c_0 \times C,$$

we get

$$h_{c_0 \times C}(x, y) = h_{c_0}(x) + O(1),$$

and similarly on the other side. This proves the first relation by Proposition 5.6 and the additivity of the height h_c in terms of c, modulo $O(1)$. By Proposition 5.2 of Chapter 4, we know that h_Δ is bounded from below on elements of $C \times C$ not lying on the diagonal. This proves the proposition.

Theorem 5.2 expressed h_δ in terms of h_θ and h_{θ^-}. Following our general pattern, we can express these heights by pull-back to the curve by using the explicit description of the intersection $\theta . C$ and $\theta^- . C$, or more generally of arbitrary translations of the theta class with C, due to Weil (generalizing classical results of Riemann over the complex numbers). We summarize these relations in a theorem. See [We 5], Theorem 20. Actually, we prove slightly more than we need now for future reference; that is, we give actual intersections of translates of the theta divisor with the curve, rather than deal only with linear equivalence classes. For this purpose, it is useful to recall some definitions.

Let $\mathfrak{a}$ be a divisor on the curve. Let:

$L(\mathfrak{a}) = $ vector space of rational functions f such that $(f) \geq -\mathfrak{a}$;

$l(\mathfrak{a}) = \dim L(\mathfrak{a})$.

Suppose that

$$\mathfrak{a} = \sum_{i=1}^{g} (P_i)$$

is a positive divisor of degree g. Let $x = S(\mathfrak{a})$. We shall say that $\mathfrak{a}$ is **non-special** if it satisfies any one of the following equivalent conditions, letting $\mathfrak{k}$ be the canonical class on C, of degree $2g - 2$.

  (i)   The points $P_1, \ldots, P_g$ are uniquely determined by x, up to a permutation.

  (ii) $l(\mathfrak{a}) = 1$.

  (iii) $x - S(\mathfrak{k}) \notin W_{g-2}^-$, where W_r is the sum of the curve r times with itself on the Jacobian, for $1 \leq r \leq g$.

The equivalence of (i) and (ii) follows from the definitions and the fact that the kernel of S_ψ in $\mathrm{Pic}_0(C)$ is trivial. As to (iii), by the Riemann–Roch theorem, we have

$$l(\mathfrak{a}) > 1 \quad \Leftrightarrow \quad l(\mathfrak{k} - \mathfrak{a}) \geq 1$$

$$\Leftrightarrow \text{ there exists a positive divisor } \mathfrak{b} \text{ such that } \mathfrak{b} \sim \mathfrak{k} - \mathfrak{a}.$$

Write $\mathfrak{b} = \sum\limits_{j=1}^{g-2} (Q_j)$. Then

$$S(\mathfrak{b}) = \sum \psi(Q_j) = S(\mathfrak{k}) - S(\mathfrak{a}),$$

or equivalently $S(\mathfrak{a}) = -S(\mathfrak{b}) + S(\mathfrak{k})$, which means $x - S(\mathfrak{k}) \in W_{g-2}^-$. This proves the equivalence between the three properties.

Any point on the Jacobian can be expressed as the sum of g points on the curve. We shall say that a point x on the Jacobian is **non-special** if it can be expressed as $x = S(\mathfrak{a})$ where $\mathfrak{a}$ is non-special. Then $\mathfrak{a}$ is uniquely determined, and will be called the **divisor on C associated with** x, or the divisor of degree g on C associated with x if we need to specify its degree. We have

$$x = S(\mathfrak{a} - g c_0).$$

Theorem 5.8. *Let $\mathfrak{k}$ be the canonical class on C. Then*

$$\Theta^- = \Theta_{-S(\mathfrak{k})}.$$

If z is a non-special point on J, and $\mathfrak{a}$ its associated divisor, then

$$\Theta^-_z . C = \mathfrak{a}.$$

For any point $z \in J$, the intersection $\theta_z . C$ has degree g, and

$$S(\theta^-_z . C) = z \qquad \text{while} \qquad S(\theta_z . C) = z + S(\mathfrak{k}).$$

This gives the intersections

$$\theta_z . C = c(z) + \mathfrak{k} - (g - 2)c_0,$$

where $c(z)$ is the element of $\mathrm{Pic}_0(C)$ corresponding to z. In particular,

$$\theta . C = \mathfrak{k} - (g - 2)c_0 \qquad \text{and} \qquad \theta^- . C = g c_0.$$

Proof. Let $x \in \Theta$ so that

$$x = P_1 + \cdots + P_{g-1},$$

where $P_1, \ldots, P_{g-1}$ are points of C. By the Riemann–Roch theorem we have

$$l\left(\mathfrak{f} - \sum_{i=1}^{g-1}(P_i)\right) \geq 1.$$

Hence there exists a positive divisor $\mathfrak{a}$ on C linearly equivalent to $\mathfrak{f} - \sum (P_i)$, and $\deg \mathfrak{a} = g - 1$. If $S(\mathfrak{a})$ denotes the image of $\mathfrak{a}$ in J (sum of the points of $\mathfrak{a}$ taken on the Jacobian), then we get

$$S(\mathfrak{a}) = S(\mathfrak{f}) - \sum P_i = S(\mathfrak{f}) - x.$$

Since $\mathfrak{a}$ is positive, of degree $g - 1$, it follows that $S(\mathfrak{a}) \in \Theta$, and hence $-x \in \Theta_{-S(\mathfrak{f})}$. This proves that $\Theta^- \subset \Theta_{-S(\mathfrak{f})}$. Since both divisors are irreducible, this proves the first relation.

The computation of $\theta^-_z.C$ is quite general and does not require the canonical class. We give the argument. Suppose $P \in C$ and $P \in \Theta^-_z$ with z generic. Then there exist points $P_1, \ldots, P_{g-1}$ in C such that

$$P = -P_1 - \cdots - P_{g-1} + z,$$

whence $z = P_1 + \cdots + P_{g-1} + P$. Taking z generic, the g points in C whose sum is z are uniquely determined up to a permutation, so the formula $S(\theta^-_z.C) = z$ follows for z generic, whence for all z. Cf. AV, Proposition 3 of Chapter II, §2. For z non-special we also get $\Theta^-_z.C = \mathfrak{a}$ as asserted. The final assertions concerning the intersections $\theta_z.C$ are then mere translations of the relation

$$S(\theta_z.C) = z + S(\mathfrak{f}),$$

obtained by subtracting from a class of degree m the multiple mc_0 to obtain a class of degree 0.

Proposition 5.9. *We have*

$$(\theta + \theta^-).C = \mathfrak{f} + 2c_0,$$

and thus for $x \in C$,

$$-h_\delta(x, x) = h_{\theta+\theta^-}(x) = h_\mathfrak{f}(x) + 2h_{c_0}(x) + O(1).$$

Proof. This comes from Theorem 5.2 together with the relations of the preceding proposition.

By Proposition 5.8, we see that the class θ is even if and only if $S(\mathfrak{f}) = 0$. Considering that we took the embedding of C into its Jacobian to be defined by

$$\psi(P) = \mathrm{Cl}(P) - c_0,$$

we see that we can make θ even by selecting c_0 such that

$$(2g - 2)c_0 = \mathfrak{f}.$$

This can always be done over a finite extension of the given field of definition of C and J, because the Jacobian is divisible. When c_0 has been selected in this fashion, we shall say that ψ is a **normalized embedding**. We then obtain:

Theorem 5.10. *Assume that ψ is normalized. Then for $x \in C$,*

$$-h_\delta(x, x) = 2gh_{c_0}(x) + O(1).$$

Since the divisors θ and θ^- are ample on J, we have already observed that the symmetric form $-h_\delta(x, y)$ is positive. We call it the **canonical form** on $J \times J$, and as such, we shall now use the standard symbols

$$\langle x, y \rangle = -h_\delta(x, y) \qquad \text{and} \qquad |x| = \sqrt{\langle x, x \rangle}.$$

to denote this scalar product. Putting together Proposition 5.7 and Theorem 5.10, we obtain **Mumford's main theorem**:

Theorem 5.11. *Assume that ψ is normalized. Then for $x, y \in C$:*

$$2g\langle x, y \rangle + 2gh_\Delta(x, y) = \langle x, x \rangle + \langle y, y \rangle + O(1).$$

If $x \neq y$ then

$$2g\langle x, y \rangle \leq |x|^2 + |y|^2 + O(1).$$

In §8 we shall use Mumford's inequality to prove that on a curve of genus ≥ 2, rational points are scattered very thinly.

§6. Definiteness Properties Over Number Fields

In this section, we suppose that all abelian varieties are defined over a number field. All points are taken in $\mathbf{Q}^a$.

Given any class $c \in \mathrm{Pic}(A)$ we let $\langle \ , \ \rangle_c$ be the symmetric bilinear form associated with the canonical height h_c, that is

$$\langle x, y \rangle_c = h_c(x + y) - h_c(x) - h_c(y).$$

We omit the index c if c is fixed throughout a discussion. If c contains a positive divisor, or an ample divisor, then we let

$$|x|_c = |x| = \sqrt{\langle x, x \rangle}.$$

Theorem 6.1. *Suppose that A is defined over a number field. Suppose that c contains an ample divisor and is even. For any point $x \in A(\mathbf{Q}^a)$ we have $|x|_c = 0$ if and only if x is a torsion point.*

Proof. For any positive integer n we have

$$|nx| = n|x|.$$

Suppose A is defined over a finite extension K of $\mathbf{Q}$ and that x is rational over K. Then nx is rational over K. There is only a finite number of points in $A(K)$ of bounded height, so the theorem follows.

Theorem 6.2. *Let W be a non-singular complete curve over a number field, and let $c \in \text{Pic}(W)$ be a divisor class. Then h_c is bounded if and only if c is of finite order. (Since here we are not on an abelian variety, h_c is only defined up to a bounded function.)*

Proof. Suppose $c \in \text{Pic}_0(W)$ so c or $-c$ is ample. Let $f(x, y) = 0$ be an affine equation for W with coefficients in the number field k. Giving x arbitrary values in k, one sees that W has infinitely many points in extensions of bounded degree over k. Then h_c is unbounded on such points by Theorem 2.6 of Chapter 3. Hence we may assume $c \in \text{Pic}_0(W)$. We have

$$h_c(P) = -h_\delta(P, S(c)) + O(1)$$

by 5.3.1, and h_δ is bilinear on $J \times J$. But J is the sum of W taken g times where g is the genus. If h_c is bounded on W then writing an arbitrary point $x \in J$ as a sum

$$x = P_1 + \cdots + P_g,$$

with $P_i \in W$, we see that the linear function

$$x \mapsto -h_\delta(x, S(c))$$

is bounded on J, hence equal to 0. Select $x = S(c)$. Then we conclude that

$$-h_\delta(S(c), S(c)) = 0.$$

But Theorem 5.2 shows that $-h_\delta(x, x) = h_{\theta + \theta^-}(x)$ and $\theta + \theta^-$ is ample. By Theorem 6.1 it follows that $S(c)$ is of finite order. This proves the theorem since the converse is trivial.

Observe that on a Jacobian variety, we have already shown that the bilinear form

$$-h_\delta(x, x)$$

is positive definite on $J \times J$.

For any abelian variety, we have an analogous statement.

Theorem 6.3. *Let A be an abelian variety over a number field, and let (A', δ) be its dual variety. Then the kernels on each side of the pairing*

$$(x, y) \mapsto h_\delta(x, y)$$

are the torsion groups A_{tor} and A'_{tor}. In other words, if y is orthogonal to all $x \in A$ then $y \in A'_{\mathrm{tor}}$, and similarly on the other side.

Proof. Let y correspond to the divisor class $c \in \mathrm{Pic}_0(A)$, that is $c = {}^t\delta(y)$. If c is not of finite order, then we let W be any non-singular curve W in A such that W generates A. [In a projective embedding, the intersection of A with a sufficiently general linear variety of suitable dimension yields such a curve.] Then $c \cdot W$ is not of finite order in $\mathrm{Pic}(W)$. This is an easy fact, since the homomorphism $J(W) \to A$ is surjective, and hence the transpose homomorphism $\mathrm{Pic}_0(A) \to \mathrm{Pic}_0(J)$ has finite kernel, and the intersection with C is injective on $\mathrm{Pic}_0(J)$. But on W we have

$$h_{c \cdot W} = h_c + O(1);$$

and on A, by Theorem 4.4,

$$h_c(x) = h_\delta(x, S(c)).$$

We can now apply Theorem 6.2 to conclude the proof.

Next we give an application of non-degeneracy to $\mathrm{Pic}(A)$.

Theorem 6.4. *Let A be defined over a number field. Then the kernel of the homomorphism $c \mapsto h_c$ from*

$$\mathrm{Pic}(A) \to \text{quadratic functions on } A(\mathbf{Q}^a)$$

consists of the elements of finite order in $\mathrm{Pic}(A)$.

Proof. The case when $c \in \mathrm{Pic}_0(A)$ is a special case of Theorem 6.3. Suppose now that c is not algebraically equivalent to 0. By AV, Corollary of Theorem 1, Chapter V, §3 we know that c is not numerically equivalent to 0. Consequently there exists a curve W in A such that $c \cdot W$ on A has degree $\neq 0$. We now have the minor technical difficulty that W may have singularities. Let

$$f: W^* \to W$$

be a birational map from a non-singular curve W^* onto W, for instance the desingularization of W itself. Then $f^*(c)$ has non-zero degree on W^*, and we have

$$h_{f^*(c)} = h_c \circ f + O(1).$$

Say $f^*(c)$ has positive degree. Then $f^*(c)$ is ample on W^*, so $h_{f^*(c)}$ is unbounded on the set of algebraic points of W^*. Hence h_c cannot be 0, since $h_c \circ f - h_{f^*(c)}$ is bounded on W^*. This concludes the proof.

There is an analogous theorem for arbitrary varieties, namely:

Theorem 6.5. *Let V be a projective non-singular variety defined over a number field. The kernel of the homomorphism*

$$c \mapsto h_c \text{ modulo bounded functions}$$

from $\mathrm{Pic}(V)$ to functions on $V(\mathbf{Q}^a)$ modulo bounded functions consists of the points of finite order in $\mathrm{Pic}(V)$.

Proof. If $c \in \mathrm{Pic}_0(V)$, then one can use the theory of the Picard variety of V, namely the existence of a morphism

$$\varphi: V \to A$$

into its Albanese variety, giving rise to a morphism

$$(\varphi, \mathrm{id}): V \times A' \to A \times A'$$

such that $(\varphi, \mathrm{id})^*(\delta)$ is a Poincaré divisor on $V \times A'$, parametrizing $\mathrm{Pic}_0(V)$. The present theorem is an easy consequence of Theorem 6.3, with an argument similar to that of Theorem 6.2.

On the other hand, if $c \notin \mathrm{Pic}_0(V)$, then one has to know the fact used in Theorem 6.4 when c does not have finite order modulo $\mathrm{Pic}_0(V)$, that there exists a curve W in V such that $c \cdot W$ has non-zero degree. One can then argue as in Theorem 6.4 to conclude the proof.

§7. Non-degenerate Heights and Euclidean Spaces

In this section we let A be an abelian variety defined over a finite extension K of the field $\mathbf{F}$, with a proper set of absolute values satisfying the product formula, so we have the theory of heights. Points of A will always be in $A(\mathbf{F}^a)$.

We recall that a positive function h on an abelian group Γ is said to be **non-degenerate** if given a number $B > 0$ there is only a finite number of elements $x \in \Gamma$ such that $h(x) \leqq B$. We shall be interested in the case of non-degenerate heights.

The fundamental fact which we exploit over number fields is:

Theorem 7.1. *If $\mathbf{F} = \mathbf{Q}$, and c contains an ample divisor, then h_c is non-degenerate on every finitely generated subgroup Γ of A.*

Indeed, $\Gamma \subset A_K$ for some finite extension K of $\mathbf{Q}$ and there is only a finite number of points on A of bounded height in such an extension.

As in the previous section, we let

$$|x|_c = |x| = \sqrt{\langle x, x \rangle_c}$$

be the norm associated with $c \in \mathrm{Pic}(A)$ containing an ample divisor.

Whether over a number field or in the general situation, we always have $|x| = 0$ if x is a torsion point. The kernel of the natural map

$$A(\mathbf{F}^a) \to \mathbf{R} \otimes A(\mathbf{F}^a)$$

consists precisely of the group of torsion points. If c is a class on A containing an ample or a positive divisor, and c is even, then we may extend the quadratic form h_c to the $\mathbf{R}$-vector space $\mathbf{R} \otimes A(\mathbf{F}^a)$ by $\mathbf{R}$-linearity on the bilinear form associated with h_c. Also note that $A(\mathbf{F}^a)$ is the union of all finitely generated subgroups, and that

$$\mathbf{R} \otimes A(\mathbf{F}^a) = \bigcup_{\Gamma} \mathbf{R} \otimes \Gamma,$$

where Γ ranges over all finitely generated subgroups of $A(\mathbf{F}^a)$. If $\mathbf{F} = \mathbf{Q}$ (the classical case), then we shall prove in the next chapter that for any number field K, the group A_K is finitely generated. We have

$$\mathbf{R} \otimes A(\mathbf{Q}^a) = \bigcup_{K} \mathbf{R} \otimes A_K,$$

and each $\mathbf{R} \otimes A_K$ is a finite dimensional vector space over $\mathbf{R}$. To test whether the extension of h_c to $\mathbf{R} \otimes A(\mathbf{F}^a)$ is positive definite, it suffices to test whether the extension of h_c to the finite dimensional space $\mathbf{R} \otimes \Gamma$ is positive definite, for every finitely generated subgroup Γ of $A(\mathbf{F}^a)$. For each Γ, we have an injection.

$$\Gamma/\Gamma_{\text{tor}} \to \mathbf{R} \otimes \Gamma,$$

mapping Γ modulo its torsion subgroup onto a lattice in $\mathbf{R} \otimes \Gamma$.

Theorem 7.2. *Let A be an abelian variety defined over a number field. Let c be a divisor class containing an ample divisor, and even. Then the extension of h_c to $\mathbf{R} \otimes A(\mathbf{Q}^a)$ is a positive definite quadratic form.*

Proof. When Néron first proved his theorem, it was believed to be automatic that the quadratic form as above is positive definite. Cassels pointed out that the matter is not quite so automatic; one needs to use Minkowski's elementary theorem that a convex set which is symmetric about the origin contains a non-zero lattice point if its volume is sufficiently large. (The reader will find a proof in almost any book on algebraic number theory.) Then the following is true.

Theorem 7.3. *Let V be a finite dimensional vector space over $\mathbf{R}$, and let L be a lattice in V. Let h be a quadratic form on V. Assume that h is positive non-degenerate on L, that is:*

(i)   *If P is in the lattice then $h(P) \geq 0$.*
(ii) *There is only a finite number of $P \in L$ with bounded $h(P)$.*

Then h is positive definite on V.

Proof. There exists a basis of V such that for any point X in V with coordinates $(x_1, \ldots, x_n)$ with respect to this basis we have

$$h(X) = \sum_{i=1}^{r} x_i^2 - \sum_{i=r+1}^{m} x_i^2, \qquad m \leq \dim V.$$

If we write $X = (X', X'', X^*)$, where X' is the projection of X on the subspace where h is positive definite, X'' the projection where h is negative definite, and X^* the projection where h is a null form, then

$$h(X) = X'^2 - X''^2.$$

Let $b, B > 0$. The inequalities

$$X'^2 < b \qquad \text{and} \qquad X''^2 < B$$

define a symmetric convex set in V. Taking b small and B large, we conclude from (i) and Minkowski's theorem that h is positive semidefinite on V. Again taking b small and allowing large coordinates for X^*, we conclude from (ii) and Minkowski's theorem that h has to be positive definite on V, as desired.

Theorem 7.3 also has the interpretation more generally over an algebraic extension of $\mathbf{F}$ that if Γ is a finitely generated subgroup of A in $A(\mathbf{F}^a)$, such that one of the canonical height functions h_c is positive non-degenerate on Γ, then the extension of h_c to the finite dimensional vector space $\mathbf{R} \otimes \Gamma$ is positive definite. This also occurs frequently in the function field case.

We shall now exploit systematically properties of positive non-degenerate heights on finitely generated subgroups of A, looking at their images in euclidean space, and using Theorem 7.3. We begin by some counting, which we base on the next quite general result.

Theorem 7.4. *Let V be a finite r-dimensional vector space over the real numbers, and let L be a lattice in V. Let q be a positive definite quadratic form on V. For each positive real number B let $L_q(B)$ be the set of elements $P \in L$ such that $q(x) \leqq B$. Then the cardinality $L_q(B)$ satisfies the asymptotic relation*

$$|L_q(B)| = \frac{\mathrm{Vol}(D)}{\mathrm{Vol}(L)} B^{r/2} + O(B^{(r-1)/2}),$$

where $\mathrm{Vol}(L)$ is the volume of a fundamental domain, and $\mathrm{Vol}(D)$ is the volume of the unit ball with respect to the quadratic form.

This theorem is a special case of Theorem 5.1, Chapter 3, giving the number of lattice points in an expanding region with smooth boundary. The set of points P such that $q(P) = 1$ is a sphere, so smooth. It is easily shown that such a submanifold is $(r - 1)$-Lischitz parametrizable as required by the theorem. The exponent $r/2$ comes instead of r because we counted by the quadratic form, and the norm on V determined by q is the square root of the form.

We apply this to get Néron's estimate for points of bounded height on abelian varieties, cf [Ne 3].

Theorem 7.5. *Let A be an abelian variety over K and let*

$$\varphi: A \to \mathbf{P}^n$$

be a projective embedding. Let c be the associated divisor class of the inverse image of a hyperplane, and

$$h_c = q_c + l_c$$

the corresponding canonical height. Assume that q_c is non-degenerate on a finitely generated subgroup Γ of $A(K)$. Let $N_{\varphi,\Gamma}(B)$ be the number of points $x \in \Gamma$ such that

$$h_\varphi(x) \leqq B.$$

Let r be the rank of Γ. There exists a constant $a_\Gamma > 0$ such that

$$N_{\varphi,\Gamma}(B) = a_\Gamma B^{r/2} + O(B^{(r-1)/2}).$$

Proof. The proof will be a simple application of Theorem 5.4, and we shall also determine explicitly the constant a_Γ. We let

$$V = \mathbf{R} \otimes \Gamma \qquad \text{and} \qquad L = \Gamma/\Gamma_{\text{tor}} \text{ in } V.$$

We have $h_\varphi = q_c + l_c + O(1)$. Let $L_\varphi(B)$ be the set of points x in L such that $h_\varphi(x) \leqq B$. Then there exist constants $C_1, C_2 > 0$ such that

$$L_q(B - C_1 B^{1/2} - C_2) \subset L_\varphi(B) \subset L_q(B + C_1 B^{1/2} + C_2),$$

because $l(x)$ is linear in x. For any torsion point t we have $q_c(x + t) = q_c(x)$. Consequently if we apply Theorem 7.4 to the quadratic form q_c we immediately deduce the desired estimate for $N_{\varphi,\Gamma}(B)$, with the constant

$$\boxed{a_\Gamma = \frac{\text{Vol}(D)w(\Gamma)}{\text{Vol}(L)}}$$

where $w(\Gamma)$ is the order of the torsion group Γ_{tor}.

Next to prove the following result I use an argument of Cassels which I reproduced incorrectly in [L 7], as many people pointed out to me.

Theorem 7.6. *Let K be a number field. Let Γ be a finitely generated subgroup of $A(K)$, and let $\{P_1, \ldots, P_r\}$ be a basis of Γ mod torsion. Let $P \in A(K)$ be a torsion point mod Γ, and let N be the smallest integer > 0 such that $NP \in \Gamma + A(K)_{\text{tor}}$. Let $|\ | = |\ |_c$ be the norm associated with the positive definite quadratic form h_c of an ample even divisor class. Then we have an inequality*

$$N \leqq \gamma^r(|P_1| + \cdots + |P_r|)^r,$$

where the constant γ depends only on A and on $[K : \mathbf{Q}]$.

Proof. Let $m_1, \ldots, m_r$ be integers and $P_0 \in \Gamma_{\text{tor}}$ be such that

$$NP = m_1 P_1 + \cdots + m_r P_r + P_0.$$

Let $1 \leq q \leq N$. Let $M = [N^{1/r}] - 1$. Cut each side of the unit cube into M segments of equal length $1/M$, so the unit cube is decomposed into M^r smaller cubes, with $M^r < N$. By Dirichlet's box principle, there are two integers q_1, q_2 with $1 \leq q_1 < q_2 \leq N$ such that

$$(q_1 m_1/N, \ldots, q_1 m_r/N) \quad \text{and} \quad (q_2 m_1/N, \ldots, q_2 m_r/N) \bmod \mathbf{Z}^r$$

lie in the same box. Let $q = q_2 - q_1$ so $1 \leq q < N$. Then there exist integers $s_1, \ldots, s_r$ such that

$$|qm_j/N - s_j| \leq 1/M \quad \text{for } j = 1, \ldots, r.$$

Let

$$n_j = qm_j - Ns_j \quad \text{so that } n_j \leq N/M.$$

Let $Q = qP - \sum_{j=1}^{r} s_j P_j$. Then

$$NQ = n_1 P_1 + \cdots + n_r P_r + P_0'$$

for some torsion point P_0'. But $Q \notin A(K)_{\text{tor}}$ (otherwise $q < N$ is a period for $P \bmod \Gamma + A(K)_{\text{tor}}$). Hence $|Q| \neq 0$, and in fact $|Q|$ is greater than some constant $C > 0$ depending only on the minimal height of points on A of degree bounded by $[K : \mathbf{Q}]$. Taking norms, we find

$$NC \leq NM^{-1} \sum_{j=1}^{r} |P_j|.$$

whence

$$[N^{1/r}] - 1 \leq M \leq C^{-1} \sum_{j=1}^{r} |P_j|.$$

Without loss of generality we can replace C^{-1} by a suitable constant γ_1 so that the right hand side is ≥ 2. The desired inequality then follows at once, with a suitable γ, depending only on A and $[K : \mathbf{Q}]$.

Finally it is a major problem to give upper bounds for a suitably selected basis of the lattice of rational points of the Mordell–Weil group, or to find generators effectively. For the convenience of the reader, I reproduce a general theorem of Hermite giving an almost orthogonalized basis for a lattice, and showing that bounds for the norms of basis elements are given in terms of the volume of a fundamental domain. For conjectures concerning explicit upper bounds for the regulator, see [L 15].

Theorem 7.7. *Let L be a lattice in a vector space V of dimension r over $\mathbf{R}$, with a positive definite quadratic form. Then there exists an orthogonal basis $\{u_1, \ldots, u_r\}$ of V, and a basis $\{e_1, \ldots, e_r\}$ of L having the following properties.*

(i) $e_1 = u_1$ *is a vector of minimal length in L.*

(ii)
$$e_i = b_{i,1} u_1 + \cdots + b_{i,i-1} u_{i-1} + u_i,$$

with $b_{i,j} \in \mathbf{R}$ for $i = 1, \ldots, r$ and $|b_{i,j}| \leq \frac{1}{2}$ for all i, j.

(iii)
$$|u_i| \leq |e_i| \leq \left(\frac{2}{\sqrt{3}}\right)^{i-1} |u_i|.$$

(iv)
$$|u_i| \leq \frac{2}{\sqrt{3}} |u_{i+1}|.$$

Remark. Before giving the proof, we point out that relations (ii) can be written more transparently in matrix form:

$$\begin{pmatrix} e_1 \\ \vdots \\ e_r \end{pmatrix} = \begin{pmatrix} 1 & & & 0 \\ & 1 & & \\ & b_{ij} & \ddots & \\ & & & 1 \end{pmatrix} \begin{pmatrix} u_1 \\ \vdots \\ u_r \end{pmatrix},$$

with $|b_{ij}| \leq \frac{1}{2}$.

Proof. Let $L_1 = L$. Let $u_1 \in L_1$ be a vector of minimal length in L_1, and let $e_1 = e_{1,1} = u_1$.

Let pr_2 be the orthogonal projection on $\langle u_1 \rangle^{\perp}$ and let $L_2 = \mathrm{pr}_2(L)$. Since V/L is compact, its projection is compact, so L_2 is a lattice in the vector space over $\mathbf{R}$ generated by L_2. Let $u_2 \in L_2$ be a vector of minimal length. Let $e_2 = e_{2,1}$ be a shortest vector in L_1 projecting on u_2 in $\langle u_1 \rangle^{\perp}$. Since the projection is norm-decreasing, it follows that

$$|u_1| \leq |e_2|$$

by the minimality of $|u_1|$ in L_1. We can write

$$e_2 = bu_1 + u_2$$

with real coefficient b satisfying $|b| \leqq \frac{1}{2}$, because we can add and subtract integral multiplies of u_1 to decrease the length if necessary. We then obtain

$$|u_1|^2 \leqq |e_2|^2 = b^2|u_1|^2 + |u_2|^2,$$

whence

$$|u_1|^2 \leqq \tfrac{4}{3}|u_2|^2$$

and therefore also

$$|u_2|^2 \leqq |e_2|^2 \leqq \tfrac{1}{4}\tfrac{4}{3}|u_2|^2 + |u_2|^2 \leqq \tfrac{4}{3}|u_2|^2.$$

This proves the desired inequalities for u_2, e_2.

Let pr_3 be the orthogonal projection on $\langle u_1, u_2 \rangle^{\perp}$, and $L_3 = \mathrm{pr}_3(L)$. Let u_3 be a vector of minimal length in L_3. Let:

$e_{3,2} = $ a shortest vector in L_2 projecting on u_3 under pr_3;

$e_{3,1} = $ a shortest vector in L_1 projecting on $e_{3,2}$ under pr_2.

We let $e_{3,1} = e_3$. Then

$$e_3 = e_{3,1} = bu_1 + e_{3,2},$$

with some real number b such that $|b| \leqq \frac{1}{2}$. The first step in the inductive argument shows that

$$|u_2|^2 \leqq |e_{3,2}|^2 \leqq \tfrac{4}{3}|u_3|^2,$$

and then

$$|u_3|^2 \leqq |e_{3,1}|^2 \leqq \tfrac{4}{3}|e_{3,2}|^2 \leqq (\tfrac{4}{3})^2|u_3|^2.$$

This settles the second step in the inductive argument. We may then continue in similar fashion, to get an orthogonal basis $\{u_1, \ldots, u_r\}$, and elements $e_{i,j-1} = $ shortest vector in L_{j-1} which projects on e_{ij} under pr_j. We let

$$e_1 = e_{1,1}, e_2 = e_{2,1}, \ldots, e_r = e_{r,1}.$$

All the desired inequalities of the theorem are then satisfied inductively. There remains to see that $\{e_1, \ldots, e_r\}$ is a basis of L. But given any vector v in L, there exists an integer k such that $v - ku_r$ projects to 0 under pr_r, and therefore also $v - ke_r$ projects to 0 under pr_r. Then $v - ke_r$ lies in $\langle u_1, \ldots, u_{r-1} \rangle$ and is also an element of L. One may then proceed by induction to conclude the proof.

A basis of the lattice as in Theorem 7.7 will be said to be **almost orthogonalized**. We recall that the **determinant** of a lattice L is defined to be the volume of a fundamental domain. If $\langle\ ,\ \rangle$ denotes the symmetric bilinear form such that

$$|P| = \sqrt{\langle P, P \rangle} \quad \text{for } P \in L,$$

then Hadamard's inequality states that

$$\det(L) \leq |P_1| \cdots |P_r|$$

for any basis $\{P_1, \ldots, P_r\}$ of L. This inequality is obvious, because the volume of a parallelepiped is no greater than the product of the lengths of the sides.

If $\{P_1, \ldots, P_r\}$ is an orthogonal basis of the lattice, then also trivially we have

$$\det(L) = |P_1| \cdots |P_r|.$$

Corollary 7.8. *Let $\{e_1, \ldots, e_r\}$ be an almost orthogonalized basis of L. Then*

(1)
$$|e_1| \cdots |e_r| \leq \left(\frac{2}{\sqrt{3}}\right)^{r(r-1)/2} \det(L).$$

(2)
$$|e_1| \leq \left(\frac{2}{\sqrt{3}}\right)^{(r-1)/2} \det(L)^{1/r}.$$

Proof. The basis $\{e_1, \ldots, e_r\}$ of L is obtained from the orthogonal basis of V by a triangular matrix all of whose diagonal elements are equal to 1. The volume of the rectangular box spanned by $\{u_1, \ldots, u_n\}$ is equal to the product of the lengths of the sides, and so

$$\det(L) = |u_1| \cdots |u_r|.$$

The first assertion follows by using (iii) in Theorem 7.7. The second follows immediately.

The vectors $u_1, \ldots, u_r$ provide an orthogonal basis for V, so that if we take an orthonormal basis in their directions, we can identify V with Euclidean space $\mathbf{R}^r$. The next corollary describes the divergence arising from taking the vectors $e_1, \ldots, e_r$ as a basis. Note that with respect to the u_i we have Pythagoras' theorem

$$\left|\sum y_i u_i\right|^2 = \sum |y_i u_i|^2.$$

Corollary 7.9. *Under the same conditions, if* $y_1, \ldots, y_r$ *are real numbers, then*

$$\gamma_1(r)^{-1} \left| \sum y_i u_i \right| \le \left| \sum y_i e_i \right| \le \gamma_2(r) \left| \sum y_i u_i \right|,$$

where

$$\gamma_1(r) = r \cdot \max(\tfrac{1}{3}\sqrt{3}^{r-1}, 1) \quad and \quad \gamma_2(r) = r \cdot \max\left(\frac{1}{2}\left(\frac{2}{\sqrt{3}}\right)^{r-1}, 1 \right).$$

Proof. We do some linear algebra. If

$$T : V \to V$$

is a linear map, we let $|T|$ be its norm, defined by

$$|Tv| \le |T||v| \quad \text{for all } v \in V,$$

and $|T|$ is the smallest real number ≥ 0 which makes this inequality valid. If $\{w_1, \ldots, w_r\}$ is an orthonormal basis of V, then any vector can be written in terms of coordinates

$$X = \sum x_i w_i,$$

and T can be represented by a matrix (t_{ij}). In that case, relative to the basis, we can define the **sup norm**

$$\|T\| = \max |t_{ij}|.$$

Then we have the inequality between the operator norm and sup norm,

$$|T| \le r\|T\|,$$

which follows easily from the Schwarz inequality applied to the components $\sum t_{ij} x_j$ of TX. We consider the linear map

$$T : V \to V \quad \text{such that } Tu_i = e_i$$

for $i = 1, \ldots, r$. We use the orthonormal basis such that

$$u_i = c_i w_i \quad \text{with } c_i > 0.$$

We have

$$T(w_i) = T(c_i^{-1} u_i) = c_i^{-1} e_i = w_i + \sum_{j=1}^{i-1} c_i^{-1} b_{ij} u_j$$

$$= w_i + \sum_{j=1}^{i-1} c_j c_i^{-1} b_{ij} w_j.$$

Then T is represented by a triangular matrix, with 1 on the diagonal, and

$$|c_j c_i^{-1} b_{ij}| \leq \frac{1}{2} \left(\frac{2}{\sqrt{3}} \right)^{i-j}.$$

Therefore

$$\|T\| \leq \max\left(\frac{1}{2} \left(\frac{2}{\sqrt{3}} \right)^{r-1}, 1 \right).$$

The right inequality of the lemma follows from the definition of T. As to the left inequality, we have

$$|\sum y_i u_i| = |\sum y_i T^{-1} e_i|$$
$$\leq |T^{-1}| |\sum y_i e_i|.$$

The corollary follows when we have estimated $|T^{-1}|$. We write

$$\text{matrix of } T = I - X \quad \text{where } X = (x_{ij}) \text{ is nilpotent,}$$

and in fact $X^r = O$. Then

$$\text{matrix of } T^{-1} = I + X + X^2 + \cdots + X^{r-1}.$$

If one is not too careful about the estimate, one then gets an estimate like $c^{r \log r}$. I am indebted to Roger Howe for the following lemma which yields a constant to the power r.

Lemma 7.10. *Let $X = (x_{ij})$ be an upper triangular matrix, so $x_{ij} = 0$ if $i \geq j$. Suppose that there are positive numbers b, c such that*

$$|x_{ij}| \leq bc^{j-i}$$

for $i < j$. Then

$$\|X + X^2 + \cdots + X^{r-1}\| \leq b(b+1)^{r-2} c^{r-1}.$$

Proof. To estimate the left hand side, we may replace x_{ij} by bc^{j-i} for $i < j$. We write

$$X = C^{-1} B C,$$

where

$$C = \begin{pmatrix} 1 & & \cdots & & 0 \\ & c & & & \\ \vdots & & c^2 & & \vdots \\ & & & \ddots & \\ 0 & & \cdots & & c^{r-1} \end{pmatrix} \quad \text{and} \quad B = \begin{pmatrix} 0 & b & \cdots & & b \\ 0 & 0 & b & & b \\ \vdots & \vdots & \ddots & \ddots & \vdots \\ 0 & 0 & \cdots & 0 & b \\ 0 & 0 & & \cdots & 0 \end{pmatrix}$$

Also let

$$
R = \begin{pmatrix}
0 & 1 & 0 & \cdots & 0 \\
0 & 0 & 1 & \cdots & 0 \\
\vdots & \vdots & \ddots & \ddots & \vdots \\
0 & 0 & \cdots & 0 & 1 \\
0 & 0 & \cdots & & 0
\end{pmatrix}
$$

have components 1 above the diagonal and zeros elsewhere. Then

$$
B = b(R + R^2 + \cdots + R^{r-1}) = bQ, \quad \text{say};
$$

and

$$
Q^k = \sum_{l=0}^{r-1} \binom{l-1}{k-1} R^l.
$$

Furthermore letting an inequality of matrices mean inequality of corresponding coefficients, we obtain:

$$
\begin{aligned}
& B + B^2 + \cdots + B^{r-1} \\
&= bQ + b^2 Q^2 + \cdots + b^{r-1} Q^{r-1} \\
&= bR + (b + b^2)R^2 + (b + 2b^2 + b^3)R^3 + \cdots + b(b+1)^{k-1} R^k + \cdots \\
&= b \begin{pmatrix}
0 & 1 & (b+1) & \cdots & (b+1)^{r-2} \\
 & \ddots & \ddots & & \vdots \\
 & & \ddots & \ddots & (b+1) \\
 & & & \ddots & 1 \\
 & & & & 0
\end{pmatrix}
\end{aligned}
$$

Conjugating by C yields the desired estimate of the lemma. We let

$$
b = \tfrac{1}{2} \quad \text{and} \quad c = \frac{2}{\sqrt{3}}.
$$

Substituting and using $|T^{-1}| \leq r\|T^{-1}\|$ immediately yields Corollary 7.9.

§8. Mumford's Theorem

Points in a finitely generated subgroup of an abelian variety are distributed essentially like lattice points in a Euclidean space. Mumford applied his estimate for the height to show that on a curve of genus ≥ 2, the points are even more thinly distributed as in the next theorem [Mu 1].

Theorem 8.1. *Let C be a curve of genus ≥ 2 and let h be the height with respect to any projective embedding. Let $\{x_n\}$ be a sequence of distinct points on C (always taken in $C(\mathbf{F}^a)$) lying in some finitely generated subgroup Γ of J, and ordered by increasing height. Assume that the associated quadratic form of h is positive non-degenerate on Γ. Then there is an integer N and a number $a > 1$ such that for all n we have*

$$h(x_{n+N}) \geq ah(x_n).$$

Proof. On a curve, two height functions h_c and $h_{c'}$ corresponding to classes of degree ≥ 1 are of the same order of magnitude, that is

$$h_c \gg\ll h_{c'},$$

as shown in Corollary 3.4 of Chapter 4. Consequently we may assume without loss of generality that h is the height corresponding to a normalized embedding of C in J as in Theorem 5.11, and we may work with the associated norm instead of h, so we have to prove

$$|x_{n+N}| \geq a|x_n|$$

with suitable a and N.

We define the cosine between points x, y with $|x| \neq 0, |y| \neq 0$ to be

$$\cos(x, y) = \frac{\langle x, y \rangle}{|x||y|}.$$

We have Mumford's inequality (Theorem 5.11)

$$2g\langle x, y \rangle \leq |x|^2 + |y|^2 + O(1).$$

We order the sequence $\{x_n\}$ by ascending norm, so

$$|x_n| \leq |x_{n+1}|,$$

and without loss of generality, we may assume that $|x_n| \neq 0$ for all n. We let V be the finite dimensional subspace in which all the x_n lie, or rather their images mod Γ_{tor}. After replacing $\{x_n\}$ by a subsequence if necessary, taking into account the fact that Γ_{tor} is finite, we may identify x_n with its image in $V = \mathbf{R} \otimes \Gamma$. We use the following lemma.

Lemma 8.2. *Given $\varepsilon > 0$, there exists an integer N such that if $v_1, \ldots, v_N$ are non-zero elements of V, then for some pair of integers $1 \leq i, j \leq N$ we have*

$$\cos(v_i, v_j) \geq 1 - \varepsilon.$$

The lemma is obvious: we can cover the unit sphere in V by small balls of very small diameter, such that any two vectors on the unit sphere which lie in the same small ball make an angle of cosine very close to 1. We let N be larger than the number of such balls. Then two vectors among $v_1, \ldots, v_N$ must lie in the same small ball.

Mumford's inequality can then be rewritten in the form

$$g \cdot \cos(x, y) \leqq \frac{1}{2} \left(\frac{|x|}{|y|} + \frac{|y|}{|x|} \right) + \frac{O(1)}{|x| \, |y|}.$$

If n is sufficiently large, then $|x_n|$ is large, and if $|x|, |y|$ are large, then the term $O(1)/|x| \, |y|$ is very small.

Let n be large, and apply the lemma to the elements x_i with

$$n \leqq i \leqq n + N.$$

There exists a pair i, j with $n \leqq i \leqq j \leqq n + N$ such that

$$g(1 - \varepsilon) \leqq g \cdot \cos(x_i, x_j) \leqq \frac{1}{2} \left(s + \frac{1}{s} \right) + O\left(\frac{1}{|x_i| \, |x_j|} \right),$$

where $s = |x_j| / |x_i| \geqq 1$. The graph of the function $f(s) = \frac{1}{2}(s + 1/s)$ is well known, and takes its minimum at $s = 1$ with $f(1) = 1$. Since $g \geqq 2$, to have the above inequality s must be large, or specifically, there exists $a > 1$ such that $s \geqq a$. This means

$$|x_j| \geqq a |x_i|.$$

This proves the theorem. For ε small and x_i large, a is approximately the number s such that $s + s^{-1} = 2g$.

Remark. The numbers a and N depend only on the embedding $C \to J$, h, $|\Gamma_{\mathrm{tor}}|$, and the rank of Γ. They do not otherwise depend on Γ.

The most important case occurs when C and J are defined over a number field K, in which case the Mordell–Weil theorem asserts that J_K is finitely generated, and the hypotheses of Theorem 8.1 are satisfied. As Mumford observes, in the concrete case of, say, the Fermat curve over the rationals

$$u^d + v^d = w^d,$$

with $d \geqq 4$, suppose we have a sequence of solutions

$$P_n = (u_n, v_n, w_n)$$

in relatively prime integers, so that

$$H(P_n) = \max(|u_n|, |v_n|, |w_n|).$$

Then there exist numbers $a_1 > 0$ and $b > 1$ such that

$$\log H(P_n) \geqq a_1 b^n,$$

so

$$H(P_n) \geqq e^{a_1 b^n}$$

grows doubly exponentially.

CHAPTER 6

The Mordell–Weil Theorem

We consider abelian varieties, defined over essentially global fields, namely, those of §2, §3, §4, Chapter 2. We shall prove an absolute result and a relative one concerning the group of rational points of an abelian variety over such fields, namely:

Theorem 1. *Let K be a finitely generated field over the prime field. Let A be an abelian variety defined over K. Then $A(K)$ is finitely generated.*

When $K = \mathbf{Q}$, $\dim A = 1$, the theorem is due to Mordell [Mo 1]. For arbitrary A over number fields, it is due to Weil [We 2], and the extension to finitely generated fields is due to Néron [Ne 1]. For a more extensive historical discussion, see the notes at the end of the chapter.

To state the second theorem, we recall a definition. Let K be a function field over the constant field k. Let A be an abelian variety defined over K. A K/k-**trace** of A is a pair (B, τ) consisting of an abelian variety B defined over k and a homomorphism

$$\tau: B \to A$$

defined over K having the following universal mapping property. Given an abelian variety C defined over k and a homomorphism $\varphi: C \to A$ defined over K, then there exists a unique homomorphism $\varphi_*: C \to B$ defined over k such that the diagram

is commutative. For the existence of the K/k-trace due to Chow, see AV, Theorem 8 of Chapter VIII, §3. We also recall that if E is an extension of k which is free from K over k, then (B, τ) is also a KE/E-trace: it does not change under constant field extensions, because K is regular over k by assumption. Furthermore, the homomorphism τ is injective.

We can now state the relative result, due to Lang–Néron [L–N].

Theorem 2. *Let K be a function field over the constant field k, and let A be an abelian variety defined over K. Let (B, τ) be a K/k-trace of A. Then $A(K)/\tau B(k)$ is finitely generated.*

For instance, if A is an elliptic curve with transcendental j-invariant over k, and $K = k(j)$, then $B = 0$ and $A(K)$ is finitely generated.

We prove the theorem in two steps. First the so-called weak Mordell–Weil theorem, where we prove that for an integer $m \geq 1$ prime to the characteristic, $A(K)/mA(K)$ is finite under the following additional assumptions:

The group of points A_m of period m is contained in $A(K)$.

In the function field case, the constant field is algebraically closed, and the dimension of K over k is equal to 1.

Indeed, we prove our result by relating our factor group $A(K)/mA(K)$ to the Galois group of certain unramified extensions of K. This is the Kummer theory, which we reproduce in §1. Using the fact that unit groups and ideal class groups are of finite type, we prove that certain unramified extensions are finite, and then that $A(K)/mA(K)$ is finite. In the function field case, these arguments certainly apply in dimension 1, and reduction steps proved in §1 show that we can limit ourselves to this case in order to prove Theorem 2. We shall also prove in §1 that we may assume that $A_m \subset A_K$, and that k is algebraically closed.

Once this is carried out, we apply properties of heights to abelian varieties and see that the logarithm of the height function is of quasi-degree 2 so that we can apply Proposition 4.1 of Chapter 3 and its corollary. In number fields, we are then finished. In function fields, we must make an analysis of sets of points of bounded height. We shall find out that they lie in a finite number of cosets of $\tau B(k)$. This will thus prove Theorem 2.

We then return to Theorem 1 for finitely generated fields over the prime field, viewing the algebraic closure k of the prime field in K as a constant field. Applying Theorem 2, the known result in number fields in characteristic 0, and the trivial fact that $B(k)$ is finite if k is finite, we conclude the proof of Theorem 1.

We can then apply Theorems 1 and 2 to get results concerning divisor classes, namely the theorem of the base (so-called by Severi).

§1. Kummer Theory

Throughout this section, we let K be a field and m an integer ≥ 1, prime to the characteristic of K.

An abelian extension E of K will be said to be of **exponent** m if $\sigma^m = 1$ for every automorphism σ of E over K.

Let A be a commutative group variety defined over K. As usual, we denote its group of rational points in K by $A(K)$. We let A_m be the subgroup of those points $a \in A$ such that $ma = 0$ (writing A additively).

We recall that the map $x \mapsto mx$ is surjective and separable (so that in particular, A_m is finite, of order equal to $\deg(m.\mathrm{id})$. Quick proof: If $f_1, \ldots, f_r$ are local parameters at the origin, then

$$f_i(mP) \equiv m \cdot f_i(P) \bmod \mathfrak{m}^2$$

where $\mathfrak{m}$ is the maximal ideal of the local ring at the origin (IAG, Proposition 1 of Chapter IX, §1). Thus $[m]$ is generically surjective and separable. If a is a given point, and x a generic point over $K(a)$, then $a - x$ is generic over K. By an m-th **root** of a we shall mean a point b such that $mb = a$. Any two such m-th roots differ by a point of A_m. If $my = x$ and $mz = a - x$ then $m(y + z) = a$. Since $K(x, y)$ and $K(x, z)$ are separable over $K(x)$ and over K, it follows that $K(y + z)$ is also separable over K.

If $a \in A(K)$, we denote by $K(1/m \cdot a)$ the field obtained by taking the composite of all fields $K(b)$ where b ranges over all m-th roots of a. Then all points of A_m are rational over that field, which is separable over K by what we have just proved. If $A_m \subset A(K)$, then this field is equal to $K(b)$ where b is any one of the m-th roots of a.

From now on, we assume that A_m is contained in $A(K)$. Let B be a subgroup of $A(K)$ containing $mA(K)$ (i.e. the set of all points ma with $a \in A(K)$). We denote by

$$E_B = K(1/m \cdot B)$$

the field obtained by taking the composite of all fields $K(1/m \cdot a)$ where a ranges over B. It is determined by B, in a given algebraic closure of K, as usual.

If σ is any isomorphism of E_B over K, then with the above notation and with $m\alpha = a$, we have

$$m(\sigma\alpha) = \sigma(m\alpha) = m\alpha.$$

Consequently $\sigma\alpha - \alpha$ is an element of A_m, say t_σ. Thus

$$\sigma\alpha = \alpha + t_\sigma.$$

Hence σ is in fact an automorphism of E_B over K, and E_B is therefore Galois. If τ is another automorphism of E_B over K, and $\tau\alpha = \alpha + t_\tau$ then clearly $\sigma\tau\alpha = \alpha + t_\sigma + t_\tau$, and hence E_B is abelian.

Let G be its Galois group. We define a bilinear map

$$G \times B \to A_m$$

as follows. Let $\sigma \in G$ and $b \in B$. Take any $\beta \in A$ such that $m\beta = b$. Then $\sigma\beta - \beta$ is clearly independent of our choice of β, and is an element of A_m, which we denote by $\langle \sigma, b \rangle$. The reader will immediately verify that this is bilinear in σ and b.

By definition, the kernel on the left consists of those σ such that $\langle \sigma, b \rangle = 1$ for all $b \in B$. As E_B is generated by elements of type $1/m \cdot b$ any such σ must be the identity.

The kernel on the right consists of those b such that $\langle \sigma, b \rangle = 1$ for all $\sigma \in G$. For any such b, the extension $K(1/m \cdot b)$ is fixed under G and hence must be K itself. This means that $1/m \cdot b$ is in fact rational over K, i.e. that b lies in $mA(K)$. We thus get a bilinear map

$$G \times B/mA(K) \to A_m$$

whose kernel on both sides is equal to the identity. Since A_m is finite, we obtain

Proposition 1.1. *Let A be a commutative group variety defined over the field K. Let m be an integer ≥ 1, prime to the characteristic. Assume that $A_m \subset A(K)$, and let B be a subgroup of $A(K)$ containing $mA(K)$. Then $K(1/m \cdot B)$ is an abelian extension of exponent m and is a finite extension if and only if $B/mA(K)$ is a finite group.*

Let us consider in greater detail the special case where A is the multiplicative group. We then write the group law multiplicatively. Our group A_m is K_m^*, the group of m-th roots of unity, and is cyclic of order m. Thus the two groups G and B/K^{*m} are dual to each other under our pairing above, and of the same order if they are finite. (We shall not need the infinite case in the sequel.)

If B_1, B_2 are two subgroups of K^* containing K^{*m} and if $B_1 \subset B_2$ then $K(B_1^{1/m}) \subset K(B_2^{1/m})$. Conversely, assume this is the case. We wish to prove $B_1 \subset B_2$. Let $b \in B_1$. Then $K(b^{1/m}) \subset K(B_2^{1/m})$ and $K(b^{1/m})$ is contained in a finitely generated subextension of $K(B_2^{1/m})$. Thus we may assume without loss of generality that B_2/K^{*m} is finitely generated. Let B_3 be the subgroup of K^* generated by B_2 and b. Then $K(B_2^{1/m}) = K(B_3^{1/m})$ and from what we have seen above, the degree of this field over K is precisely

$$(B_2 : K^{*m}) \quad \text{or} \quad (B_3 : K^{*m}).$$

Thus these two indices are equal, and $B_2 = B_3$. Summarizing:

Theorem 1.2. *Let K be a field, m an integer ≥ 1 prime to its characteristic. Assume that the m-th roots of unity lie in K. Then there is a bijective* *correspondence between abelian extensions E of K of exponent m and subgroups B of K* containing K*^m, given by E = K(B^{1/m}), and*

$$[E:K] = (B:K^{*m}),$$

in the sense that if one index is finite, so is the other, and they are equal.

Proof. We have proved everything except that every abelian extension of K of exponent m is of type $K(B^{1/m})$ for some B. But such an extension is a composite of finite cyclic extensions of exponents m. Our assertion follows then from Galois theory (Hilbert's theorem 90).

We shall say that the abelian extension $K(B^{1/m})$ and B **belong** to each other. Theorem 1.2 is said to express the Kummer theory, and extensions of this type are called **Kummer extensions**.

We wish now to determine when a Kummer extension is unramified for a discrete valuation.

Proposition 1.3. *Let v be a discrete absolute value on K, and $a \in K^*$. Assume that m is prime to the characteristic of the residue class field of v. Then v is unramified in $K(a^{1/m})$ if and only if $\mathrm{ord}_v\, a$ is divisible by m.*

Proof. Assume first that the order of a is divisible by m. Then we can write $a = \pi^{rm}u$ where π is a prime element, r an integer, and u a unit. We see that $K(a^{1/m}) = K(u^{1/m})$ and that $u^{1/m}$ is a unit, satisfying $X^m - u = 0$. Since m is assumed to be prime to the characteristic, we can apply the criterion of Proposition 3.4 of Chapter 1 to conclude that $K(a^{1/m})$ is unramified at any extension of v. Conversely, assume this the case. Then if $a = \pi^s u$ with some integer s, it must be so that m divides s, otherwise, the value group of our extension $K(a^{1/m})$ would obviously be unequal to the value group $v(K^*)$, and so the extension could not be unramified.

Let K have a proper set of absolute values M_K and assume that all the non-archimedean absolute values of M_K are discrete. We shall say that an algebraic extension E of K is **unramified** over M_K, or that M_K is **unramified** in E, if every non-archimedean $v \in M_K$ is unramified in E.

We continue to assume that the m-th roots of unity lie in K, and we let E be a Kummer extension of K, of exponent m, belonging to the subgroup B of K^*. Using Proposition 1.3, we see that M_K is unramified in E if and only if, for each $b \in B$, there exists an M_K-ideal $\mathfrak{b}$ such that

$$(b) = m\mathfrak{b}.$$

Hence:

Theorem 1.4. *Let K be a field with a proper set of absolute values M_K. Assume all the non-archimedean ones are discrete. Let m be an integer* prime to the characteristics of the residue class fields of the discrete $v \in M_K$, and assume all m-th roots of unity lie in K. Let B_u be the subgroup of K* belonging to the maximal abelian extension of K, of exponent m, and unramified over M_K. Then B_u consists of those elements $b \in K^*$ for which there exists an M_K-ideal $\mathfrak{b}$ such that $(b) = m\mathfrak{b}$. If C is the group of M_K-ideal classes, C_m those of period m, and U the group of M_K-units, then we have an exact sequence

$$0 \to U/U^m \to B_u/K^{*m} \to C_m \to 0.$$

Proof. If $b \in B_u$, and we have $(b) = m\mathfrak{b}$ as above, then we map b on the ideal class of $\mathfrak{b}$, which is obviously in C_m. The map is surjective in view of Proposition 1.3, and K^{*m} is obviously contained in the kernel. Thus we get the map of B_u/K^{*m} onto C_m. The reader will immediately verify that its kernel consists of the units modulo $U^m = U \cap K^{*m}$.

Corollary 1.5. *If in the above theorem, both C_m and U/U^m are finite, then so is the maximal abelian extension of K, of exponent m, and unramified over M_K.*

One can often reduce the study of ramified extensions to that of un-ramified extensions, over a large ground field, as follows:

Lemma 1.6. *Let K be a field, m an integer ≥ 1 prime to its characteristic, and E an abelian extension of exponent m. Let v be a discrete valuation on K, and π a prime element. Assume that the m-th roots of unity lie in K and that m is prime to the characteristic of the residue class field of v. Let $L = K(\pi^{1/m})$. Then every absolute value w of L extending v is un-ramified in EL.*

Proof. This is immediate from Proposition 1.3.

We recall that an unramified extension stays unramified under translation (Proposition 3.5 of Chapter 1). Thus repeated application of Lemma 1.6 kills the ramification at any given finite number of absolute values of K, by means of a finite extension. We apply this to get the fundamental result:

Theorem 1.7. *Let K be a number field, or a function field in one variable over the algebraically closed constant field k. Let m be an integer ≥ 1, and prime to the characteristic. Let M_K be the canonical set of absolute values of K, as in §2 and §3 of Chapter 2, and let S be a finite subset of M_K. Then there is only a finite number of abelian extension of K, of exponent m, which are unramified over $M_K - S$.*

Proof. We may assume S contains the divisors of m. Let E be the composite of all abelian extension of K, of exponent m, unramified over $M_K - S$. Let L be obtained by adjoining the m-th roots of unity to K, and also $\pi_1^{1/m}, \ldots, \pi_s^{1/m}$ where $\pi_1, \ldots, \pi_s$ are prime elements at the discrete valuations of S. Then EL, viewed as an extension of L, is unramified over M_L by our lemma. We now apply Corollary 1.5 the finiteness of the class number of Theorem 6.3, Chapter 2 and the unit theorem in number fields. In the function field case, the units are merely the non-zero constants, and the existence of the Jacobian guarantees that the divisor classes of exponent m form a finite group. Thus our corollary also works in this case.

§2. The Weak Mordell–Weil Theorem

If A is an abelian variety defined over a field K, and v is a discrete valuation on K, then we can define the **reduction** of A with respect to this valuation [Sh]. We denote it by A_v. In general, A_v is merely a cycle.

If A_v has one component, with multiplicity 1, which is an abelian variety whose law of composition is obtained by reducing that of A, then we shall say that the reduction is **non-degenerate** or **good**. The following lemma can be proved by a systematic procedure which follows the one at the end of [Sh].

Lemma 2.1. *Let K be a field with a proper set M_K of discrete valuations. Let A be an abelian variety defined over K. Then for all but a finite number of $v \in M_K$, the reduction A_v is non-degenerate.*

Proof. We leave it to the reader. It consists in looking separately at each geometric property entering into the definition of an abelian variety and its reduction: the fact that the variety remains non-singular, that certain maps remain morphisms, that the associative law holds, etc. In each case, one proves that there exists an element $a \in K^*$ such that if $|a|_v = 1$, then the reduction A_v has the corresponding geometric property.

Lemma 2.2. *Let K be a field with a discrete valuation v. Let A be an abelian variety defined over K, and having a non-degenerate reduction A_v. Let m be an integer $\geqq 1$, prime to the characteristic of K and to the characteristic of the residue class field of v. Assume $A_m \subset A(K)$. Then v is unramified in the extension $K(1/m \cdot A(K))$.*

Proof. Since our statement is local in v, we may assume that K is complete under our valuation. Furthermore, it suffices to prove that for each $a \in A(K)$, the extension $K(1/m \cdot a)$ is unramified over K. Let P be a point on A such that $mP = a$. The automorphisms of $K(P)$ over K are given by translations as in §1, and if

$$n = [K(P) : K]$$

then there are n elements $a_1, \ldots, a_n$ of A_m such that the automorphisms σ_i ($i = 1, \ldots, n$) satisfying

$$\sigma_i P = P + a_i$$

give all automorphisms of $K(P)$ over K. Let a_i' be the reduced points. Since m is prime to the characteristic of the residue class field, it follows that $a_1', \ldots, a_n'$ are distinct. If φ is any place of $K(P)$ extending the canonical place of v, then $\varphi\sigma_i$ are also places of $K(P)$ and

$$\varphi\sigma_i(P) = (P + a_i)' = P' + a_i'.$$

Hence the $\varphi\sigma_i$ are distinct, and by Proposition 3.2 of Chapter 1 we conclude that $K(P)$ is unramified over K, as desired.

From the two lemmas, Proposition 1.1 and Theorem 1.7 we get:

Proposition 2.3. *Let K be a number field, or a function field in one variable over an algebraically closed constant field k. Let m be an integer ≥ 1, prime to the characteristic of K, and A an abelian variety defined over K such that $A_m \subset A(K)$. Then $A(K)/mA(K)$ is finite.*

Of course, in the function field case, we shall later eliminate the restriction that the function field is of dimension 1. For the moment, we turn to the full Mordell–Weil theorem.

§3. The Infinite Descent

Let F be a field with a proper set of absolute values M_F satisfying the product formula, and let K be a finite extension of F, with its set of absolute values M_K. Let A be an abelian variety defined over K, which we assume to be embedded in projective space by means of the complete linear system $\mathscr{L}$. Let $X \in \mathscr{L}$, let m be an integer > 1, and $a \in A(K)$. Let

$$\varphi : A \rightarrow A$$

be the isogeny $\varphi u = mu + a$. Then

$$\varphi^{-1}(X) = [m]^{-1} X_{-a} \approx m^2 X_{-a} \approx m^2 X,$$

where $\approx$ is algebraic equivalence. By Proposition 3.3 of Chapter 5, we conclude that $h \circ \varphi$ is quasi-equivalent to $m^2 h$, where h is the height associated with the given projective embedding. In particular, h is of quasi-degree 2 on $A(K)$ in the sense of Chapter 3, §4 and we can apply Corollary 4.2 of that chapter to conclude the proof of the Mordell–Weil theorem over number fields, since there is only a finite number of points of bounded height in a given number field.

Of course, for the infinite descent, one could select an even ample class c, and work with the positive definite canonical quadratic form h_c and the associated norm $| \ |_c$ instead of the function of quasi-degree 2. Readers can suit their tastes as to which heights to use.

In the function field case, we need additional arguments to determine more precisely the nature of a set of points of bounded height. This is carried out in §5, after some reduction steps.

§4. Reduction Steps

The propositions which we prove in this section will be used to reduce our main theorem in function fields to special cases. Throughout this section, K will denote a finitely generated regular extension of a field k.

We first show that to prove our main theorem we may extend our function field by a finite separable extension.

Proposition 4.1. *Let* $L \supset K$ *be a finite separable extension of* K, *also regular over* k. *Let* A *be an abelian variety defined over* K. *Let* $(A^{K/k}, \tau_K)$ *be its* K/k-trace, *and* $(A^{L/k}, \tau_L)$ *its* L/k-trace. *Then the factor group*

$$[\tau_L A^{L/k}(k) \cap A(K)]/\tau_K A^{K/k}(k)$$

is finite.

Proof. Let Γ_L be the graph of τ_L. It is defined over L. Let us take the intersection

$$H = \bigcap \Gamma_L^\sigma$$

of Γ_L and its conjugates over K, where σ ranges over the distinct isomorphisms of L over k. Then H is K-closed, and is an algebraic subgroup of $A^{L/k} \times A$. Its connected component H_0 is therefore defined over K by Chow's theorem (AV, Theorem 5, Chapter II, §1).

We have

$$\Gamma_L \cap [A^{L/k}(k) \times A(K)] = H \cap [A^{L/k}(k) \times A(K)]$$

and this group contains $H_0 \cap [A^{L/k}(k) \times A(K)]$ as a subgroup of finite index since H_0 is of finite index in H.

We have a surjective homomorphism

$$\Gamma_L \cap [A^{L/k}(k) \times A(K)] \to \tau_L A^{L/k}(k) \cap A(K)$$

by projection on the second factor. We contend that the inverse image of $\tau_K A^{K/k}(k)$ contains $H_0 \cap [(A^{L/k}(k) \times A(K)]$. From this it will be clear that the desired factor group is finite.

Since H_0 is contained in the graph of a homomorphism of $A^{L/k}$ into A, it is itself the graph of a homomorphism β of its projection B on the first factor into A. Furthermore, B is contained in $A^{L/k}$ and is defined over K, and hence over k by Chow's theorem. By the universal mapping of the K/k-trace there exists a commutative diagram

with β_* defined over k, hence $\beta B(k)$ is contained in $\tau_K A^{K/k}(k)$. This proves our contention and concludes the proof of our proposition.

Corollary 4.2. *If Theorem 2 is true for L/k, then it is true for K/k, i.e. if $A_L/\tau_L A^{L/k}(k)$ is finitely generated, so is $A_K/\tau_K A^{K/k}(k)$.*

Proof. This follows immediately from the proposition and the fact that we have an injection

$$0 \to A_K/[\tau_L A^{L/k}(k) \cap A_K] \to A_L/\tau_L A^{L/k}(k).$$

Next we show that we may extend the constant field.

Proposition 4.3. *Let k' be any extension of k which is independent of K over k, and let $K' = Kk'$. Let A be an abelian variety defined over K and let (B, τ) be its K/k-trace. Then*

$$A(K) \cap \tau B(k') = \tau B(k).$$

Proof. We know that (B, τ) is also a K'/k'-trace of A. The inclusion $\tau B(k) \subset A(K) \cap \tau B(k')$ is obvious. Conversely, let b be a point of $B(k')$ such that τb is rational over K. If we knew that τ is regular, i.e. that it is birational between B and its image in A, then in view of the fact that τ is defined over K, we would see immediately that $\tau^{-1}\tau b$ is rational over k. As we do not know that τ is regular, we use Chow's regularity theorem of AV, Chapter IX.

We can write $K = k(u)$ for suitable parameters u, and $A = A_u$, $\tau = \tau_u$. Let $u_1, \ldots, u_m$ be independent generic specializations of u over k'. For m large, the map

$$x \to (\tau_{u_1} x, \ldots, \tau_{u_m} x)$$

of B into $A_{u_1} \times A_{u_2} \times \cdots \times A_{u_m}$ is regular, and thus birational between B and its image. If τb is rational over K, then $(\tau_{u_1} b, \ldots, \tau_{u_m} b)$ is rational over $k(u_1, \ldots, u_m)$; hence b is rational over

$$k(u_1, \ldots, u_m).$$

Since we took $k(u_1, \ldots, u_m)$ free from k' over k, and since b is rational over k', it follows that b is rational over k. This concludes the proof.

Finally, we give the reduction to the case where K is of dimension 1 over k.

Proposition 4.4. *Let $K \supset E \supset k$ be a tower of fields such that K is regular over E and E regular over k. If Theorem 2 is true for K/E and E/k, then it is true for K/k.*

Proof. Let A be an abelian variety defined over K, and let $(A^{K/E}, \tau_{K/E})$ be a K/E-trace of A. Let $(A^{E/k}, \tau_{E/k})$ be an E/k-trace of $A^{K/E}$. By the universal property, there is an injective homomorphism $\beta : A^{E/k} \to A^{K/k}$ defined over k such that the following diagram is commutative:

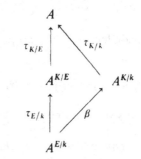

All the homomorphisms are injective. We have

$$A_K \supset \tau_{K/E} A^{K/E}(E) \supset \tau_{K/E}\tau_{E/k} A^{E/k}(k) = \tau_{K/k}\beta A^{E/k}(k).$$

If we assume that $A^{K/E}(E)$ is of finite type modulo $\tau_{E/k} A^{E/k}(k)$, it follows that A_K modulo $\tau_{K/k}\beta A^{E/k}(k)$ is also of finite type. But this last group is contained in $\tau_{K/k} A^{K/k}(k)$. We see, therefore, that it suffices to prove Theorem 2 for each step K/E and E/k of the tower.

It is well known and easy to show that one can always construct a tower $E_0 = k \subset E_1 \subset \cdots \subset E_n = K$ such that each step E_i/E_{i-1} is regular and of transcendence degree 1. Such a construction comes from the lemma of Bertini's theorem, which states that if x, y are two elements of K, algebraically independent over k, and such that $y \notin K^p k$, then for all but a finite number of constants $c \in k$, K is regular over $k(x + cy)$. (Cf. IAG, Chapter VI.) If k is finite, one needs a small additional argument. However, in view of Proposition 4.3, for our purposes, we may assume that k is infinite.

The above reduction steps have thus reduced the proof of Theorem 2 to the case where $A(K)$ contains A_m, k is algebraically closed, and K is of transcendence degree 1 over k.

To conclude our proof of Theorem 2, we must investigate the nature of sets of points of bounded height.

§5. Points of Bounded Height

Let $W \subset \mathbf{P}^r$ be a projective variety, non-singular in codimension 1, defined over a field k. Let w be a generic point of W over k, and $K = k(w)$ a function field for W over k. Let $A = A_w$ be an abelian variety defined over K, and embedded in projective space $\mathbf{P}^n$ over K. Let u be a generic point of A over K. We shall denote by $W \circ A$ the locus of (w, u) over k. Then $W \circ A$ is the graph of the algebraic family parametrized by W, of which A is a generic member. If we put $Z = W \circ A$, then

$$A = Z(w) = \mathrm{pr}_2[Z \cdot (w \times \mathbf{P}^n)].$$

Given a rational point P of A over K, we shall denote by W_P the locus of (w, P) over k. Then

$$W_P \subset W \circ A \subset W \times \mathbf{P}^n \subset \mathbf{P}^r \times \mathbf{P}^n.$$

Note that W_P and W are birationally equivalent under the projection pr_1.

There is a projective embedding

$$\varphi: \mathbf{P}^r \times \mathbf{P}^n \to \mathbf{P}^{(r+1)(n+1)-1}$$

which to each product of points with homogeneous coordinates

$$(x_0, \ldots, x_r) \times (y_0, \ldots, y_n)$$

assigns the point with homogeneous coordinates $(x_i y_j)$. One verifies immediately that for any $P \in A_K$ we have

$$h_\varphi(w, P) \leq h(w) + h(P) = \deg W + h(P).$$

Here, of course, $h = h_{M_K}$ is the height, taken relative to our projective embeddings and the canonical set of absolute values on K obtained from the prime rational divisors of W over k.

Let c_3 be a number > 0 and S a subset of A_K such that $H(P) \leq c_3$ for all $P \in S$. Then by Theorem 3.6 of Chapter 3 there exists a number c_4 such that $\deg \varphi(W_P) \leq c_4$ and hence, by the theory of Chow coordinates, W_P can belong only to a finite number of algebraic families on $W \circ A$ for $P \in S$. We shall be through with the proof of Theorem 2 once we have proved:

Theorem 5.1. *Let W be a projective variety, non-singular in codimension 1, defined over a field k. Let w be a generic point of W over k, and $K = k(w)$ a function field of W over k. Let A be an abelian variety defined over K, and (B, τ) be its K/k-trace. For each point $P \in A_K$, let W_P be the locus of (w, P) over k. If $P, P' \in A(K)$ are such that W_P and $W_{P'}$ are in the same algebraic family on $W \circ A$, then P and P' are congruent modulo $\tau B(k)$ (i.e. $P - P' \in \tau B(k)$).*

Proof. In view of Proposition 4.3, we may assume that k is algebraically closed. We shall prove our theorem by going from W_P to $W_{P'}$ passing through a generic member of the family. Our first step is to show that a generic member is of the same type as W_P.

Lemma 5.2. *Let U be a subvariety of $W \circ A$, defined over an extension k_1' of k independent of K over k, and such that W_P is a specialization of U over k. Put $K_1' = Kk_1'$. Then there exists a rational point P_1 of A in K_1 such that $U = W_{P_1}$.*

Proof. We have $\mathrm{pr}_1 W_P = W$. Hence by the compatibility of projections with specializations, we must have $\mathrm{pr}_1 U = W$. But $\dim U = \dim W$, and hence a generic point of U over k_1 is of type (w, Q) for some $Q \in \mathbf{P}^n$, algebraic over $k_1(w)$. We must show that Q is rational over $k_1(w)$, and that Q is in A.

On $W \times \mathbf{P}^n$ the intersection $U \cdot (w \times \mathbf{P}^n)$ is obviously defined, and in view of the above remarks, we have

$$U \cdot (w \times \mathbf{P}^n) = \sum_{i=1}^{d} (w, Q_i)$$

for suitable points Q_i. Taking the projection on W, we see that $d = 1$. Since $U \cdot (w \times \mathbf{P}^n)$ is rational over $k_1(w)$, so is $Q_1 = P_1$. Finally, we have $P_1 \in A$ because

$$w \times P_1 = U \cdot (w \times \mathbf{P}^n) \subset (W \circ A) \cdot (w \times \mathbf{P}^n) = w \times A.$$

Note that in the lemma, we have made no use of the fact that A is an abelian variety: the same statement holds for any algebraic system of varieties.

It will suffice to prove Theorem 5.1 for W_P and W_{P_1}. Indeed, given W_P and $W_{P'}$ as in the theorem, there exists a subvariety W_{P_1} of $W \circ A$, defined over some k_1, such that both W_P and $W_{P'}$ are specializations of W_{P_1}. If $P - P_1 \in \tau B(k_1)$ and $P' - P_1 \in \tau B(k_1)$ then $P - P' \in \tau B(k_1)$, and by Proposition 4.3 we get $P - P' \in \tau B(k)$.

The field k_1 may be assumed to be of transcendence degree 1 over k. Indeed, there exists a variety T, a generic point t_1, and a point t of T such that W_P is the unique specialization of W_{P_1} over the specialization $t_1 \to t$ (over k). The specialization $t_1 \to t$ over k can be extended to a place of $k(t_1)$ corresponding to a discrete valuation ring by blowing up our point t. Thus there exists a field k_1 such that $k \subset k_1 \subset k(t_1)$, $k(t_1)$ has dimension 1 over k_1, and the specialization $t_1 \to t$ is in fact over k_1. Using Proposition 4.3 we may replace k by k_1, and in fact may go over to the algebraic closure of our constant field, without loss of generality for the proof of Theorem 5.1, and we may assume that T is nonsingular.

We have the following diagram of fields:

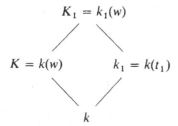

$$K_1 = k_1(w)$$

$$K = k(w) \qquad k_1 = k(t_1)$$

$$k$$

Let $\mathrm{Alb}(T)$ be the Albanese variety of T, defined over k, and let

$$f: T \to \mathrm{Alb}(T)$$

be a canonical map, defined over k. Let

$$g: T \to A$$

be the rational map defined over K by the expression $g(t_1) = P_1$. Then there is a homomorphism $\alpha: \mathrm{Alb}(T) \to A$ and a point $a \in A$ such that the following diagram is commutative:

$$T \xrightarrow{\ f\ } \mathrm{Alb}(T)$$

$$g \downarrow \quad \swarrow_{\alpha + a}$$

$$A$$

By the definition of the K/k-trace, there exists a homomorphism

$$\beta: \mathrm{Alb}(T) \to B$$

defined over k such that the following diagram is commutative:

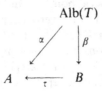

Since t is simple on T, f is defined at t. Furthermore, since W_P is the unique specialization of W_{P_1} over $t_1 \to t$ (relative to k or K), it follows that P is the unique specialization of P_1 over $t_1 \to t$ (relative to K). Hence $g(t) = P$, and

$$P_1 - P = g(t_1) - g(t) \, \alpha f(t_1) + a - \alpha f(t) - a$$
$$= \tau \beta f(t_1) - \tau \beta(t)$$
$$= \tau b$$

with $b = \beta[f(t_1) - f(t)] \in B(k_1)$. This proves our theorem.

The next theorem provides the properties of non-degeneracy in the function field case, analogous to the number field case.

Theorem 5.3. *Let A be projectively embedded in the above theorem, and let the height be taken relative to this embedding and the canonical set of absolute values $M_{k(W)}$. Let S be a subset of $A(K)$ of bounded height. Then the points of S lie in a finite number of cosets of $\tau B(k)$.*

Proof. As we saw in our discussion preceding the theorem, points P of bounded height are such that the varieties W_P have bounded degree, and thus we can apply Theorem 5.2.

We are now in a position to transpose the theorems of Chapter 5, §6 and §7 to the present case.

Theorem 5.4. *Let c be a divisor class containing an ample divisor and even. We view h_c as a function on $A(K^a)$. For simplicity let k be algebraically closed.*

5.4.1. *For any point $x \in A(K^a)$ we have $|x|_c = 0$ if and only if*

$$x \in A_{\text{tor}} + \tau B(k).$$

5.4.2. *Let (A', δ) be the dual variety. Then the kernels on each side of the pairing*

$$(x, y) \mapsto h_\delta(x, y)$$

are $A_{\text{tor}} + \tau B(k)$ and $A'_{\text{tor}} + \tau' B'(k)$.

5.4.3.   *The kernel of the homomorphism $c \mapsto h_c$ from*

$$\text{Pic}(A) \rightarrow \textit{quadratic functions on } A(K^a)$$

consists of $\text{Pic}(A)_{\text{tor}} + \tau'B'(k)$.

5.4.4.   *Let Γ be a finitely generated subgroup of*

$$A(K)/(\tau B(k) + A(K)_{\text{tor}}).$$

Then the canonical height h_c is non-degegenerate on Γ, and its extension to $\mathbf{R} \otimes (A(K^a)/\tau B(k))$ is a positive definite quadratic form.

The proofs are the same as over number fields, since the only extra ingredient to the general facts is the non-degeneracy provided by Theorem 5.3 in the function field case.

§6. Theorem of the Base

Let V be a projective variety, non-singular in codimension 1, and defined over the algebraically closed field k which we may take to be a universal domain. The group of divisors $D(V)_k$ on V, rational over k, contains the usual subgroups of divisors which are algebraically equivalent to 0 and linearly equivalent to 0, respectively.

$$D(V)_k \supset D_a(V)_k \supset D_l(V)_k.$$

The Picard group $D_a(V)_k/D_l(V)_k$ has the structure of an abelian variety, the Picard variety. On the other hand we have the **Néron–Severi group**

$$NS(V)_k = D(V)_k/D_a(V)_k.$$

Theorem 6.1 (Néron, Theorem of the Base.) *Let V be a projective variety, non-singular in codimension 1 over an algebraically closed field k. Then $NS(V)_k$ is finitely generated.*

We shall show below how this theorem can be reduced to Theorem 2. First, let us prove a corollary.

Corollary 6.2. *Let K be a field of finite type over the prime field, and let V be a projective variety, non-singular in codimension 1, and defined over K. Let $D(V)_K$ be the group of divisors on V rational over K, and $D_l(V)_K$ those* *which are linearly equivalent to zero and rational over K. Then the factor group*

$$D(V)_K/D_l(V)_K$$

is of finite type.

Proof. To begin with, we may extend K by a finite extension so that V has a rational point. Then $D_a(V)_K/D_l(V)_K$ is represented by the group $A'(K)$ of the rational points on the Picard variety, over K. If k is the algebraic closure of the prime field in K, then we just apply Theorem 2 and the known theorem over number fields to conclude that $A'(K)$ is finitely generated. We can now use the theorem of the base to finish the proof.

Let us now prove Theorem 6.1. Let L_u by a generic linear variety over k, such that $V \cdot L_u = C_u$ is a generic curve (cf. IAG, Chapter 7). Let J be the Jacobian of C_u, defined over the field $k(u)$. Let $D_0(V)_k$ be the subgroup of $D(V)_k$ consisting of those divisors $X \in D(V)_k$ such that $X \cdot C_u$ is of degree 0. Then $D(V)_k/D_0(V)_k$ is infinite cyclic, and it will suffice to prove that $D_0(V)_k/D_a(V)_k$ is of finite type.

Let $\varphi: Z_0(C_u) \to Z_0(J)$ be the canonical map of the 0-cycles of degree 0 on C_u into the 0-cycles on the Jacobian. Then

$$\mathfrak{a} \mapsto S(\varphi(\mathfrak{a}))$$

maps $Z_0(C_u)$ into J, and preserves rationality (i.e. if $\mathfrak{a}$ is rational over $k(u)$, so is $S(\varphi(\mathfrak{a}))$). We get a homomorphism

$$\lambda: D_0(V)_k \to J(k(u))$$

given by the formula

$$\lambda(X) = S(\varphi(X \cdot C_u)).$$

According to an elementary equivalence criterion [We 4], one knows that the kernel E of λ, which contains $D_l(V)_k$, is of finite type modulo $D_l(V)_k$. One can in fact prove that this kernel is precisely $D_l(V)_k$ if we have chosen our projective embedding suitably. This is irrelevant here, and we see in any case that $[E + D_l(V)_k]/D_l(V)_k$ is of finite type. To prove the theorem of the base, it will therefore suffice to prove that

$$D_0(V)_k/[E + D_a(V)_k]$$

is of finite type.

The inverse image $\lambda^{-1}(\lambda(D_a(V)_k))$ is precisely equal to

$$E + D_a(V)_k.$$

We have therefore an injection

$$0 \to D_0(V)_k/[E + D_a(V)_k] \to J(k(u))/\lambda D_a(V)_k.$$

Now let $\psi: V \to A$ be a canonical map of V into its Albanese variety, defined over k. For a suitable constant $b \in A$, we have a commutative diagram

$$
\begin{array}{ccc}
C_u & \xrightarrow{\ i\ } & V \\
\varphi \downarrow & & \downarrow \psi \\
J & \xrightarrow[i_* + b]{} & A
\end{array}
$$

where i_* is induced by the inclusion.

Let us look at the inverse images of divisors in $D_a(V)_k$ under the composite maps $(i_* + b) \circ \varphi$ and $\psi \circ i$. The formalism $(f \circ g)^{-1} = g^{-1} \circ f^{-1}$ can be applied to the first according to Appendix 1 of AV since J and A are non-singular. A direct verification using intersection theory (associativity and the definition of $C_u = V \cdot L_u$) shows that it can also be applied to the second. According to the definition of the Picard variety, which we know is obtained by pull back from that of the Albanese variety, we see that the map

$$\lambda: D_a(V)_k \to J(k(u))$$

induces the rational homomorphism $'i_*$ on $A'(k) = D_a(V)_k/D_l(V)_k$ if we denote as usual by the upper index t the transpose of i_* on the Picard varieties. Consequently we have an injection

$$0 \to D_0(V)_k/[E + D_a(V)_k] \to J(k(u))/'i_* A'(k).$$

By Chow's theory of the $k(u)/k$-trace, one knows that $(A', 'i_*)$ is a $k(u)/k$-trace of $J' = J$ (AV, Chapter 8, Theorem 12). Consequently we can now apply Theorem 2, and the theorem is proved.

Historical Note and Other Comments

Poincaré [Po] in 1901 first defined the rank both of an elliptic curve, and also of "groups of p points on a curve of genus p." He wrote as if it was a matter of course that the rank is finite, that is, the group of rational points is finitely generated. It was not a matter of course, and Mordell [Mo 1] proved what amounted to a conjecture that the group of rational points of an elliptic curve over $\mathbf{Q}$ is finitely generated. In his paper, Mordell also conjectures that the set of integral points is finite, and that the set of rational points on a curve of genus ≥ 2 is finite. The conjecture about integral points was proved by Siegel [Si], but the other conjecture is still unproved.

Weil [We 2] proved Poincaré's conjecture in higher dimension, and also dealt with abelian varieties (actually Jacobians) over arbitrary number fields.

The extension to number fields and abelian varieties offered serious problems, because at that time, among other things, it was not clear how to estimate the size of a point. In Mordell's case, over the rationals, there was no problem. For this purpose, Weil [We 2] invented his decomposition theorem. Siegel [Si] introduced the height. Actually, as was shown by Lang-Néron [L-N], one does not need the full theory to carry out the infinite descent: the definition of heights and the standard properties suffice. However, the principle of the proof has remained the same since the early days of the theory, first the finiteness of $A(K)/mA(K)$, and second the infinite descent. As was said above, this was done by Weil in his thesis [We 2].

I have given the proof of Lang-Tate for the first. Another method will be found in Roquette [Ro], but one should note that the finite generation of the unit group and divisor class group and the nondegeneracy of a reduction for almost all p are the essential ingredients of these proofs.

As the techniques of algebraic geometry became better, and properties or heights more organized, the infinite descent could be somewhat slicked up, cf. the progression [We 2], [Ne 1], [L-N].

Some time ago, Severi noticed an analogy between the theorem of the base on surfaces and the Mordell-Weil theorem. Making the analogy more precise, Néron in his thesis [Ne 1] proved Theorem 1 (i.e. the Mordell-Weil theorem for fields of finite type) and also the theorem of the base, in any characteristic. He had to overcome technical difficulties in algebraic geometry, because the Picard variety was not yet available. The fact that one can extract a relative Mordell-Weil theorem concerning only abelian varieties was proved by Lang-Néron [L-N], whose exposition for this and the theorem of the base we have followed here. The theorem concerning sets of bounded height is also taken from [L-N].

As usual, our results are purely qualitative. It is of considerable interest to try to push the theory in the quantitative direction, and to give an explicit determination for the set of generators of the Mordell-Weil group of rational points. In the geometric case, there is a result of Igusa [Ig] which ties it up with the second Betti number (he deals with a pencil on a surface), and with the group of principal homogeneous spaces of the abelian variety under consideration. This seems to be one of the main directions in which the theory is developing, and shows that one can expect some hard questions to arise. We may recall the following exact sequence, as in Lang-Tate [L-T].

Let A be an abelian variety (or even a commutative group variety) defined over the field K, and denote by A also its group of points rational over the separable closure of K. Let G be the Galois group of K^a over K.

Then G operates on A, and if we let the cohomology be the limit cohomology from finite separable extensions, we have the exact sequence

$$0 \to A_m \to A \xrightarrow{m} A \to 0$$

for each integer $m > 0$ prime to the characteristic of K, giving rise to the exact cohomology sequence

$$0 \to A_m \cap A_K \to A_K \to A_K \to H^1(G, A_m) \to H^1(G, A) \xrightarrow{m} H^1(G, A) \to$$

the beginning which we may rewrite in the following form:

$$0 \to A_K/mA_K \to H^1(G, A_m) \to H^1(G, A)_m \to 0$$

the group on the right being the kernel in $H^1(G, A)$ of the multiplication by m. Thus $H^1(G, A)_m$ is simply the group of principal homogeneous spaces of A over K, of period m. If $A_m \subset A_K$, then we get back the Kummer theory.

Special cases when A is an elliptic curve with complex multiplication have been treated by Cassels [Ca 1]. Local results for elliptic curves have been obtained by Shafarevitch [Sh 2] and by Tate [Ta 1], who also gives a complete duality theory for abelian varieties, in preparation for global dualities. One of the main problems here is the **conjecture of Shafarevitch–Tate**, to the effect that those elements of $H^1(G, A)$ which split at all primes form a finite group.

As in the case of number fields, the order of the Shafarevich–Tate group occurs linked with the regulator of the Mordell–Weil group in the Birch–Swinnerton–Dyer conjecture, see [Ta 2], [Ta 3] and [L 15] for conjectures concerning its connection with explicit bounds for generators of the Mordell–Weil group.

Concerning Mordell's conjectures for integral points and rational points on curves of higher genus, see the comments of Chapter 8.

CHAPTER 7

The Thue–Siegel–Roth Theorem

We shall give an exposition of this important theorem under axioms which are valid in number fields and function fields, namely, the product formula and a weak Riemann–Roch condition. We also need characteristic 0. In §1 we state the theorem. In §2 we reformulate it by a formal argument in a more manageable form (because we consider approximations at several absolute values). This should not be confused with the deep problem of simultaneous approximations at one absolute value solved by Schmidt. In §3 we give the proof, basing it on two propositions. We then prove Proposition 3.1 in §4 and §5, and Proposition 3.2 in §6, §7, §8 and §9. Finally, in §10, we give a geometric formulation adapted to the applications to algebraic geometry and possible generalizations to varieties of arbitrary dimensions.

The proof of Theorem 1.1 is self contained and elementary (the use of the Gauss lemma of Chapter 3 obviously fits this description).

§1. Statement of the Theorem

Throughout this chapter, we shall assume that K is a field with a proper set of absolute values M_K.

(As we deal with a fixed field K, we could omit the condition that the absolute values are well behaved: it will never be used.)

We denote by S_∞ a fixed finite, non-empty subset of M_K, containing all the archimedean absolute values, if there are any. If S is a finite subset of M_K, we denote by I_S the subset of elements $x \in K$ such that

$$|x|_v \leqq 1 \quad \text{for all } v \notin S$$

and call this the S-**integers**. If $S \supset S_\infty$ then I_S is a ring, taking into account the non-archimedean nature of the absolute values outside S. If $S = S_\infty$ we also write I_∞ or I instead of I_{S_∞} and call this simply the **integers**. We shall always assume that K is the quotient field of I_∞.

If all absolute values of M_K are non-archimedean, then the set of elements $x \in K$ such that

$$|x|_v \leqq 1 \quad \text{for all } v \in M_K$$

is a field, and all its non-zero elements have in fact absolute value 1 for all $v \in M_K$. One sees this immediately from the non-archimedean property and the product formula. This field will be called the **constant field** and will be denoted by k. Given an M_K-divisor $\mathfrak{d}$ one sees that the set $L(\mathfrak{d})$ consisting of those elements $x \in K$ satisfying $|x|_v \leqq \mathfrak{d}(v)$ is a vector space over k, again by the non-archimedean nature of the absolute values.

The following two axioms which we shall consider are very weak forms of **Riemann–Roch theorems** (see the appendix).

Axiom 1a. *There exist numbers $C_1, C_2, B_0 > 0$ depending only on K and S_∞ such that for all $B \geq B_0$, the set of elements $x \in I_\infty$ such that*

$$|x|_v \leqq B \quad \text{all } v \in S_\infty$$

is finite, and the number of elements $\lambda(B)$ in the set satisfies the inequalities

$$C_1 B^N \leqq \lambda(B) \leqq C_2 B^N,$$

where $N = \sum_{v \in S_\infty} N_v$.

Axiom 1b. *All absolute values of M_K are non-archimedean. There exist numbers $C_1, C_2, B_0 > 0$ and an integer $d > 0$ depending only on K and S_∞ such that for all integers $v > 0$ the set of elements $x \in I_\infty$ such that*

$$|x|_v \leqq B_0^v \quad \text{all } v \in S_\infty$$

is a finite dimensional vector space over the constant field k, whose dimension $l(B_0^v)$ satisfies the inequalities

$$C_1 v^d \leqq l(B_0^v) \leqq C_2 v^d$$

We note that Axiom 1a is a consequence of Theorem 6.6, Chapter 2 whenever K is a number field, and S_∞ the set of archimedean absolute values. If K is a function field in one variable over the constant field k, then Axiom 1b is a consequence of the Riemann–Roch theorem, which asserts something much more precise. In that case, if k is algebraically closed, the multiplicities turn out to be equal to 1 with suitably normalized absolute values, namely,

$$|x|_v = (1/e)^{\operatorname{ord}_v x}$$

e being the constant $2.718 \ldots$.

Axiom 1b also applies to higher dimensional projective, normal varieties. One selects an absolutely irreducible hyperplane D to determine the discrete valuation at infinity. After taking a high multiple $D_0 = v_0 D$, one knows (from elementary facts) that $l(vD_0)$ is a polynomial of degree d in v, so that Axiom 1b is satisfied. We shall use Axiom 1b to prove the next theorem. In the applications, one can also reduce the higher dimensional case to that of curves by induction (cf. [L 4]). In our axiomatic setup here, we may as well deal simultaneously with all cases, especially in view of the very elementary proofs which can be given for our axioms in the special cases of interest to us.

Theorem 1.1. *Let K be a field of characteristic 0, with a proper set of absolute values M_K satisfying the product formula with multiplicities $N_v \geq 1$. Let S_∞ be a finite, non-empty subset of M_K containing all the archimedean absolute values. Assume that K is the quotient field of I_∞ and that Axiom 1a or 1b is satisfied. Let S be a finite subset of M_K. For each $v \in S$ let α_v be algebraic over K, and assume that v is extended to K^a in some way. Let κ be real > 2. Then the elements $\beta \in K$ satisfying the approximation condition*

$$\prod_{v \in S} \inf(1, \|\alpha_v - \beta\|_v) \leq \frac{1}{H(\beta)^\kappa}$$

have bounded height.

As in Chapter 3, we use the height

$$H_K(\beta) = H(\beta) = \prod_{v \in M_K} \sup(1, \|\beta\|_v)$$

and $\|\beta\|_v = |\beta|_v^{N_v}$.

It is useful to make a few remarks concerning our statement of the theorem.

(i)   The theorem is strengthened the smaller we make the left side and the larger we make the right side of our inequality. In particular, the smaller we make κ, the better the theorem. (It is known, however, that 2 is a best possible result.) The larger we make the set S, the stronger the theorem.

(ii)   The right-hand side $1/H(\beta)^\kappa$ could be replaced by $C/H(\beta)^\kappa$ where C is a fixed number. Indeed, if there were β's satisfying the modified inequality with C, whose height tends to infinity, then writing $\kappa = \kappa_1 + \xi$ with $\xi > 0$ and $\kappa_1 > 2$ we have

$$\frac{C}{H(\beta)^\kappa} = \frac{1}{H(\beta)^{\kappa_1}} \cdot \frac{C}{H(\beta)^\xi}$$

and for large $H(\beta)$, we see that $C/H(\beta)^\xi \leqq 1$. It is a very interesting question to estimate in number fields the number of solutions of the inequality as a function of C, asymptotically for $C \to \infty$.

(iii)   Suppose that β ranges through a sequence of elements of K such that $\|\alpha_v - \beta\|_v$ tends to 0. We know from Chapter 1 that the extension of v to K^a corresponds to an embedding of K^a into K_v^a. In particular, viewing α_v as embedded in K_v^a and thus algebraic over K_v, if α_v is approximated very closely by elements β of K, then α_v must in fact lie in K_v.

(iv)   If α, α' are two distinct elements of K_v for some v, and if β approximates α, then β stays away from α'. As β approaches α, its distance from α' approaches the distance between α and α'. Hence it would add no greater generality to our statement if we took a product over several α_v for each v.

(v)   There is no reason not to let β approach infinity, and we could add to the approximation condition on the left-hand side a product

$$\prod_{v \in S} \inf(1, \|1/\beta\|_v).$$

This version of the theorem can be reduced to the other by making a linear projective transformation $T(\beta) = (a\beta + b)/(c\beta + d)$ such that the transform of β approaches elements at a finite distance for all $v \in S$, if it converges at all. It is trivially checked that if T is a non-singular transformation with coefficients in K, then $H(T(\beta))$ is equivalent to $H(\beta)$, i.e. each is less than a constant multiple of the other.

(vi)   If α is in K, and S contains one absolute value, then the theorem is trivial, and with $\kappa > 1$ instead of $\kappa > 2$. Indeed, one can even replace the finite product on the left by the complete product over all absolute values of M_K, in which case one gets the reciprocal of the height of $(\alpha - \beta)$ which is equivalent to the height of β. From this the assertion is clear.

It is known that if one puts conditions of a multiplicative nature on the β, then it is possible to lower the value of κ. This is the way one can interpret some known variants of the theorem, for instance as in [Ri 2]. Actually, these are immediate consequences of the theorem as we have stated it. As an illustration, we give the proof of a theorem of Mahler, transported to our general situation.

Corollary 1.2. *Use the same assumptions as in Theorem* 1.3, *and assume in addition that the elements β of K are subject to the condition*

$$\|\beta\|_v = 1$$

for all $v \notin S$ and $\alpha_v \neq 0$. Then the conclusion of the theorem is valid if $\kappa > 0$ (instead of $\kappa > 2$).

Proof. In view of our additional assumption, we have

$$H(\beta) = \prod_{v \in S} \sup(1, \|\beta\|_v) = \prod_{v \in S} \inf(1, \|1/\beta\|_v)^{-1}.$$

Similarly, using $H(\beta) = H(1/\beta)$, we have

$$H(\beta) = \prod_{v \in S} \inf(1, \|\beta\|_v)^{-1}.$$

Using Remarks (iv) and (v), we know that if $\xi > 2$ then the elements β of K satisfying

$$\prod_{v \in S} \min(1, \|\alpha_v - \beta\|_v) \prod_{v \in S} \min(1, \|\beta\|_v) \prod_{v \in S} \min(1, \|1/\beta\|_v) \leqq \frac{1}{H(\beta)^{\xi}}$$

have bounded height. Using the equalities just derived, our assertion is now obvious.

Finally, before starting the proof, it may be illuminating to consider the special case where $K = \mathbf{Q}$ and the set of absolute values is the ordinary set of absolute values. Then the multiplicities are equal to 1, and Axiom 1a is obviously satisfied.

When only one absolute value is considered, say the ordinary one, then the inequality has frequently been written

$$\left| \alpha - \frac{m}{n} \right| \leqq \frac{1}{|n|^{\kappa}}$$

taking m, n relatively prime integers. Since m/n comes very close to α, the two integers $|m|, |n|$ must have the same order of magnitude, so that instead of taking $1/|n|^{\kappa}$ for the right-hand side, one can replace $|n|$ by $\sup(|m|, |n|)$ which is precisely the height of $\beta = m/n$. This identifies the most classical form of the theorem with ours, taking into account Remark (iii), the completion being the real numbers.

Suppose we consider a p-adic absolute value. Then the term corresponding to p is sometimes written in the form

$$\min(1, |n\alpha - m|_p)$$

following Mahler. If $|\alpha - (m/n)|_p$ becomes very small, then $|m/n|_p$ cannot tend to infinity. It is bounded from above by a number depending on α. In particular, $|m|_p$ and $|n|_p$ lie between two fixed bounds depending on α, and thus having

$$\min(1, |n\alpha - m|_p)$$

is no improvement to having

$$\min\left(1, \left|\alpha - \frac{m}{n}\right|_p\right)$$

in the product.

We have thus seen how our formulation covers the classical formulations and their variants, and we are now ready to prove the theorem.

§2. Reduction to Simultaneous Approximations

Rather than deal with the approximation condition in the form in which we have given it, i.e. with a product, it is easier to deal with a simultaneous set of approximations for all $v \in S$. Thus we reformulate our theorem in the following manner.

Theorem 2.1. *Let the hypotheses be as in Theorem 1.1. For each $v \in S$, suppose a real number $\lambda_v \geq 0$ is given such that*

$$\sum_{v \in S} \lambda_v = 1.$$

Then the elements β in K satisfying the simultaneous system of inequalities

$$\inf(1, \|\alpha_v - \beta\|_v) \leq \frac{1}{H(\beta)^{\kappa \lambda_v}}$$

for all $v \in S$ have bounded height.

It is obvious that Theorem 1 implies Theorem 2.1 (take the product). We shall now show that Theorem 2.1 implies Theorem 1.1. Suppose we have a sequence of solutions β to the approximation condition of Theorem 1, whose height tends to infinity. For such β, write

$$\inf(1, \|\alpha_v - \beta\|_v) = \frac{1}{H(\beta)^{\kappa \xi_v(\beta)}}$$

with a real number $\xi_v(\beta) \geq 0$. Then by hypothesis,

$$\sum_{v \in S} \xi_v(\beta) \geq 1.$$

Now select κ' such that $\kappa > \kappa' > 2$. Choose a positive integer A such that

$$A\left(\frac{\kappa}{\kappa'} - 1\right) > s$$

where s is the number of elements of S. Using induction, and the obvious fact that $[x + y] \leq [x] + [y] + 1$ (the bracket being the largest integer $\leq$) we get

$$A + s \leq A \frac{\kappa}{\kappa'} \leq \left[\sum_{v \in S} A \frac{\kappa}{\kappa'} \xi_v(\beta) \right] + 1 \leq \sum_{v \in S} \left[A \frac{\kappa}{\kappa'} \xi_v(\beta) \right] + s,$$

whence

$$A \leq \sum_{v \in S} \left[A \frac{\kappa}{\kappa'} \xi_v(\beta) \right].$$

Consequently there exist integers $a_v(\beta) \geq 0$ such that

$$a_v(\beta) \leq \left[A \frac{\kappa}{\kappa'} \xi_v(\beta) \right] \leq A \frac{\kappa}{\kappa'} \xi_v(\beta).$$

and $\sum a_v(\beta) = A$. From this we see that there is only a finite number of possible distributions of such integers $a_v(\beta)$, and hence, restricting our attention to a subsequence of our elements β if necessary, we can assume that the $a_v(\beta)$ are the same for all β. We write them a_v. We then put

$$\lambda_v = a_v / A$$

so that $0 \leq \lambda_v \leq 1$ and $\sum \lambda_v = 1$. For each β in our subsequence we have $\kappa' \lambda_v \leq \kappa \xi_v(\beta)$, and hence these β satisfy the simultaneous system of inequalities

$$\inf(1, \|\alpha_v - \beta\|_v) \leq \frac{1}{H(\beta)^{\kappa' \lambda_v}}.$$

We can therefore apply Theorem 2.1, and have therefore shown what we wanted.

We make one more remark which is technically useful in the proof. We may assume without loss of generality that all the α_v are integral over I_∞. Indeed, the α_v satisfy an equation

$$a_n Y^n + \cdots + a_0 = 0$$

with $a_i \in K$. (Take the product of their irreducible equations over K.) Using our assumption that K is the quotient field of I_∞ we can clear denominators, and thus assume that all a_i lie in I_∞. Multiplying this equation by a_n^{n-1}, we see that $a_n \alpha_v$ is integral over I_∞ for each v. Now if β approaches α_v, then $a_n \beta$ approaches $a_n \alpha_v$ essentially as well, and we can use a trick as in

Remark (ii) to show that we may deal with $a_n\alpha_v$ instead of α_v. This proves our assertion. Thus we may assume $S \supset S_\infty$, and we let

$$N = \sum_{v \in S} N_v.$$

§3. Basic Steps of the Proof

Let K be a field, and $F(X) \in K[X_1, \ldots, X_m] = K[X]$ a polynomial in several variables with coefficients in K. Let $r_1, \ldots, r_m$ be integers (in $\mathbf{Z}$) ≥ 0 and assume that the degree of F in X_j is $\leq r_j$. For any variable t, we have a Taylor expansion

$$F(X_1, \ldots, X_m) = \sum F^{(i)}(t, \ldots, t)(t - X_1)^{i_1} \cdots (t - X_m)^{i_m}$$

the sum being taken over $(i) = (i_1, \ldots, i_m)$, that is, m-tuples of integers satisfying $0 \leq i_1 \leq r_1, \ldots, 0 \leq i_m \leq r_m$. Each $F^{(i)}$ is a polynomial, namely

$$F^{(i)} = \frac{1}{i_1! \cdots i_m!} \, \partial_1^{i_1} \cdots \partial_m^{i_m} F$$

and $\partial_1, \ldots, \partial_m$ are the partial derivatives. Thus the coefficients of $F^{(i)}$ are integral multiples of the coefficients of F, since the factorials in the denominator divide the integers which will occur as a result of repeated partial differentiation. Furthermore, such integral multiples will be bounded by the product of the binomial coefficients

$$\binom{r_1}{i_1} \cdots \binom{r_m}{i_m}$$

which is itself bounded by $2^{r_1 + \cdots + r_m}$. This remark will be of use later when we estimate the absolute value of F. We also observe that the number of terms in our sum is certainly bounded by

$$(r_1 + 1) \cdots (r_m + 1)$$

which is also bounded by $2^{r_1 + \cdots + r_m}$.

We shall now state two results, to be proved in the following sections. We shall then indicate how Theorem 2.1 follows from them.

If we have K, S_∞, I_∞ as described in §1, and a polynomial $G(X_1, \ldots, X_m)$ in $I_\infty[X]$ then we write

$$B_\infty(G) = \sup |G|_v$$

the sup being taken over all $v \in S_\infty$. It is a bound for the absolute values of the coefficients of G in S_∞.

We shall meet in the statements and proofs which follow certain constants depending on the original data. We number them consecutively and always indicate what they depend on.

Proposition 3.1. *Let K be a field with a proper set of absolute values M_K. Let S_∞ be a finite non-empty subset of M_K containing all the archimedean absolute values. Let I_∞ be its integers. Assume Axiom 1a or 1b is satisfied. Let $f(Y)$ be a polynomial in one variable in $I_\infty[Y]$ with leading coefficient 1 and of degree n. Let $r_1, r_2, \ldots$ be integers > 0, and let $0 < \varepsilon < \frac{1}{2}$. Then for all $m > (2n/\varepsilon)^2$ (or $m > (2n/C_1\varepsilon)^2$ in case of Axiom 1b) there exists a polynomial*

$$G(X_1, \ldots, X_m) \neq 0$$

in $I_\infty[X]$ satisfying:

(1) $\qquad\qquad\qquad\qquad \deg G \quad in \quad X_j \quad is \quad \leqq r_j.$

(2) *If* $\qquad\qquad\qquad \dfrac{i_1}{r_1} + \cdots + \dfrac{i_m}{r_m} \leqq m\left(\dfrac{1}{2} - \varepsilon\right),$

then for any root α of f, we have

$$G^{(i)}(\alpha, \ldots, \alpha) = 0.$$

(3) $\qquad\qquad B_\infty(G) \leqq C_3^{r_1 + \cdots + r_m} \quad where \quad C_3 = C_3(K, S_\infty, f).$

Remark. From now on, all constants $C_4, C_5, \ldots$ depend on K, S_∞, f only, unless otherwise specified. In each case, they can be easily evaluated.

The second proposition will give us a means of preventing a polynomial from vanishing at a point $(\beta_1, \ldots, \beta_m)$.

It will be convenient to use both multiplicative and logarithmic heights. Thus we denote by H the multiplicative height, and $h = \log H$. Both will be taken relative to a given field K with a proper set of absolute values satisfying the product formula.

Proposition 3.2. *Let K be a field of characteristic 0, with a proper set of absolute values M_K satisfying the product formula with multiplicities $N_v \geqq 1$. Let m be an integer > 0. Let $0 < \delta < 1/16^{m+N}$. Let*

$$G(X_1, \ldots, X_m) \neq 0$$

be in $I_\infty[X]$ of degree $\leq r_j$ in X_j. Assume that the r_j satisfy

(4) $$\delta r_1 > r_2, \ldots, \delta r_m > 10,$$

that is $\delta r_{j-1} > r_j$ for $j < m$ and $\delta r_m > 10$. We say that $r_1, \ldots, r_m$ are δ-decreasing.

Let $\beta_1, \ldots, \beta_m$ be elements of heights $h_1, \ldots, h_m$ respectively satisfying

(5) $$N \log 4 + 4m \leq \delta h_1,$$

(6) $$r_1 h_1 \leq r_j h_j \quad \text{for } j = 1, \ldots, m.$$

Assume also that G satisfies

(7) $$\log B_\infty(G) \leq \delta r_1 h_1.$$

Then there exists an m-tuple $(v_1, \ldots, v_m)$ such that

(8) $$v_1/r_1 + \cdots + v_m/r_m \leq (20)^m \delta^{(1/2)^m}$$

and $G^{(v)}(\beta_1, \ldots, \beta_m) \neq 0$.

Using Propositions 3.1 and 3.2, we show how to prove Theorem 2.1.

We assume that we have elements β in K satisfying our simultaneous inequalities whose heights tend to infinity, and show that given ε, we must have $\kappa < 2 + \varepsilon'$ where ε' is small when ε is small. We can of course assume that $\varepsilon < \frac{1}{10}$.

We choose $m, \delta, \beta_1, \ldots, \beta_m, r_1, \ldots, r_m$ in that order, subject to certain conditions, as follows.

We first select $m > (2n/\varepsilon)^2$ or $> (2n/C_1\varepsilon)^2$ so that we can apply Proposition 3.1. We then select $\delta < \varepsilon$ so that $(20)^m \delta^{(1/2)^m}$ is also less than ε.

We now select β_1 of height h_1 so large that (5) is satisfied, and that

$$\delta h_1 > m \log C_3$$

(the constant of (3)).

We select $\beta_2, \ldots, \beta_m$ of heights $h_2, \ldots, h_m$ respectively such that they increase δ-rapidly, that is to say

$$\delta h_2 > 2h_1,$$

$$\cdots$$

$$\delta h_m > 2h_{m-1}.$$

We choose an integer r_1 so large that

$$r_1 \delta h_1 > 10 h_j \quad \text{for all } j = 2, \ldots, m$$

and then integers $r_2, \ldots, r_m$ such that

$$r_1 \frac{h_1}{h_j} \leqq r_j < 1 + r_1 \frac{h_1}{h_j}.$$

It follows immediately from the above inequalities that the r_j satisfy the condition

(9) $r_1 h_1 \leqq r_j h_j \leqq (1 + \varepsilon) r_1 h_1$

even with $[1 + (\frac{1}{10})\delta]$ instead of $(1 + \varepsilon)$ on the right-hand side. From this one concludes that (4) is satisfied, i.e. the r_j are δ-decreasing, namely,

$$r_j/r_{j+1} = (r_j/r_1)(r_1/r_{j+1})$$

and using (9) we get (4) at once.

We now apply Proposition 3.1. We had seen at the end of §2 that we could assume the α_v to be integral, satisfying a polynomial of degree n over K with leading coefficient 1. We find a polynomial G as in that proposition. By condition (3) we conclude that

$$\log B_\infty(G) \leqq \delta r_1 h_1.$$

By Proposition 3.2, some derivative $G^{(v)} = F$ is such that $F(\beta) \neq 0$ and $v_1/r_1 + \cdots + v_m/r_m \leqq \varepsilon$. Consequently, by (2) it follows that if

$$i_1/r_1 + \cdots + i_m/r_m \leqq m(\tfrac{1}{2} - \varepsilon) - \varepsilon$$

then for any root α of f (i.e. any α_v) we have

$$F^{(i)}(\alpha, \ldots \alpha) = 0.$$

We estimate $\|F(\beta)\|_v$ for $v \in S$ such that $\lambda_v \neq 0$. In the expansion

$$F(\beta) = \sum F^{(i)}(\alpha_v, \ldots, \alpha_v)(\alpha_v - \beta_1)^{i_1} \cdots (\alpha_v - \beta_m)^{i_m}$$

all the terms will be 0 except those belonging to an (i) such that

$$i_1/r_1 + \cdots + i_m/r_m \geqq m(\tfrac{1}{2} - 2\varepsilon).$$

The total number of terms in the sum is bounded by

$$2^{r_1 + \cdots + r_m} \leqq 2^{m r_1} \leqq H_1^{\delta r_1 m}$$

since the r_j are decreasing. An upper bound for $|F^{(i)}(\alpha_v, \ldots, \alpha_v)|_v$ is certainly

$$2^{mr_1} 2^{mr_1} C_3^{mr_1} B_\infty(G) \leqq H_1^{2\delta r_1 m}$$

using the fact that we chose H_1 large. The constant C_3 need merely be a bound for the value of α_v at v. Thus putting everything together, we get

$$\|F(\beta)\|_v \leqq H_1^{3\delta mr_1 N_v} \sup \|\alpha_v - \beta_1\|_v^{i_1} \cdots \|\alpha_v - \beta_m\|_v^{i_m}$$

the sup being taken for those (i) satisfying the condition mentioned above. Using the hypotheses of Theorem 2.1, we get

$$\log\|F(\beta)\|_v \leqq 3\delta mr_1 N_v h_1 - \kappa\lambda_v(i_1 h_1 + \cdots + i_m h_m).$$

We now replace i_1 by $r_1 i_1/r_1, \ldots, i_m$ by $r_m i_m/r_m$. We then replace $r_j h_j$ by $r_1 h_1$ using (9), making the denominator smaller and the fraction bigger. We then replace $i_1/r_1 + \cdots + i_m/r_m$ by $m(\frac{1}{2} - 2\varepsilon)$ thus finally obtaining

$$\log\|F(\beta)\|_v \leqq 3\delta mr_1 N_v - m(\frac{1}{2} - 2\varepsilon)\kappa\lambda_v r_1 h_1.$$

If $v \in S$ and $\lambda_v = 0$, we estimate $\|F(\beta)\|_v$ naively, and get

$$\|F(\beta)\|_v \leqq H_1^{3\delta mr_1 N_v} \sup(1, \|\beta_1\|_v)^{r_1} \cdots \sup(1, \|\beta_m\|_v)^{r_m}$$

using the same kind of crude estimates as before for the number of terms, coefficients, etc.

If $v \notin S$, so that v is non-archimedean, then we simply get

$$\log\|F(\beta)\|_v \leqq \log \sup(1, \|\beta_1\|_v)^{r_1} \cdots \sup(1, \|\beta_m\|_v)^{r_m}.$$

We now take the sum over all $v \in M_K$. We almost get the heights of the β_j on the right, up to a finite number of terms. We make the right-hand side only bigger if we add these terms. Having done so, we use (9) which guarantees

$$r_j h_j \leqq r_1 h_1(1 + \varepsilon),$$

and thus obtain

$$\sum_{v \in M_K} \log\|F(\beta)\|_v \leqq 6\delta mNr_1 h_1 + mr_1 h_1(1 + \varepsilon) - mr_1 h_1(\frac{1}{2} - 2\varepsilon)\kappa.$$

Since $F(\beta) \neq 0$, the left-hand side is equal to 0, we cancel $mr_1 h_1$ and get

$$\kappa(\frac{1}{2} - 2\varepsilon) \leqq 6\delta N + 1 + \varepsilon.$$

From this we immediately get $\kappa \leqq 2 + \varepsilon'$ as desired.

§4. A Combinatorial Lemma

Let $r_1,\ldots,r_m$ be integers ≥ 1 and let $\varepsilon > 0$. The number of sets of integers $(i_1,\ldots,i_m)$ satisfying

$$0 \leq i_1 \leq r_1,\ldots,0 \leq i_m \leq r_m$$

and

$$\frac{i_1}{r_1} + \cdots + \frac{i_m}{r_m} \leq m(\tfrac{1}{2} - \varepsilon)$$

does not exceed

$$\frac{1}{\varepsilon m^{1/2}} (r_1 + 1) \cdots (r_m + 1)$$

Proof. By induction on m. The assertion is trivial if $m = 1$. Take $m > 1$. We write

$$m(\tfrac{1}{2} - \varepsilon) = \tfrac{1}{2}(m - \lambda)$$

so that $m\varepsilon = \lambda/2$. In terms of λ our upper bound reads

$$\frac{2m^{1/2}}{\lambda} (r_1 + 1) \cdots (r_m + 1).$$

Our assertion is trivial if $\lambda \leq 2m^{1/2}$. We may therefore assume that $\lambda > 2m^{1/2}$. For each i_m, r_m fixed, we consider the solutions of

$$\frac{i_1}{r_1} + \cdots + \frac{i_{m-1}}{r_{m-1}} \leq \tfrac{1}{2}(m - \lambda) - \frac{i_m}{r_m}$$

$$\leq \frac{m - 1}{2} - \frac{\lambda - 1 + 2i_m/r_m}{2}.$$

We apply the induction hypothesis and sum over i_m. For this, we first make a computation:

$$\sum_{i=0}^{r} \frac{2}{\lambda - 1 + 2i/r} = \sum_{i=0}^{r} \left[\frac{1}{\lambda - 1 + 2i/r} + \frac{1}{\lambda + 1 - 2i/r} \right]$$

$$= \sum_{i=0}^{r} \frac{2\lambda}{\lambda^2 - (1 - 2i/r)^2}$$

$$\leq \sum_{i=0}^{r} \frac{2\lambda}{\lambda^2 - 1} = (r + 1) \frac{2\lambda}{\lambda^2 - 1}.$$

We now let $i = i_m$ and $r = r_m$. Our number e is therefore bounded by

$$\sum \frac{2(m - 1)^{1/2}}{\lambda - 1 + 2i_m/r_m} (r_1 + 1) \cdots (r_{m-1} + 1)$$

$$\leq \frac{2\lambda(m - 1)^{1/2}}{\lambda^2 - 1} (r_1 + 1) \cdots (r_m + 1).$$

We must therefore check whether

$$\frac{2\lambda(m - 1)^{1/2}}{\lambda^2 - 1} \leq \frac{2m^{1/2}}{\lambda},$$

or equivalently, whether

$$[(m - 1)/m]^{1/2} \leq (\lambda^2 - 1)/\lambda^2.$$

Observe that $(1 - 1/m)^{1/2} \leq 1 - \frac{1}{2}m$ (square the right-hand side). It suffices therefore that $1 - \frac{1}{2}m \leq 1 - 1/\lambda^2$. But this is true in view of our original hypothesis on λ, and our lemma is proved.

§5. Proof of Proposition 3.1

We have to use either Axiom 1a or 1b. Let us start with 1a, and indicate afterwards the modifications of language necessary to use Axiom 1b instead.

The number of polynomials $F(X_1, \ldots, X_m)$ such that

(10) $\deg F$ in X_j is $\leq r_j$

(11) $F \in I_\infty[X]$ and $B_\infty(F) \leq B$

is at least $(C_1 B)^{N(r_1 + 1) \cdots (r_m + 1)}$. For each set of integers (i) satisfying the condition of (2), we let $RF^{(i)}$ be the remainder of the division of $F^{(i)}(Y, \ldots, Y)$ by $f(Y)$. The degree of $F^{(i)}(Y, \ldots, Y)$ (which is a polynomial in one variable) is at most $r_1 + \cdots + r_m$. Using the remarks at the beginning of §3, and long division, we find by an inductive estimate that

(12) $B_\infty(RF^{(i)}) \leq C_5^{r_1 + \cdots + r_m} B_\infty(F),$

where C_5 depends on f and S_∞ only. In this way, each polynomial satisfying (10) and (11) gives rise to a set of e remainders

$$(RF^{(i)_1}, \ldots, RF^{(i)_e}),$$

where e is bounded as in the preceding section. Each remainder is of degree $< n$.

By Axiom 1a, the number of such sets of remainders is at most

$$(13) \qquad [C_5^{r_1 + \cdots + r_m} B]^{Nne}.$$

We want two distinct polynomials with the same set of remainders, say F, G. Then $F - G$ will have remainders 0 for all (i) satisfying (2), and thus the conclusion of (2) will be satisfied. Furthermore, $B_\infty(F - G)$ will be at most $2B$, and thus (3) will also be satisfied. Condition (1) is trivially satisfied since we started with (10).

We shall certainly get our two distinct polynomials if

$$(14) \qquad (C_1 B^N)^{(r_1 + 1) \cdots + (r_m + 1)} > [C_5^{r_1 + \cdots + r_m} B]^{Nne},$$

and this will certainly be true if we replace e by the upper estimate derived in the preceding section. The product $(r_1 + 1) \cdots (r_m + 1)$ in the exponent of both sides can then be cancelled. Let $C_6 = C_5^N$. It will suffice to get

$$(15) \qquad C_1 B^N > [C_6^{r_1 + \cdots + r_m} B^N]^{n/\varepsilon m^{1/2}}.$$

If now $m > (2n/\varepsilon)^2$, then it will suffice to get

$$(16) \qquad C_1 B^N > [C_6^{r_1 + \cdots + r_m} B^N]^{1/2},$$

and thus it will in fact suffice to get

$$(17) \qquad B^{N/2} > C_7^{r_1 + \cdots + r_m},$$

with a suitable C_7. Letting $C_8 = C_7^{2/N}$ we get a condition of the type desired to satisfy (3). This concludes the proof, using Axiom 1a.

To use Axiom 1b, we have to count linear dimensions.

The polynomials $F(X)$ satisfying (10) and

$$(11b) \qquad F \in I_\infty[X] \qquad \text{and} \qquad B_\infty(F) \leqq B_0^v$$

are a vector space over the constants k of dimension at least equal to $(r_1 + 1) \cdots (r_m + 1) C_1 v^d$.

For each set of integers (i) satisfying the condition of (2), we find the same estimate (12) (without bothering to take into account the fact that we are dealing with non-archimedean absolute values). The map

$$F \longmapsto (RF^{(i)_1}, \ldots, RF^{(i)_e})$$

is now a k-linear map. By Axiom 1b, the dimension over k of the vector space which is the image of this map is at most

(13b) $$ne[v_0(r_1 + \cdots + r_m) + v]^d.$$

The integer v_0 is determined so that $C_5 \leqq B_0^{v_0}$. We now want the kernel of our k-linear map to be larger than 0, and for this it will suffice to have the inequality of dimensions:

(14b) $(r_1 + 1) \cdots (r_m + 1)C_1 v^d$

$$> n \frac{1}{\varepsilon m^{1/2}} (r_1 + 1) \cdots (r_m + 1) \cdot [v_0(r_1 + \cdots + r_m) + v]^d.$$

We cancel the product of the $(r_j + 1)$, and take $m > (2n/C_1\varepsilon)^2$. We then see that it suffices to achieve

$$v^d > \tfrac{1}{2}[v_0(r_1 + \cdots + r_m) + v]^d$$

and taking a d-th root, it is enough to get

$$v\left(1 - \frac{1}{2^{1/d}}\right) > v_0(r_1 + \cdots + r_m).$$

Putting this condition in terms of $B_\infty(F)$, we see that our proposition is proved.

§6. Wronskians

Let R be a ring. A **derivation** d of R is a linear map of R into itself satisfying

$$d(xy) = x \cdot dy + y \cdot dx.$$

Those elements x of R such that $dx = 0$ form a subring (because $d(1) = 0$) containing the prime ring, and are called the ring of **constants**. This ring of constants is a field if R is a field. A **differential ring** is a ring together with a finite set of derivations $d_1, \ldots, d_n$ which commute with each other (as operators on R). Its subring of constants is by definition the intersection of the rings of constants of each d_j.

Example. Let k be a field, and R the polynomial ring $k[X_1, \ldots, X_n]$. Let d_j be the partial derivative with respect to X_j, taken formally. Then R is a differential ring, and if k has characteristic 0, it is the field of constants.

If R is a differential ring, and $d_1, \ldots, d_n$ its derivations, we   can form monomials

$$d_1^{i_1} \cdots d_n^{i_n},$$

where i_ν are integers ≥ 0. These monomials can be viewed as linear operators on R in the obvious way. The **degree** of the above monomial is by definition $i_1 + \cdots + i_n$. We denote by $D(r)$ the set of monomials of degree $< r$, where r is an integer ≥ 1.

The following proposition will be applied to the case of polynomials in several variables, as above. I am indebted to Kolchin for the formulation and proof in its present degree of generality.

Proposition 6.1. *Let F be a differential field, k the field of constants for the derivations $d_1, \ldots, d_n$ of F. Let $D(r)$ be the monomials of degree $< r$ in the derivations. Let $x_1, \ldots, x_n$ be elements of F. If they are linearly dependent over k, then*

$$\det(\theta_i x_j) = 0$$

for all operators $\theta_1, \ldots, \theta_n \in D(r)$ and any r. Conversely, if this determinant is 0 for all choices of $\theta_i \in D(i)$, then they are linear dependent over k.

Proof. The first assertion is trivial. We prove the second by induction on n, and the assertion is trivial for $n = 1$. We assume $n > 1$. Observe that $\theta_1 = 1$. We consider determinants:

$$\begin{vmatrix} \theta_1 x_1 & \theta_1 x_2 & \cdots & \theta_1 x_n \\ & \cdots & \\ \theta_n x_1 & \theta_n x_2 & \cdots & \theta_n x_n \end{vmatrix}, \qquad \theta_i \in D(i).$$

If all the subdeterminants of size $n - 1$ above the last line are 0 for all choices of $\theta_i \in D(i)$, then any $n - 1$ among the x's are linearly dependent by induction, and we are done. Thus we may assume that some determinant of size $n - 1$ above the last line is not 0, say

$$\begin{vmatrix} x_1 & x_2 & \cdots & x_{n-1} \\ \theta_2' x_1 & \theta_2' x_2 & \cdots & \theta_2' x_{n-1} \\ & \cdots & \\ \theta_{n-1}' x_1 & \cdots & & \theta_{n-1}' x_{n-1} \end{vmatrix} \neq 0$$

for some choice of $\theta_i' \in D(i)$, $i = 1, \ldots, n - 1$.

We complete this determinant to one with rows having n elements. Then these rows are linearly independent over F since the smaller rows are already linearly independent over F. Hence any row

$$(\theta x_1, \ldots, \theta x_n), \qquad \theta \in D(n)$$

is linearly dependent over F on our expanded $n - 1$ rows, and thus the matrix

$$\begin{bmatrix} x_1 & \cdots & x_n \\ \theta'_2 x_1 & \cdots & \theta'_2 x_n \\ & \cdots & \\ \theta'_{n-1} x_{n-1} & \cdots & \theta'_{n-1} x_n \\ \theta x_1 & \cdots & \theta x_n \\ \varphi x_1 & \cdots & \varphi x_n \\ \cdots & \cdots & \cdots \end{bmatrix}$$

has rank $n - 1$ for any choice of $\theta, \varphi, \ldots$ in $D(n)$. It therefore has column rank equal to $n - 1$, and hence there exist elements $a_1, \ldots, a_n \in F$ with $a_n = 1$ such that for all $\theta \in D(n)$ we have

(18) $$a_1 \theta x_1 + \cdots + a_n \theta x_n = 0$$

because the last column can be expressed as a linear combination of the first $(n - 1)$ columns. Let d be any one of the derivations d_i. Applying d to (18) we get

$$a_1(d\theta x_1) + \cdots + a_n(d\theta x_n) + (da_1)\theta x_1 + \cdots + (da_n)\theta x_n = 0.$$

In this last expression, let $A(\theta)$ and $B(\theta)$ denote the first sum involving the $d\theta$, and the second sum respectively. We have $A(\theta) + B(\theta) = 0$. If θ happens to lie in $D(n - 1)$ then $d\theta$ lies in $D(n)$ and hence $A(\theta) = 0$ by (18). This holds in particular for everyone of the derivations θ'_j and gives $B(\theta'_j) = 0$ for all j. This is a set of $n - 1$ equations involving $n - 1$ variables

$$da_1, \ldots, da_{n-1}$$

because $a_n = 1$ and $da_n = 0$. The determinant of the coefficients is not zero, and hence $da_1 = \cdots = da_{n-1} = 0$. Since d was any one of the derivations d_i it follows that $a_1, \ldots, a_{n-1}$ are constants. Letting $\theta = 1$ in (18) proves what we wanted.

§7. Factorization of a Polynomial

Although it is not true that a polynomial in several variables factorizes into a product of polynomials in fewer variables, we shall see, nevertheless, that a suitable derivative of it does. This will allow us to carry out an induction argument later.

Let K be a field of characteristic 0. We consider the polynomial ring $K[X_1,\ldots,X_m] = K[X]$ with the standard partial derivatives. By a **differential operator** D we mean

$$D = \frac{1}{i_1!\cdots i_m!}\,\partial_1^{i_1}\cdots\partial_m^{i_m}$$

and we call $i_1 + \cdots + i_m$ the **degree** of the operator. Applied to any polynomial G, it yields another polynomial whose coefficients are integral multiples of those of G.

Lemma 7.1. *Let K be a field of characteristic 0. Let $G(X) = G(X_1,\ldots,X_m)$ be a non-zero polynomial in $K[X]$ in m variables, $m \geq 2$, of degree $\leq r_j$ in X_j for each $j = 1,\ldots,m$. Then there exists an integer l satisfying*

(19) $$1 \leq l \leq r_m + 1$$

and differential operators $D_0,\ldots,D_{l-1}$, on the variables $X_1,\ldots,X_{m-1}$, of degrees $\leq 0,\ldots,l-1$, respectively, such that if

$$F(X_1,\ldots,X_m) = \det\left(D_\nu\,\frac{1}{\mu!}\,\partial_m^\mu G\right) \qquad (\nu,\mu = 0,\ldots,l-1),$$

then $F \neq 0$, and

$$F(X_1,\ldots,X_m) = U(X_1,\ldots,X_{m-1})V(X_m),$$

where U, V are polynomials, U is of degree at most lr_j in X_j for

$$j = 1,\ldots,m-1$$

and V is of degree at most lr_m in X_m.

Proof. We consider all representations of G in the form

$$G(X) = \varphi_0(X_m)\psi_0(X_1,\ldots,X_{m-1}) + \cdots + \varphi_{l-1}(X_m)\psi_{l-1}(X_1,\ldots,X_{m-1}),$$

where the φ_ν and ψ_μ are polynomials with coefficients in K, of degree at most r_j in X_j. Such a representation is possible, for instance, with $l - 1 = r_m$ and $\psi_\nu(X_m) = X_m^\nu$. From all such representations, we select the one for which l is least. Then

$$\varphi_0(X_m),\ldots,\varphi_{l-1}(X_m)$$

are linearly independent over K, for, if not, one would immediately deduce another representation of G as above with a smaller l. Similarly,

$$\psi_0(X_1, \ldots, X_{m-1}), \ldots, \psi_{l-1}(X_1, \ldots, X_{m-1})$$

are linearly independent over K. Also $1 \leq l \leq r_m + 1$.

We now apply Proposition 6.1 to these two sets of polynomials. The θ's in Proposition 6.1 were taken with coefficient 1. Since we are in characteristic 0, putting a factorial in the denominator results in no loss of generality, and hence there exist differential operators $D_0, \ldots, D_{l-1}$, in the variables $X_1, \ldots, X_{m-1}$, of degrees $\leq 0, \ldots, l-1$, respectively, such that

$$U(X_1, \ldots, X_{m-1}) = \det(D_\mu \psi_\rho) \qquad (\mu, \rho = 0, \ldots, l-1)$$

is not 0. Similarly,

$$V(X_m) = \det\left(\frac{1}{\nu!} \partial_m^\nu \psi_\rho\right) \qquad (\nu, \rho = 0, \ldots, l-1)$$

is not 0. Multiplying these two determinants, we see that $UV = F$. The degrees of U, V in the X_j obviously satisfy the required condition, since the determinants are $l \times l$.

We now estimate the coefficients of F in terms of those of G.

Lemma 7.2. *Let the notation be as in the previous lemma. Let v be an absolute value on K. If v is non-archimedean then $|F|_v \leq |G|_v^l$. If v induces the ordinary absolute value on $\mathbf{Q}$, then*

$$|F|_v \leq [(r_1 + 1) \cdots (r_m + 1)]^l l! \, 2^{l(r_1 + \cdots + r_m)} |G|_v^l.$$

Proof. If v is non-archimedean, our assertion is obvious. If v induces the ordinary absolute value on $\mathbf{Q}$, we regard G as a sum of $(r_1 + 1) \cdots (r_m + 1)$ terms, each of the form

$$a_{(i)} X_1^{i_1} \cdots X_m^{i_m},$$

where $|a_{(i)}| \leq |G|$. We can develop our determinant into a sum of

$$[(r_1 + 1) \cdots (r_m + 1)]^l$$

determinants, the general element in each such determinant being of the form

$$a_{(i)} D_\mu \frac{1}{\nu!} \partial_m^\nu X_1^{i_1} \cdots X_m^{i_m}.$$

By the remarks at the beginning of §3, we see that the effect of the differential operators on the estimates is at most to multiply them by $2^{r_1 + \cdots + r_m}$. Hence the coefficients of each of the $l!$ terms in the expansion of an individual determinant have absolute values not exceeding

$$[|G|2^{r_1 + \cdots + r_m}]^l$$

and our assertion follows.

§8. The Index

Let K be a field, $F(X_1, \ldots, X_m)$ a polynomial in $K[X]$, and $F \neq 0$. Let $r_1, \ldots, r_m$ be numbers > 0, and $\beta_1, \ldots, \beta_m$ elements of K. To simplify the notation, we denote by $r_{(m)}$ and $\beta_{(m)}$ the vectors of the r's and β's respectively.

Let t be a new variable. Consider the expansion

$$F(\beta_1 + X_1 t^{1/r_1}, \ldots, \beta_m + X_m t^{1/r_m})$$
$$= \sum F^{(i)}(\beta_1, \ldots, \beta_m) X_1^{i_1} \cdots X_m^{i_m} t^{i_1/r_1 + \cdots + i_m/r_m}.$$

We can view the right-hand side as a polynomial in fractional powers of t. We let

$$\theta = \text{ind}_{r_{(m)}, \beta_{(m)}} F$$

be the smallest number occurring as an exponent of t on the right-hand side such that t^θ has a non-zero coefficient, and call it the **index** of F at $\beta_{(m)}$ relative to $r_{(m)}$. Then θ is also the minimum of the numbers $(i_1/r_1 + \cdots + i_m/r_m)$ such that $F^{(i)}(\beta_1, \ldots, \beta_m) \neq 0$.

We note that the index is always ≥ 0, and equals 0 if and only if $F(\beta_1, \ldots, \beta_m) \neq 0$.

It follows immediately from the definitions that the index satisfies the following properties:

Ind. 1. *If F, G are two polynomials, neither of which is 0, then*

$$\text{ind}(F + G) \geq \min(\text{ind } F, \text{ind } G)$$
$$\text{ind}(FG) = \text{ind } F + \text{ind } G$$

(because the degree of a product is equal to the sum of the degrees). *If F is a polynomial in $X_1, \ldots, X_{m-1}$ and G is a polynomial in X_m, then*

$$\text{ind}_{r_{(m)}, \beta_{(m)}}(FG) = \text{ind}_{r_{(m-1)} \beta_{(m-1)}}(F) + \text{ind}_{r_m, \beta_m}(G).$$

Ind. 2. *The index of*

$$\partial_1^{v_1} \cdots \partial_m^{v_m} F$$

at $\beta_{(m)}$ relative to $r_{(m)}$ is at least equal to

$$\operatorname{ind}_{r_{(m)}, \beta_{(m)}}(F) - v_1/r_1 - \cdots - v_m/r_m,$$

for any integers $v_1, \dots, v_m \geqq 0$, provided the derived polynomial is not identically 0.

Ind. 3. *Let l be a number > 0, and $lr_{(m)} = (lr_1, \dots, lr_m)$. Then*

$$\operatorname{ind}_{lr_{(m)}, \beta_{(m)}}(F) = \frac{1}{l} \operatorname{ind}_{r_{(m)}, \beta_{(m)}}(F).$$

We now wish to compare the indices of G and F in Lemma 1.

Lemma 8.1. *Let the notation be as in Lemma 7.1. Let $0 < \delta \leqq 1$. Assume that $r_1, \dots, r_m$ are δ-decreasing, that is to say*

(20) $$\delta r_1 \geqq r_2, \dots, \delta r_m \geqq 10.$$

Let $\beta_1, \dots, \beta_m$ be elements of K, and put $\operatorname{ind} = \operatorname{ind}_{r_{(m)}, \beta_{(m)}}$. Let $\theta = \operatorname{ind}(G)$. Then

$$\theta^2 \text{ or } \theta \leqq \frac{4}{l} \operatorname{ind}(F) + d\delta,$$

Proof. Consider a differential operator of the form

$$D_\mu = \frac{1}{i_1! \cdots i_{m-1}!} \partial_1^{i_1} \cdots \partial_{m-1}^{i_{m-1}}$$

on $X_1, \dots, X_{m-1}$ of degree $\leqq l - 1$. If the polynomial

$$D_\mu \frac{1}{v!} \partial_m^v G$$

does not vanish identically, its index at $\beta_{(m)}$ relative to $r_{(m)}$ is at least

$$\theta - i_1/r_1 - \cdots - i_{m-1}/r_{m-1} - v/r_m.$$

We make this number smaller if we replace the r_j $(j = 1, \ldots, m - 1)$ by r_{m-1}. Hence this index is at least

$$\theta - \frac{l - 1}{r_{m-1}} - \frac{v}{r_m} \geq \theta - \frac{r_m}{r_{m-1}} - \frac{v}{r_m}$$

Since $r_m/r_{m-1} \leq \delta$, and since the index is never negative, it must be at least

$$\max(0, \theta - v/r_m) - \delta.$$

If we expand the determinant giving the expression for F, we obtain for F a sum of $l!$ terms, the typical term being of the form

$$\pm (D_{\mu_0} G)\left(D_{\mu_1} \frac{1}{1!} \partial_m G\right) \cdots \left(D_{\mu_{l-1}} \frac{1}{(l - 1)!} \partial_m^{l-1} G\right),$$

where the D_μ have degree $\leq l - 1$. By one of the basic properties of the index, if such a term does not vanish identically its index is at least

$$\sigma = \sum_{v=0}^{l-1} \max(0, \theta - v/r_m) - l\delta.$$

Since F is a sum of such terms, and is not 0, it follows that $\mathrm{ind}(F) \geq \sigma$. We must now solve for θ in terms of $\mathrm{ind}(F)$. We can assume $\theta r_m > 10$, for if not, we have

$$\theta \leq 10/r_m \leq \delta$$

and our inequality is certainly satisfied.

If $\theta r_m < l$ then we have

$$\sum_{v=0}^{l-1} \max(0, \theta - v/r_m) = 1/r_m \sum_{v=0}^{l-1} \max(0, \theta r_m - v)$$

$$\geq 1/r_m \sum_{0 \leq v \leq [\theta r_m]} ([\theta r_m] - v)$$

$$\geq \frac{[\theta r_m]([\theta r_m] + 1)}{2r_m}$$

$$\geq \frac{[\theta r_m]^2}{2r_m}.$$

If a is a number > 10, then $(a - 1)^2/2 > a^2/3$. Consequently, since $\theta r_m > 10$, we have by a trivial computation

$$[\theta r_m]^2/2r_m \geq (\theta r_m)^2/3r_m = \theta^2 r_m/3.$$

Thus in this case, $\text{ind}(F) \geqq \theta^2 r_m/3 - l\delta$ and

$$\theta^2 \leqq \frac{l}{r_m} \frac{3}{l} \text{ind}(F) + \frac{3l}{r_m} \delta \leqq \frac{4}{l} \text{ind}(F) + 4\delta,$$

since $r_m \geqq 10$ and $l \leqq r_m + 1$.

On the other hand, if $\theta r_m \geqq l$ we have

$$\sum_{v=0}^{l-1} \max(0, \theta - v/r_m) = \sum_{v=0}^{l-1} (\theta - v/r_m) \geqq l\theta/2.$$

From this our assertion is again obvious, and this concludes the proof.

§9. Proof of Proposition 3.2

We come to the proof of Proposition 3.2. We shall in fact prove a slightly stronger version of it. If K is a field with a proper set of absolute values M_K satisfying the product formula with multiplicities $N_v \geqq 1$, and

$$F(X_1, \ldots, X_m) \in K[X]$$

is a polynomial, then we define as usual

$$H(F) = \prod_{v \in M_K} |F|_v^{N_v} = \prod_{v \in M_K} \|F\|_v, \qquad \text{and} \qquad h(F) = \log H(F).$$

Proposition 9.1. *Let K be a field of characteristic 0, with a proper set of absolute values M_K satisfying the product formula with multiplicities $N_v \geqq 1$. Let S be a finite subset of M_K containing the archimedean absolute values, and let $N = \sum_{v \in S} N_v$. Let*

(21) $$0 < \delta < 1/16^{m+N}.$$

Let r_j $(j = 1, \ldots, m)$ be integers δ-decreasing, that is

(4) $$\delta r_1 > r_2, \ldots, \delta r_m > 10.$$

Let $G(X_1, \ldots, X_m)$ be a non-zero polynomial in $I_K[X]$ of degree $\leqq r_j$ in X_j and let β_j $(j = 1, \ldots, m)$ be elements of K with $h(\beta_j) = h_j$. Assume that:

(22) $$N \log 4 + 4m \leqq \delta h_1,$$

(23) $$r_1 h_1 \leqq r_j h_j,$$

(24) $$h(G) \leqq N\delta r_1 h_1.$$

Then

$$\text{ind}_{r_{(m)}, \beta_{(m)}}(G) \leq (20)^m \delta^{(1/2)^m}.$$

Remark. What we want, of course, is that if δ is small, the coefficients of G are small, and the r_j are δ-decreasing, then the index of G is small. The hypotheses are made precise in order to load the induction.

Proof. We prove Proposition 9.1 by induction on m. Suppose first $m = 1$. Write $G(X) = (X - \beta)^e F(X)$ and suppose $\deg G \leq r$. By Gauss' lemma we have

$$H[(X - \beta)^e]H(F) \leq H(G)4^{rN}.$$

For each absolute value in M_K, it is clear that

$$|(X - \beta)|^e \leq |(X - \beta)^e|$$

and hence $H(\beta)^e \leq H[(X - \beta)^e]$. From this and the fact $H(F) \geq 1$ we get $H(\beta)^e \leq H(G)4^{rN}$, whence

$$eh(\beta) \leq h(G) + rN \log 4.$$

By hypothesis and conditions (24) and (22) we get

$$e \leq N\delta r + r\delta$$

and consequently the index $\text{ind}_{r, \beta}(G) = e/r$ satisfies

$$e/r \leq \delta(N + 1) \leq 6\delta^{1/2},$$

if we use (21). Thus the case $m = 1$ is proved.

Let $m > 1$. We use Lemma 7.1 and get a polynomial $F = UV$ as in this lemma. The estimate for its coefficients obtained in Lemma 7.2 can now be transformed as follows, using (4):

$$|F|_v \leq 2^{mr_1 l} 2^{r_1 l} 2^{mr_1 l} |G|_v^l$$

replacing $(r_j + 1)$ by 2^{r_1} and $l!$ by $l^l \leq 2^{r_1 l}$. Using (22) and (24) this yields

$$h(F) \leq Nmr_1 l \log 8 + h(G) \leq 2\delta N r_1 h_1 l.$$

We also know that the degree of F in X_j is $\leq lr_j$.

Since $F = UV$ and the variables are separated, one can verify immediately that $H(F) = H(U)H(V)$, and hence we get the same estimate for $H(U)$ and $H(V)$ that we have just obtained for $H(F)$, because the height is always

≥ 1. Both U and V have degrees $\leq lr_j$ in X_j. In (24) we can replace G by U or V and δ by 2δ, so that it remains valid (using (23) for V). In (4) we can also replace δ by 2δ and r_j by lr_j so that it remains valid. Thus the induction hypothesis is satisfied for each one of U, V with lr_j and 2δ.

Using Property 1 of the index, we get by induction

$$\mathrm{ind}_{lr_{(m-1)},\,\beta_{(m-1)}}(U) \leq (20)^{m-1}(2\delta)^{(1/2)^{m-1}}$$
$$\mathrm{ind}_{lr_m,\,\beta_m}(V) \leq (20)(2\delta)^{(1/2)}.$$

The index for V is certainly at most equal to the index for U. The index of F at $\beta_{(m)}$ relative to $lr_{(m)}$ is the sum of these two. Using Lemma 8.1 and the fact that $4\delta \leq 4(2\delta)^{1/2}$ we get

$$\theta^2 \text{ or } \theta \leq (4 \cdot 3 \cdot \sqrt{2})(20)^{m-1}(2\delta)^{(1/2)m-1}$$

from which we get

$$\theta \leq (20)^m \delta^{(1/2)^m}$$

thereby proving our proposition, and Roth's theorem.

§10. A Geometric Formulation of Roth's Theorem

Throughout this section, K is a field satisfying the hypotheses of Theorem 1.1 so that this theorem is valid for it.

We reformulate Theorem 1.1 slightly, as an intermediate step to the final statement of Theorem 10.1 below.

Let $G(Y)$ be a polynomial in $K[Y]$ (one variable) and assume that the multiplicity of its roots is at most r for some integer $r > 0$. Let S be a finite subset of M_K. Let $C > 0$ be a number, and $\kappa > 2$. Then the solutions β in K of the inequality

$$\prod_{v \in S} \inf(1, \|G(\beta)\|_v) \leq \frac{C}{H(\beta)^{\kappa r}}$$

have bounded height.

To prove this, we may assume that G has leading coefficient 1, and say

$$G(Y) = \prod_{i=1}^{d} (Y - \alpha_i)^{e_i}$$

is a factorization in K^a. We extend our absolute values of S to K^a in any way. The expression on the left-hand side of our inequality is greater or equal to

$$\prod_{v \in S} \prod_{i=1}^{d} \inf(1, \|\beta - \alpha_i\|_v)^{e_i},$$

which is itself greater or equal to the expression obtained by replacing e_i by r for each i. Now we are in the situation of Theorem 1.1, taking into account the remark (iv) following it, namely, the solutions β of the inequality

$$\prod_{v \in S} \prod_{i=1}^{d} \inf(1, \|\beta - \alpha_i\|_v)^r \le \frac{1}{H(\beta)^{\kappa r}}$$

have bounded height, hence so do the solutions of our original inequality.

We observe that Theorem 1.1 is in fact equivalent to this theorem, for if α is algebraic over K, we take for $G(Y)$ its irreducible equation over K. Its roots occur with multiplicity 1, and if β approximates α, then it stays away from the other roots, at a distance greater than some fixed constant, once it is sufficiently close to α. We can do the same for each α_v, and thus we see that the implication is trivial.

Theorem 10.1. *Let W be a complete non-singular curve defined over K. Let z and y be two non-constant functions in $K(W)$, and let r be the largest of the multiplicities of the zeros of z. Assume that y is defined at all the zeros and poles of z and gives an injective mapping of this set of zeros into K^a. Let κ be a number > 2 and let $C > 0$. Then the points $Q \in W(K)$ which are not poles or zeros of z and are such that*

$$\prod_{v \in S} \inf(1, \|z(Q)\|_v) \le \frac{C}{H(y(Q))^{\kappa r}}$$

have bounded height H_y.

Proof. Without loss of generality, we may assume that $K(W) = K(z, y)$ (otherwise, replace W by a non-singular model of $K(z, y)$). We may also assume that the absolute values

$$|y(Q)|_v$$

are bounded for all $v \in S$. Indeed, if there is an infinite sequence of points Q whose height $H_y(Q) = H(y(Q))$ tends to infinity and which satisfies the above inequality, but with $|y(Q)|_v$ unbounded for some $v \in S$, we then consider a non-singular linear transformation on y, say

$$y' = (ay + b)/(cy + d),$$

with coefficients in K, which has the same property as y in the statement of the theorem, but such that $|y'(Q)|_v$ is bounded for all $v \in S$, and a subsequence of points Q. $H_{y'}$ is then equivalent to H_y (trivial by direct verification, or use Proposition 2.1 of Chapter 4). If S has only one absolute value v, then we can actually take $1/y$ instead of y.

Let Φ be the set of zeros of z. Since y has no pole in Φ, it is integral over the local ring $\mathfrak{o}$ of the point $z = 0$ in the function field $K(z)$. Let $F(Y)$ be its irreducible equation over $\mathfrak{o}$. Then

$$F(Y) \equiv G(Y) \pmod{z},$$

where $G(Y)$ is a polynomial with coefficients in K with leading coefficient 1 and mod z means modulo the maximal ideal of $\mathfrak{o}$ generated by z.

By hypothesis, y induces an injection $Q \mapsto y(Q)$ of Q into K^a. The multiplicity of a root of $G(Y)$ is thus $\leqq$ the multiplicity of a point on W in the inverse image of the point $z = 0$, since it is the multiplicity of a zero of z. One can see this formally for instance as follows. Let $(y^{(1)})$ be an affine generic point of w over K all of whose coordinates are integral over $\mathfrak{o}$. If $(y^{(1)}, \ldots, y^{(m)})$ is a complete set of conjugates of $(y^{(1)})$ over $K(z)$, then the cycle on W which is the inverse image of $z = 0$ consists of a specialization $(\bar{y}^{(1)}, \ldots, \bar{y}^{(m)})$ of $(y^{(1)}, \ldots, y^{(m)})$ over $z \to 0$. The conjugates of y correspond to the conjugates $(y^{(1)}, \ldots, y^{(m)})$, and one can then use IAG, Theorem 2 of Chapter I, §4 applied to the polynomial $F(Y)$.

We can write

$$F(Y) = G(Y) + zA(z, Y),$$

where $A(z, Y)$ is a polynomial in Y with coefficients in $\mathfrak{o}$. Since $A(0, Y)$ is defined, so is $A(z(Q), Y)$ for small values of $\|z(Q)\|_v$, which is all that we need to consider. Thus the values

$$\|A(z(Q), y(Q))\|_v$$

remain bounded since we may assume that $\|y(Q)\|_v$ remains bounded for all $v \in S$. Since $F(y) = 0$, we get an estimate for $G(y(Q))$, namely,

$$\|G(y(Q))\|_v \leqq C'\|z(Q)\|_v$$

for some constant C' and all $v \in S$. Taking the usual product, we get

$$\prod_{v \in S} \inf(1, \|G(y(Q))\|_v) \leqq C'' \prod_{v \in S} \inf(1, \|z(Q)\|_v)$$

$$\leqq \frac{C''}{H(y(Q))^{\kappa r}},$$

which leaves us in the situation discussed at the beginning of this section, and concludes the proof.

Actually, it will be even more convenient to make use of the projection property of Chapter 4, §4. Of course, in proving these properties of heights, we worked with a ground field **F** and absolute values which were well behaved in order to state our theorems for the absolute heights. Here, it does not matter since we shall work over a fixed field K (we incorporate the multiplicities N_v directly into the data). Properties of heights clearly remain valid under our present set up. Thus from the projection property, we get:

Theorem 10.2. *Let W be a projective non-singular curve defined over K. Let ψ be a non-constant function in $K(W)$ and let r be the largest of the multiplicities of the poles of ψ. Let κ be a number > 2, C a number > 0. Let S be a finite subset of M_K. Then the points $Q \in W_K$ which are not among the poles of ψ, and are such that*

$$\prod_{v \in S} \sup(1, \|\psi(Q)\|_v) \geq C \cdot H(Q)^{\kappa r}$$

have bounded height.

Proof. We let Φ be the set of zeros and poles of ψ. We make a projection to get a function y so that we can apply the preceding theorem. We have also found it convenient to invert the formula taking $z = 1/\psi$.

In the applications, we could either apply the formula directly, or work with only one absolute value in S. For simplicity, we shall follow this second alternative in the next chapter.

Historical Note

We refer to Schneider's book for a complete historical note on the Liouville–Thue–Siegel–Mahler–Dyson–Schneider–Roth theorem. Mahler's contribution was to have included several primes (i.e. taking the product). Schneider already worked with several variables and the index in his Crelle paper of 1936. Roth's greatest contribution (aside from giving a very clear picture of the whole structure of the proof) lies in Proposition 3.2 and its proof.

The proof follows that given by Roth without any essential change.

Schmidt [Schm 1], [Schm 2] has extended Roth's theorem to simultaneous approximations, and linear forms in several variables. Baker [Ba] gives a very terse and very good exposition having all the advantages of terseness (the main ideas can be seen most clearly), but the disadvantages that readers must be well acquainted with the techniques to take care of details which are obvious to the technician but may not be obvious to the novice.

In [L 10] I defined the notion of **type** of approximation, where instead of having $2 + \varepsilon$ in the exponent, one uses a function of the height and one looks at inequalities

$$|\alpha - \beta| \leqq \frac{1}{H(\beta)^2 f(H(\beta))}.$$

It is a problem to determine the type for algebraic numbers and the classical transcendental numbers, known only for e and a few others related to e, cf. [L 10]. With such finer measures of approximation, one gets asymptotic quantitative results with canonical error terms involving the type. These give more structure than the measure theoretic results of Schmidt [Schm 3], Theorem 2.

It is also a problem to extend Schmidt's results and give them a geometric formulation on projective non-singular varieties V. The condition $\kappa > 2$ should be replaced by a similar one (instead of 2, perhaps $1 + n$ where $n = \dim V$), and the multiplicity r by a multiplicity which must be a bi-holomorphic invariant, so that it remains the same in unramified coverings. This is essential if one is to apply the theorem to isogenies of abelian varieties, cf. the comments at the end of Chapter 8.

The pattern of proof in Schmidt's theorems is the same as in Roth's theorem, although of course more complicated because of the higher dimension. This means that the proofs are still not effective, and it is a standard problem to find a new structure which makes them effective, let alone get a better type than H^ε. Bombieri has obtained some effective results, but far from the Roth exponent in [Bo].

CHAPTER 8

Siegel's Theorem and Integral Points

If C is an affine curve defined over a ring R finitely generated over $\mathbf{Z}$, and if its genus is ≥ 1, then C has only a finite number of points in R. This is the central result of the chapter. We shall also give a relative formulation of it for a curve defined over a ring which is a finitely generated algebra over an arbitrary field k of characteristic 0. In that case, the presence of infinitely many points in R implies that the curve actually comes from a curve defined over the constant field and that its points are of a special nature (excluding possibly a finite number).

In §1, we put together Roth's theorem and the weak Mordell–Weil theorem to show that integral points are of bounded height. It is actually more convenient (although not really more general) to deal with a non-constant function φ whose values $\varphi(P)$ lie in a ring of the above type.

In §2, we then consider the two standard examples and prove the statements of absolute and relative finiteness.

We shall also give Siegel's theorems classifying all the curves of genus 0 which have infinitely many integral points.

In §6 and §7 we investigate (irreducible) polynomial equations

$$f(X, Y) = 0$$

such that there are infinitely many roots of unity satisfying $f(\zeta', \zeta'') = 0$. We shall find that f has a very simple form, essentially $cX^r + dY^s$ or $cX^rY^s + d$. More generally, let Γ_0 be a finitely generated subgroup of complex numbers, and let $\Gamma = \Gamma_0^{1/\infty}$ be its division group in $\mathbf{C}^*$ (the group of all elements such that some non-zero integral power lies in Γ_0). If $f(x, y) = 0$ for infinitely many $x, y \in \Gamma$ then again $f(X, Y)$ has the above shape. The notes and comments at the end of the chapter will relate this situation with the Mordell conjecture and various problems on abelian varieties.

It is not out of place here to recall a folklore anecdote about Chevalley and Zariski, who once had a discussion concerning curves, and neither seemed able to understand the other. In desperation, Chevalley finally asked Zariski: "What do you mean by a curve?" They were in front of a blackboard, and Zariski said: "Well, of course I mean this!" And he drew the following picture.

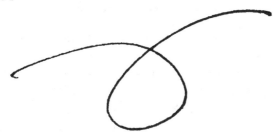

He continued: "And what do *you* mean by a curve?" Chevalley answered: "I don't mean this at all. I mean $f(x, y) = 0$."

From §3 to §7 we adopt Chevalley's meaning.

§1. Height of Integral Points

Throughout this section, we shall deal with a field **F** *with a proper set of absolute values M_F satisfying the product formula. Let K be a finite extension of* **F**, *with its set of absolute values M_K. We let $H = H_K$ be the relative height.*

Let $v \in M_K$. We shall say that *K* **satisfies Roth's theorem for** v if the following assertion is true:

Given a projective non-singular curve U defined over K, let $\psi \in K(U)$ be a non-constant function, and let r be the largest of the orders of the poles of ψ. Let ρ be a number > 2, and $c > 0$. Then the points $Q \in U(K)$ which are not poles of ψ, and are such that

$$|\psi(Q)|_v \geq cH(Q)^{\kappa r}$$

have bounded height.

Using the weak Mordell–Weil theorem, our purpose is to improve this statement for curves of genus ≥ 1. We shall prove:

Theorem 1.1. *Let K be as above, satisfying Roth's theorem for one of the absolute values $v \in M_K$. Let C be a complete non-singular curve of genus ≥ 1 defined over K, let J be its Jacobian, also defined over K, let m be an integer > 1 prime to the characteristic such that $J(K)/mJ(K)$ is finite. Let φ be a non-constant function in $K(C)$, and let ρ, c be numbers > 0. Then the height of points P in $C(K)$ which are not poles of φ and such that*

$$|\varphi(P)|_v \geq cH(P)^\rho$$

is bounded.

In view of Proposition 3.3 of Chapter 4, we observe that if a set of points has bounded height in one projective embedding, then it has bounded height in every projective embedding. Therefore we need not specify the embedding in statements such as the above.

If C is a complete non-singular curve defined over K, of genus ≥ 1, and with a rational point, then its Jacobian J is also defined over K, and we may assume that C is contained in it under a canonical mapping. We shall take J with a fixed projective embedding, and we shall take on C the induced embedding. The height $H = H_K$ is taken relative to this embedding.

Proposition 1.2. *Let m be an integer > 0, unequal to the characteristic of K, and assume that $J(K)/mJ(K)$ is finite. Let $\mathfrak{S}$ be an infinite set of rational points of C in K. Then there exists an unramified covering $\omega: U \to C$ defined over K, an infinite set of rational points $\mathfrak{S}'$ of U in K, such that ω induces an injection of $\mathfrak{S}'$ into $\mathfrak{S}$, and a projective embedding of U over K such that $H \circ \omega$ is quasi-equivalent to H^{m^2}. (Of course, in $H \circ \omega$ the H refers to the height on C, while in H^{m^2} it refers to the height on U.)*

Proof. Let $a_1, \ldots, a_n$ be representatives of cosets of $J(K)/mJ(K)$. Infinitely many $P \in \mathfrak{S}$ lie in the same coset, and so there exists one point, say a_1, and infinitely many points $Q \in J_K$ such that $mQ + a_1$ lies in $\mathfrak{S}$. We let $\mathfrak{S}'$ be this infinite set of points Q. The covering $\omega: J \to J$ given by $\omega u = mu + a_1$ is unramified, and its restriction U to C also gives an unramified covering of C, which is an irreducible covering of the same degree as ω. The inverse image of a point in C lies in U. Thus $\mathfrak{S}'$ is actually a subset of $U(K)$. Restricting one's attention to a subset of $\mathfrak{S}'$ guarantees that ω induces an injection on this subset.

To prove the relation concerning the height, we may work on the Jacobian itself since we take U in the projective embedding induced by that of J. If X is a hyperplane section of J, then

$$\omega^{-1}(X) = (m\delta)^{-1} X_{-a_1} \equiv m^2 X,$$

where $\equiv$ is algebraic equivalence. But $H_{\omega^{-1}(X)} \sim H_X \circ \omega$ by functoriality and $H_{\omega^{-1}(X)}$ is quasi-equivalent to $H_{m^2X} \sim H_X^{m^2}$ by Proposition 3.3 of Chapter 4. Our proposition is therefore proved.

Although it is not difficult to prove that the restriction of our covering ω to C is of the same degree as ω (cf. [L 1]) it is actually irrelevant here, since we could just as well use one of the components of this restriction that contains infinitely many points of S'. Such a component must then be defined over K.

We note that $\varphi(P) = \varphi(\omega(Q))$. Let ψ be the function $\varphi \circ \omega$ on U. Let κ be a number > 2, and let m be large enough so that $m^2\rho > \kappa r$. Then for a suitable constant $c > 0$, our inequality becomes

$$|\psi(Q)|_v \geq cH(Q)^{\kappa r}$$

and we can apply Roth's theorem, because the orders of the poles of ψ on U are the same as those of φ on C, since our covering is unramified. This concludes the proof of Theorem 1.1.

Theorem 1.3. *Let K, C, φ be as in Theorem 1.1. Let S be a finite, non-empty subset of M_K, containing all the archimedean primes and assume that K satisfies Roth's theorem for each $v \in S$. Let R be a subring of K all of whose elements are integral for each valuation $v \notin S$, and let $\mathfrak{R}$ be the subset of $C(K)$ consisting of those points P which are not poles of φ, and such that $\varphi(P) \in R$. Then the points of $\mathfrak{R}$ have bounded height.*

Proof. Assume the contrary, and let $\mathfrak{R}_1$ be a subsequence of points of $\mathfrak{R}$ whose height tends to infinity. By assumption, we have $|\xi|_v \leq 1$ for $v \notin S$ and $\xi \in R$. Hence for all $P \in \mathfrak{R}_1$ we get

$$H(\varphi(P)) = \prod_{v \in S} \sup(1, |\varphi(P)|_v)^{N_v}.$$

Let $N = [K : F]$ and let s be the number of elements of S. Since $N_v \leq N$, we have at most Ns terms in our product, of type

$$\sup(1, |\varphi(P)|_v).$$

Consequently, for each $P \in \mathfrak{R}_1$ there exists one v in S such that

$$|\varphi(P)|_v \geq H(\varphi(P))^{1/Ns}.$$

Hence there exists an infinite subset $\mathfrak{R}_2$ of $\mathfrak{R}_1$ such that for some $v \in S$ and all points $P \in \mathfrak{R}_2$ we have

$$|\varphi(P)|_v \geq H(\varphi(P))^{1/Ns}.$$

In view of Corollary 3.4 of Chapter 4, we can compare $H(\varphi(P))$ and $H((P))$. If d is the degree of C in its given projective embedding, and φ is viewed as a map into $\mathbf{P}^1$, then there is a number $c_1 > 0$ depending on ε such that for P in $C(K)$ we have

$$H(P)^{r/d - \varepsilon} \leq c_1 H(\varphi(P)).$$

Combining this with our previous inequality, we see that there is a number $\rho > 0$ such that for some $v \in S$ and all $P \in \mathfrak{R}_2$ we have for suitable $c_2 > 0$:

$$|\varphi(P)|_v \geq c_2 H(P)^\rho$$

and we can apply Theorem 1.1 to get our contradiction.

The special cases of interest of Theorem 1.3 are number fields and function fields. In the case of number fields, we take for R a finitely generated ring over $\mathbf{Z}$, and then we conclude that the set of points P such that $\varphi(P)$ lies in R, is actually finite. This will be generalized to rings of finite type over $\mathbf{Z}$, including transcendental elements in the next section.

In the case of function fields, we shall consider a function field K over a constant field k of characteristic 0. The set of absolute values is that of Chapter 2, §3. All the conditions of Theorem 1.3 are satisfied, except that $J(K)/mJ(K)$ need not be finite. However, in view of the fact that the height is geometric, i.e. does not depend on the constant field (Chapter 3, Proposition 3.2) we may go over to the function field Kk^a over k^a, i.e. we may without loss of generality assume that k is algebraically closed. In that case, we can apply Theorem 10.2 of Chapter 7, taking into account that $B(k)$ (and hence $\tau B(k)$) is divisible, and thus that $J(K)/mJ(K)$ is indeed finite.

In the next section, we shall give a more precise description of our sets of bounded height and sets of integral points.

§2. Finiteness Theorems

If K is a function field over a constant field k we shall pick a model as in Chapter 2 §3 to compute heights. If C is a complete non-singular curve defined over K, and a subset of points of $C(K)$ has bounded height in some projective embedding, then we recall that it has bounded height in every projective embedding.

> **Proposition 2.1.** *Let K be a function field over a constant field k of characteristic 0. Let C be a complete non-singular curve of genus ≥ 1 defined over K. Let $\mathfrak{R}$ be an infinite subset of $C(K)$ of bounded height. Regard C as embedded in its Jacobian J over K, and let (B, τ) be a K/k-trace of J. Then τ is an isomorphism. The points of $\mathfrak{R}$ lie in a finite number of cosets of $\tau(B(k))$. If infinitely many of them lie in one coset, and so are of type $a + \tau b$ where a is some point of $J(K)$ and b ranges over an infinite subset of $B(k)$, then $C_0 = \tau^{-1}(C_{-a})$ is defined over k, and τ induces an isomorphism of C_0 onto C_{-a}.*

Proof. We may assume that J and B are embedded in projective space. We see from Theorem 5.3 of Chapter 6 that the points of $\mathfrak{R}$ lie in a finite number of cosets of $\tau B(k)$. Assume that infinitely many lie in the coset $a + \tau B(k)$. We know that τ establishes an isomorphism between B and τB by AV, Corollary 2 of Theorem 9, Chapter VIII. Since infinitely many points of $\tau B(k)$ lie in C_{-a}, it follows that their K-closure in τB or J is precisely C_{-a}. Put $C_0 = \tau^{-1}(C_{-a})$. Then C_0 is a curve contained in B, and τ induces an isomorphism τ_0 of C_0 onto C_{-a}. Furthermore, C_0 contains infinitely many points b of B rational over k. It is then a trivial matter to conclude that C_0 is defined over k, because these infinitely many points are both k- and K-dense in C_0. Since τB contains a translation of C, it follows that $\tau B = J$ is the Jacobian, i.e. τ is an isomorphism.

Remark. If we do not assume characteristic 0 in the preceding proposition, then τ is merely bijective, and C_0 may be defined over a purely inseparable extension of k.

Corollary 2.2. *Let K, k be as above. Let C be a complete non-singular curve of genus ≥ 2 defined over K, and let $\mathfrak{R}$ be an infinite subset of $C(K)$ consisting of points of bounded height. Then there exists a curve C_0 defined over k and a birational transformation $T: C_0 \to C$ defined over K such that all but a finite number of points of $\mathfrak{R}$ are images under T of rational points of C_0 in k.*

Proof. Since the genus is ≥ 2, the curve cannot be equal to any translation of itself in its Jacobian. (If it were, so would the divisor Θ, and it isn't, even up to linear equivalence, by AV, Theorem 3 of Chapter VI, §3.) Hence there can only be one coset having infinitely many points of C, and we apply the proposition. Combining our corollary with the results obtained in the previous section, we get:

Theorem 2.3. *Let K be a function field over a constant field k of characteristic 0. Let R be a subring of K of finite type over k. Let C be a complete non-singular curve of genus ≥ 1 defined over K, and φ a non-constant function on C, also defined over K. Let $\mathfrak{R}$ be the subset of $C(K)$ consisting of those points P such that $\varphi(P) \in R$. If $\mathfrak{R}$ is infinite, then there exists a curve C_0 defined over k and a birational transformation $T: C_0 \to C$ defined over K. If the genus is ≥ 2, then all but a finite number of points of $\mathfrak{R}$ are images under T of points of $C_0(k)$. If the genus is 1, then the points of $\mathfrak{R}$ lie in a finite number of cosets of $T(C_0(k))$.*

Our next theorem gives the absolute result.

Theorem 2.4. *Let K be a field of finite type over $\mathbf{Q}$, and R a subring of K of finite type over $\mathbf{Z}$. Let C be a projective non-singular curve of genus ≥ 1 defined over K, and let φ be a non-constant function in $K(C)$. Then there is only a finite number of points $P \in C(K)$ which are not poles of φ and such that $\varphi(P) \in R$.*

Proof. If R is algebraic over $\mathbf{Z}$, this is simply Theorem 1.3 together with the finiteness of points of bounded height in a given number field. We shall reduce the general case to this one by a specialization argument, made possible by the previous theorem.

Let $\mathfrak{R}$ be the subset of points $P \in C(K)$ such that $\varphi(P) \in \mathfrak{R}$. Suppose that $\mathfrak{R}$ is infinite. Let k be the algebraic closure of $\mathbf{Q}$ in K. Then by Theorem 2.3, C is birationally equivalent over K to a curve C_0 defined over k. If the genus of C is 1, we restrict our attention to an infinite subset of points of $\mathfrak{R}$ which lie in the same coset of $T(C_0(k))$. Then without loss of generality, we may assume $C = C_0$, and that we have infinitely many points of C in K such that $\varphi(P) \in R$, where φ is a function defined over K. We shall now prove our theorem by induction on the dimension of K over $\mathbf{Q}$.

Let F be a subfield of K containing k such that the dimension of K over F is 1. There exists a discrete valuation ring $\mathfrak{o}$ of K containing F and R whose residue class field $E = \mathfrak{o}/\mathfrak{m}$ is finite over F and such that the reduction φ' of φ mod $\mathfrak{m}$ is a non-constant function $\varphi' : C \to \mathbf{P}^1$ (of the same degree as φ). For any point Q of $C(K)$, we get a specialized point Q' in $C(E)$ and $\varphi'(Q') = \varphi(Q)'$ using the compatibility of intersections and reductions, i.e. formally, using the graphs:

$$[\Gamma_\varphi \cdot (Q \times \mathbf{P}^1)]' = \Gamma'_\varphi \cdot (Q' \times \mathbf{P}^1)$$

the left-hand side being $(Q \times \varphi(Q))'$ and the right-hand side being $Q' \times \varphi'(Q')$. This yields infinitely many points Q' of $C(E)$ such that $\varphi'(Q')$ lies in the ring R', the image of R in the homomorphism $\mathfrak{o} \to \mathfrak{o}/\mathfrak{m}$. Since E is of finite type over $\mathbf{Q}$, of dimension one less than that of K, and since R' is still of finite type over $\mathbf{Z}$, this concludes the proof.

§3. The Curve $ax + by = 1$

Theorem 3.1. *Consider the equation*

$$ax + by = 1$$

with coefficients $a, b \neq 0$ in a field K of characteristic 0. Let Γ be a finitely generated multiplicative subgroup of K^*. Then there is only a finite number of solutions of the above equation with $x, y \in \Gamma$.

Proof. Let n be an integer ≥ 3. If there were infinitely many, then there would be infinitely many in one coset of Γ/Γ^n, which we could write in the form

$$x = a_1 \xi^n, \quad \text{and} \quad y = b_1 \eta^n.$$

This would give infinitely many solutions in $\Gamma \times \Gamma$ of the equation

$$aa_1 \xi^n + bb_1 \eta^n = 1$$

which has genus ≥ 1. We can view these solutions as lying in the ring finitely generated by Γ over $\mathbf{Z}$, thereby contradicting Theorem 2.4.

By a **linear torus** (**torus** for short) we shall mean a group variety which is isomorphic to a product of multiplicative groups after a possible extension of the base field. We note that given a finitely generated subgroup of a commutative group variety, there always exists a field of finite type over the prime field over which all the points of this subgroup are rational.

Theorem 3.2. *Let G be a torus in characteristic 0. Let Γ be a finitely generated subgroup of G, and let C be a curve contained in G. If $C \cap \Gamma$ is infinite, then C is the translation of a subtorus.*

Proof. Let K be a field of finite type over $\mathbf{Q}$ such that G is isomorphic to a product of multiplicative groups over K, and such that all points of Γ are rational over K. Let $(x_1, \ldots, x_n)$ be a generic point of C over K, each x_i lying in the multiplicative group. In view of Theorem 2.4, C must have genus 0, and hence $K(x_1, \ldots, x_n) = K(t)$ for some parameter t. Thus each x_i is a rational function of t, say

$$x_i = \varphi_i(t).$$

Let m be an integer ≥ 1. Infinitely many points of C must lie in some coset of Γ/Γ^m. Thus there exist elements $a_i \in K^*$ such that if we put $x_i = a_i \xi_i^m$, then the curve C' whose generic point is $(\xi_1, \ldots, \xi_n)$ over possibly a finite extension of K has infinitely many points in Γ. This implies that this curve also has genus 0. From this, we deduce that each function $\varphi_i(t)$ has at most two singularities (zero or pole). Indeed, a curve defined by an equation

$$\xi^m = a\varphi(t), \qquad a \in K^*,$$

where $\varphi(t)$ is a rational function of t has genus ≥ 1 as soon as m is sufficiently large, and relatively prime to the order of the zeros and poles of φ, provided that φ has at least 3 singularities. We see this from the genus formula

$$2g' - 2 = -2m + \sum (e_P - 1)$$

each ramification index being equal to m in this case. Thus the term $\sum (e_P - 1)$ grows at least like $3(m - 1)$ as m tends to infinity, and the genus does become large.

Choosing our parameter t properly, we may write say for $i = 1$,

$$x_1 = b_1 t^{r_1}$$

for some integer $r_1 \neq 0$. We now contend that each x_i can be written $b_i t^{r_i}$. This will prove our theorem.

Our covering C' of C contains as an intermediate covering the curve defined by the equation

$$\xi^m = b_1 t^{r_1}(b_i \varphi_i(t))^{\pm 1}$$

which must be of genus 0. This implies that $t^{r_1}\varphi_i(t)^{\pm 1}$ has at most two singularities, and hence clearly that $\varphi_i(t) = b_i t^{r_i}$ for some integer r_i. Our theorem is proved.

We remark that the translation of a subtorus has the generic point

$$(a_1 t^{r_1}, \ldots, a_n t^{r_n})$$

with integers $r_1, \ldots, r_n$.

§4. The Thue–Siegel Curve

Let R be a finitely generated ring over $\mathbf{Z}$, and let K be its quotient field. We also assume that R is integrally closed. By Chapter 2, §7 we know that the group of units R^* is finitely generated, and so is the group of divisor classes.

Theorem 4.1. *Let $\alpha_1, \ldots, \alpha_n, \lambda$ be elements of K. Assume that at least three elements among the α_i are distinct, say $\alpha_1, \alpha_2, \alpha_3$. Then the equation*

$$\prod_{i=1}^{n} (X - \alpha_i Y) = \lambda$$

has only a finite number of solutions in R.

Proof. We shall follow Siegel [Sie 1] Part 2, §1 where Siegel shows how to reduce solving this equation in integers to a finite number of equations

$$au + bu' = 1$$

in units u, u' as follows.

To begin with, without loss of generality, we may assume that α_i, λ are integral, since we may adjoin them to R to obtain another finitely generated ring, and then use Proposition 4.1 of Chapter 2 to get a ring which is integrally closed but still finitely generated.

For any integral solution x, y the factors $x - \alpha_i y$ then divide λ, and consequently lie in a finite number of cosets of the units. We now use the Siegel identity

$$\frac{\alpha_3 - \alpha_1}{\alpha_2 - \alpha_1} \frac{x - \alpha_2 y}{x - \alpha_3 y} + \frac{\alpha_2 - \alpha_3}{\alpha_2 - \alpha_1} \frac{x - \alpha_1 y}{x - \alpha_3 y} = 1.$$

Each ratio

$$\frac{x - \alpha_i y}{x - \alpha_j y}$$

lies in only a finite number of cosets of the units, and therefore the original equation gives rise to a finite number of equations of type

$$au + bu' = 1$$

in units. If we know that these have only a finite number of solutions, then we conclude that the ratios above are finite in number. For a fixed pair of elements (γ, γ') in K, the equations

$$\frac{x - \alpha_2 y}{x - \alpha_3 y} = \gamma \quad \text{and} \quad \frac{x - \alpha_1 y}{x - \alpha_3 y} = \gamma'$$

then determine the pair (x, y) up to a proportionality factor. Thus all solutions of the original equation corresponding to a fixed pair (u, u') are of type

$$(x, y) = \rho(x_0, y_0), \qquad \rho \in K,$$

where x_0, y_0 are fixed. Thus the original equation yields

$$\rho^n \prod_{i=1}^{n} (x_0 - \alpha_i y_0) = \lambda.$$

This shows that ρ can have only a finite number of values and concludes the proof of the reduction of the Thue–Siegel equation to the unit equation treated in §3.

Remark. It is instructive to see how Siegel's argument over number fields goes over at once to the more general case. However, one can also see the theorem as a special case of Theorem 2.4, since the Thue–Siegel curve has genus ≥ 1.

§5. Curves of Genus 0

In this section we give Siegel's classification of curves of genus 0 which admit infinitely many integral points.

Theorem 5.1. *Let K be a field of finite type over $\mathbf{Q}$, and R a subring of finite type over $\mathbf{Z}$. Let C be a complete non-singular curve of genus 0, defined over K, and let φ be a non-constant function in $K(C)$. Suppose that* *there are infinitely many points of $C(K)$ such that $\varphi(P)$ lies in R. Then φ has at most two distinct poles.*

Proof. First we may assume without loss of generality that R is integrally closed (Proposition 4.1 of Chapter 2), and that K is the quotient field of R as in the last section (adjoin a finite number of elements to R). We shall use Chapter 2, §7 constantly.

Suppose φ has at least three distinct poles. We can select a model of C such that none of the poles or zeros of φ lie at infinity. This means that $K(C) = K(t)$ for some transcendental element t over K, and that

$$\varphi(t) = c\,\frac{f(t)}{g(t)},$$

where c is a leading coefficient, and

$$f(t) = \prod_{i=1}^{n}(t - \beta_i), \qquad g(t) = \prod_{i=1}^{n}(t - \alpha_i)$$

are polynomials of the same degree and relatively prime. Note that one can also write

$$\varphi(t) = c\,\frac{\prod(1 - \beta_i/t)}{\prod(1 - \alpha_i/t)}$$

since f, g have the same degree. Making a finite extension of K if necessary, we may assume that the roots α_i, β_i lie in K.

Also we can extend R so that $\mathrm{Pic}(R)$ is trivial, that is every divisor of R is principal, by Corollary 7.7 of Chapter 2. Thus any non-zero element t in K can be written as a quotient of integral elements

$$t = x/y,$$

such that x, y are relatively prime (that is, their divisors have no component in common). Let us write

$$g(x, y) = \prod (x - \alpha_i y),$$
$$f(x, y) = \prod (x - \beta_i y).$$

Since f, g are relatively prime, there exist relations

$$A_1(x, y)f(x, y) + B_1(x, y)g(x, y) = y^n,$$
$$A_2(x, y)f(x, y) + B_2(x, y)g(x, y) = x^n,$$

where A_1, A_2, B_1, B_2 are (homogeneous) polynomials. If x, y are obtained from a solution t such that $\varphi(t)$ is integral, then dividing by $g(x, y)$ shows that $g(x, y)$ divides $y^n d$ and $x^n d$ where d is a denominator for the coefficients of A_1, A_2, B_1, B_2. Since x, y are relatively prime, it follows that

$$g(x, y) = \lambda u,$$

where λ ranges over a finite number of elements of K, and u ranges over units.

Replacing K by the finite extension obtained by adjoining n-th roots of units, we may write $u = w^n$ for some unit w. Since $g(x, y)$ is homogeneous of degree n, the above equation to be solved in integral x, y is equivalent to the equation

$$g(x, y) = \lambda$$

to be solved in integral x, y. This reduces the case of genus 0 to the case treated in the preceding section.

Knowing that our function φ has only two poles, we can then describe it further. On the one hand, these two poles may be both finite, determined by $\alpha, \bar{\alpha}$ where α is a quadratic irrationality over K. Thus

$$\varphi(t) = f(t)/g(t),$$

where $g(t)$ is a power of $(t - \alpha)(t - \bar{\alpha})$, essentially a norm. If we make a quadratic extension of K in this case, or if all the roots of g are in K, then we can choose our parametrization of C by t in such a way that one of the poles is at infinity, and thus $\varphi(t)$ takes the form

$$\varphi(t) = f(t)/t^m,$$

where $f(t)$ is a polynomial and m is an integer ≥ 0. Thus the only possibility for integral points are the obvious ones.

We could also analyze the situation first in the relative case, in order to reduce it afterwards to the absolute case or the case of number fields. Let us just state one more result.

Theorem 5.2. *Let K be a function field over the constant field k of characteristic 0, and let R be a finitely generated subalgebra of K over k. Let C be a complete non-singular curve of genus 0 defined over K, and φ a non-constant function in $K(C)$. If φ has at least three distinct poles, then the points P of $C(K)$ such that $\varphi(P) \in R$ have bounded height.*

Proof. Just as in the preceding theorem, we end up with an equation $ax + by = 1$ and solutions in R^* (after possibly extending R to a larger ring). We can then proceed as before in the reduction to curves of genus $\geqq 1$ to get what we want.

We leave it to the reader to show that elements of bounded height in K^* lie in a finite number of cosets of k^*. Our theorem reduces the study of integral points on C to such points of bounded height.

Remark. Instead of the computations involving factorizations of polynomials in the proof of Theorem 5.1, and also in Theorem 4.1, one could use Weil's decomposition theorem to be proved in Chapter 10, Theorem 3.7. The entire proof of Theorem 5.1 can then be replaced by the following much shorter argument.

Suppose there are infinitely many points $P \in C(K)$ such that $\varphi(P) \in R$. Then for such points P the Weil function $\lambda_{\alpha_i}(P, v)$ is bounded by some constant $\gamma(v)$ for each i. This means that the function

$$\frac{t - \alpha_i}{t - \alpha_j}$$

for $i \neq j$, which has no zero or pole except α_i and α_j, is such that for any value t' of t in K, $t' \neq \alpha_i$ or α_j, the M_R-divisor of $(t' - \alpha_i)/(t' - \alpha_j)$ lies between two fixed bounds. Excluding the primes occurring in a finite set (Proposition 7.1 of Chapter 2), and using Theorem 7.4 of Chapter 2, we see that $(t' - \alpha_i)/(t' - \alpha_j)$ lies in a finitely generated group. Now we apply Siegel's identity

$$\frac{\alpha_2 - \alpha_1}{\alpha_3 - \alpha_1} \frac{t - \alpha_3}{t - \alpha_2} + \frac{\alpha_3 - \alpha_2}{\alpha_3 - \alpha_1} \frac{t - \alpha_1}{t - \alpha_2} = 1.$$

Thus from the assumptions we get infinitely many points in a finitely generated multiplicative group, and we can apply the result of §3 to conclude the proof.

§6. Torsion Points on Curves

Let $f(X, Y) = 0$ be the equation of a curve over the complex numbers. We assume that f is absolutely irreducible. We may view this equation as defining a curve in the product of multiplicative groups $\mathbf{G}_m \times \mathbf{G}_m$. In this section we are interested in the intersection of the curve with the torsion points, which in this case are the roots of unity. Thus we are interested in the roots of unity ζ', ζ'' such that $f(\zeta', \zeta'') = 0$.

Theorem 6.1. *Let $f(X, Y)$ be an irreducible polynomial over **C**. If there exist infinitely many roots of unity (ζ', ζ'') such that $f(\zeta', \zeta'') = 0$ then f is a polynomial of the form*

$$aX^m + bY^n \qquad or \qquad cX^nY^m + d.$$

In geometric language, one may say that

if the curve has an infinite intersection with the group of torsion points on a torus, then the curve is the translation of a subtorus.

After making a translation, we may assume without loss of generality that $f(1, 1) = 0$, that is the curve passes through the origin. In that case the curve is itself a subtorus, i.e. a subgroup variety.

For the context of this conjecture and its relation with problems on abelian varieties, see the notes and comments at the end of the chapter. I had conjectured Theorem 6.1 and a proof was originally shown to me by Ihara, Serre and Tate [L 9]. I shall now reproduce Tate's proof. (Serre's was similar.)

After dividing f by some coefficient, we may assume without loss of generality that some coefficient of f is equal to 1. Suppose there are infinitely many pairs of roots of unity $\zeta = (\zeta', \zeta'')$ such that $f(\zeta) = 0$. Then the co-efficients of f lie in some field $\mathbf{Q}(\zeta_m) = k$ generated over $\mathbf{Q}$ by a primitive m-th root of unity. One sees this by considering the field $F = \mathbf{Q}(\boldsymbol{\mu})$ obtained by adjoining all roots of unity to $\mathbf{Q}$. Let K be the field obtained from F by adjoining the coefficients of f. If $K \neq F$ there exists a conjugate σ of K over F such that $f^\sigma \neq f$ and then f^σ, f have infinitely many zeros in common. Since they are irreducible, each divides the other by Bezout's theorem, and since f has some coefficient equal to 1, it follows that $f = f^\sigma$. This proves that the coefficients of f lie in $\mathbf{Q}(\boldsymbol{\mu})$, after we have normalized these coefficients so that one of them is equal to 1.

Let n be the period of ζ, i.e. the least common multiple of the periods of ζ', ζ''. Let d be a positive integer prime to n. There exists an automorphism σ of $\mathbf{Q}(\zeta)$ such that $\sigma\zeta = \zeta^d$. If in addition $d \equiv 1 \pmod{m}$, then σ can be extended to an automorphism of $k(\zeta)$ inducing the identity on k. Then $f(\zeta^d) = 0$, so that ζ is a zero of $f(X, Y)$ and also of $f(X^d, Y^d)$. But

$$[k(\zeta) : k] \geq \phi(n)/m.$$

Applying any automorphism τ of $k(\zeta)$ over k, we find that $\tau\zeta$ is also a common zero of these two polynomials, which have therefore at least $\phi(n)/m$ zeros in common. However, by Bezout's theorem, these polynomials have at most $(\deg f)^2 d$ common zeros, unless $f(X, Y)$ divides $f(X^d, Y^d)$. As soon as n is large enough, we can use the prime number theorem giving the existence of primes in arithmetic progressions to find a prime number d satisfying the above conditions, such that d is much smaller than $\phi(n)$. Hence we conclude that $f(X, Y)$ divides $f(X^d, Y^d)$, and it is then an exercise to show that $f(X, Y) = 0$ defines a subgroup variety of $\mathbf{C}^* \times \mathbf{C}^*$. This concludes the proof, which also shows that n is bounded in terms of $\deg f$ and m.

Note that we can avoid the congruence condition $d \equiv 1 \bmod m$ by using the following variation of Tate's argument. Let $r = [k : \mathbf{Q}]$. We extend σ to an automorphism of $k(\zeta)$. Then σ^r induces the identity on k, and

$$\sigma^r \zeta = \zeta^{d^r}.$$

Then $f(\zeta^{d^r}) = 0$, so that ζ is a zero of $f(X, Y)$ and also of $f(X^{d^r}, Y^{d^r})$. We can then argue as before, using only the lemma:

Lemma 6.2. *Given an integer s, there exists an integer n_0 such that for all $n > n_0$, there exists a prime number p not dividing n, such that $p^s \leq n$.*

Proof. The worst case occurs when n is a product of distinct primes, in which case the assertion is an immediate consequence of the fact that $\pi(N)$ is of the order of magnitude of $N/\log N$.

For the analogue on abelian varieties, cf. the notes and comments at the end of the chapter.

Concerning the exercise about $f(X^d, Y^d)$ one has the more general statement as in [L 9]:

Lemma 6.3. *Let G be a commutative group variety in characteristic 0. Let V be a curve on G, passing through the origin. Assume that there exists an integer $d > 1$ such that for all $x \in V$ the point dx also lies in V. (We write the group law additively.) Then V is a subgroup of G.*

Proof. Let p be a prime number dividing d. Let k_0 be a field of definition for G and V, finitely generated over the rationals. We can embed k_0 in a finite extension k of the p-adic field $\mathbf{Q}_p$. If x is a point of $V(k)$ sufficiently close to the origin, then dx lies in $V(k)$, and the points $d^n x$ approach 0 as n tends to infinity (for the p-adic topology on $V(k)$). Taking the inverse image by the exponential map of a sufficiently small neighborhood U of 0 in $V(k)$, we find on the tangent space at the origin that $\exp^{-1}(U)$ has an infinite intersection with a straight line, having 0 as point of accumulation. This implies that $\exp^{-1}(U)$ contains a small (infinite) subgroup, and hence that U contains a small subgroup. Since V is a curve, this small subgroup is Zariski dense in V, and hence V is a group, as was to be shown.

I shall now give a more precise version of Theorem 6.1 due to Liardet [Li 1], [Li 2].

Theorem 6.4. *Let $f(X, Y)$ be a polynomial of degree m in X and n in Y, $mn \neq 0$, and with coefficients in a subfield K of $\mathbf{C}$. Let $K_0 = K \cap \mathbf{Q}^a$. There exists a constant $C = C(m, n, [K_0 : \mathbf{Q}])$ such that if $f(\zeta_a, \zeta_b) = 0$ with primitive roots of unity of order a, b respectively, and if $a > C$, then $f(X, Y)$ has an irreducible factor of the form*

$$X^r Y^s - \zeta \qquad or \qquad \zeta X^r - Y^s$$

with some root of unity ζ, integers r, s and $s > 0$.

The proof occupies the rest of this section. First we remark that if a is sufficiently large, then $f(\zeta_a, Y) \neq 0$. Otherwise, $f(X, Y)$ is divisible by $X - \zeta_a$ and hence by $X - \zeta_a^\sigma$ for all $\sigma \in \mathrm{Gal}(K(\zeta_a)/K)$ so $[K(\zeta_a) : K] \leq m$. But

$$[K(\zeta_a) : K] = [K_0(\zeta_a) : K_0]$$
$$\geq [\mathbf{Q}(\zeta_a) : \mathbf{Q}]/[K_0 : \mathbf{Q}],$$

so $f(\zeta_a, Y) \neq 0$ provided $\phi(a) > m[K_0 : \mathbf{Q}]$, which we now assume.

Next we need a lemma which will tell us what r and s are.

Lemma 6.5. *Let u be an element of a cyclic group G of order d, and let v be a generator. Then there exist integers r, s with*

$$1 \leq s \leq \sqrt{d} \qquad and \qquad |r| \leq \frac{d}{[\sqrt{d}]} \leq 2\sqrt{d}.$$

such that $u^s = v^r$.

Proof. Let N be an integer with $0 < N \leq \sqrt{d}$. Write $u = v^q$ with some integer q. Consider the $N + 1$ integers

$$\langle jq \rangle, \qquad j = 0, \ldots, N,$$

where $\langle t \rangle$ is the smallest integer ≥ 0 representing the class of t in $\mathbf{Z}/d\mathbf{Z}$. By Dirichlet's box principle, there exists an integer s with $1 \leq s \leq N$ and an integer i such that

$$|sq - id| \leq d/N.$$

We take $N = [\sqrt{d}]$. Let $r = sq - id$. Then $0 \leqq |r| \leqq d/N$ and

$$u^s = v^{sq} = v^r.$$

This concludes the proof.

We reformulate the lemma in the context of roots of unity.

Lemma 6.6. *Let a, b be positive integers and ζ_a, ζ_b primitive roots of unity of order a, b respectively. Let $d = (a, b)$ and write*

$$a = da_0 a', \qquad b = db_0 b',$$

where a_0, b_0 have only prime factors dividing d, while a', b' are relatively prime to d. Then there is an expression

$$\zeta_b^{b_0 s} = \zeta \zeta_a^r,$$

where ζ is a b'-root of unity, and

$$0 < s \leqq \sqrt{d}, \qquad |r| < 2\sqrt{d}.$$

Proof. Let G be the group $\boldsymbol{\mu}_{ab'}/\boldsymbol{\mu}_{b'}$. Then $\zeta_b^{b_0}$ lies in the cyclic group generated by ζ_a mod $\boldsymbol{\mu}_{b'}$, when we view these elements in G. We can then apply Lemma 6.5 to conclude the proof.

We return to the proof of the theorem. By Lemma 6.6, the two curves

$$f(X, Y) = 0 \qquad \text{and} \qquad \zeta X^r - Y^{b_0 s} = 0$$

have the common zero (ζ_a, ζ_b), and have therefore at least as many common zeros as the number of conjugates of ζ_a over $K(\zeta)$, that is

$$[K(\zeta, \zeta_a) : K(\zeta)] = [K_0(\zeta, \zeta_a) : K_0(\zeta)]$$

because K and $\mathbf{Q}^a$ are linearly disjoint over K_0. We then have

$$[K_0(\zeta, \zeta_a) : K_0(\zeta)] \geqq [\mathbf{Q}(\zeta, \zeta_a) : \mathbf{Q}(\zeta)]/[K_0(\zeta) : \mathbf{Q}(\zeta)]$$

$$\geqq \phi(a)/[K_0 : \mathbf{Q}].$$

But the projective degrees of the two curves are respectively

$$m + n \qquad \text{and} \qquad |r| + b_0 s.$$

For the curves to have a component in common, it suffices that

$$\phi(a) > [K_0 : \mathbf{Q}](m + n)(|r| + b_0 s).$$

By the inequality of Lemma 6.6, it suffices that

(1) $$\phi(a) > [K_0 : \mathbf{Q}](m + n)(2 + b_0)\sqrt{a}.$$

By the definition of b_0 as having only prime factors dividing d, we find the upper bound

$$
\begin{aligned}
b_0 = [\mathbf{Q}(\zeta_{db_0}) : \mathbf{Q}(\zeta_d)] &\leq [\mathbf{Q}(\zeta_b) : \mathbf{Q}(\zeta_d)] \\
&= [\mathbf{Q}(\zeta_a, \zeta_b) : \mathbf{Q}(\zeta_a)] \\
&\leq [K_0(\zeta_a, \zeta_b) : K_0(\zeta_a)][K_0 : \mathbf{Q}]. \\
&\leq n[K_0 : \mathbf{Q}]
\end{aligned}
$$

because $f(\zeta_a, Y) \neq 0$ and $f(\zeta_a, \zeta_b) = 0$. Using (1) it suffices that

(2) $$\phi(a) > [K_0 : \mathbf{Q}]^2(m + n)3n\sqrt{a},$$

where the factor of 3 is there so that $2 + b_0 \leq 3b_0$. We have now given a lower bound for a entirely in terms of $[K_0 : \mathbf{Q}]$, m, n. By elementary number theory, one knows that $\phi(a) \gg a/\log \log a$, so (2) is practically a bound of the form

$$\sqrt{a} > [K_0 : \mathbf{Q}]^2(m + n)3n$$

up to some log log factor. This is unimportant here.

After replacing X by X^{-1} we may assume that $f(X, Y)$ and $\zeta X^r - Y^s$ have a common factor with r, s non-negative integers, $s > 0$. Let $t = (r, s)$. Then the factorization into irreducible factors over $\mathbf{C}$ of $\zeta X^r - Y^s$ is given by

$$\zeta X^r - Y^s = \prod_{z^t = \zeta} (z X^{r/t} - Y^{s/t}).$$

Each factor is of the desired type, thereby proving the theorem.

Liardet's Theorem 6.4 is the first case of still another generalization which will be given in the next section.

§7. Division Points on Curves

Let K be a finitely generated field over $\mathbf{Q}$ and let Γ_0 be a finitely generated subgroup of K^*. We are interested in the **division group** Γ of Γ_0 in $\mathbf{C}^*$, that is the group of all elements $z \in \mathbf{C}^*$ such that $z^n \in \Gamma_0$ for some positive integer n. For every finite extension L of K we let

$$\Gamma_L = \Gamma \cap L,$$

so Γ_K is the division group of Γ_0 in K.

Lemma 7.1. *The group Γ_K is finitely generated.*

Proof. Let R be the ring generated over $\mathbf{Z}$ by Γ_0 and let R' be its integral closure in K. Then Γ_K is a subgroup of the units in R' and is therefore finitely generated by Corollary 7.5 of Chapter 2.

If K is a number field, algebraic number theory gives explicit bounds for the ranks and heights of generators of these groups, which are S-units for a suitable finite set of primes S. In fact one has the following explicit bound. The method used already occurs in Stark [St]. The proof is identical with that of the analogous theorem on abelian varieties, Theorem 7.6 of Chapter 5.

Theorem 7.2. *Let K be a number field. Let Γ_0 be a finitely generated subgroup of K^*, and let $\{u_1,\ldots,u_r\}$ be a basis of Γ_0 mod torsion. Let $u \in K^*$ be a torsion point mod Γ_0, and let N be the smallest integer > 0 such that $u^N \in \Gamma_0\mu(K)$, where $\mu(K)$ is the group of roots of unity in K. Then there exists a number γ depending only on $[K:\mathbf{Q}]$ such that*

$$N \leqq \gamma^r(h_1 + \cdots + h_r)^r,$$

where $h_i = h(u_i)$.

Proof. Let $m_1,\ldots,m_r$ be integers and $u_0 \in \mu(K)$ be such that

$$u^N = u_1^{m_1} \cdots u_r^{m_r} u_0.$$

Let $1 \leqq q \leqq N$. Let $M = [N^{1/r}] - 1$. Cut each side of the unit cube into M segments of equal length $1/M$, so the unit cube is decomposed into M^r smaller cubes, with $M^r < N$. By Dirichlet's box principle, there are two integers q_1, q_2 with $1 \leqq q_1 < q_2 \leqq N$ such that

$$(q_1 m_1/N, \ldots, q_1 m_r/N) \quad \text{and} \quad (q_2 m_1/N, \ldots, q_2 m_r/N) \bmod \mathbf{Z}^r$$

lie in the same box. Let $q = q_2 - q_1$ so $1 \leqq q < N$. Then there exist integers $s_1,\ldots,s_r$ such that

$$|qm_j/N - s_j| \leqq 1/M \quad \text{for } j = 1,\ldots,r.$$

Let

$$n_j = qm_j - Ns_j \quad \text{so that } |n_j| \leqq N/M.$$

Let

$$w = u^q \prod_{j=1}^{r} u_j^{-s_j}.$$

Then

$$w^N = u_1^{n_1} \cdots u_r^{n_r} u_0'$$

for some root of unity u_0'. But $w \notin \mu(K)$ (otherwise $q < N$ is a period for $u \bmod \Gamma_0 \mu(K)$). Hence $h(w) \neq 0$ by Proposition 1.1 of Chapter 3, and so $h(w)$ is greater than some constant $C > 0$ depending only on the minimal height of algebraic numbers of degree bounded by $[K : \mathbf{Q}]$. Using the sub-additivity of the height, we find

$$NC \leqq NM^{-1} \sum_{j=1}^{r} h_j$$

which proves the theorem.

To unify Theorem 3.2 dealing with the intersection of a curve and a finitely generated group, and Theorem 6.1 dealing with the intersection of a curve and the torsion group, I conjectured the following statement in [L 9] and [L 16].

Theorem 7.3. Let Γ_0 be a *finitely generated multiplicative group of complex numbers, and let Γ be its division group, that is the group of complex numbers z such that $z^n \in \Gamma_0$ for some integer $n \neq 0$ (depending on z). Let $f(X, Y) = 0$ be a curve in the plane (absolutely irreducible). If there exist infinitely many elements $x, y \in \Gamma$ such that $f(x, y) = 0$, then f is a polynomial of the form*

$$aX^m + bY^n = 0 \qquad or \qquad cX^nY^m + d = 0.$$

In other words:

If a curve has an infinite intersection with the division group of a finitely generated group, then it is the translation of a subtorus.

Liardet [Li 1], [Li 2] proved this conjecture, even in a more precise form as follows.

Let K be a subfield of $\mathbf{C}$ and let $x \in \mathbf{C}^*$. We define:

$M_K(x) =$ smallest positive integer r such that $x^r \in K$ if some non-zero power of x lies in K;

$M_K(x) = 0$ otherwise.

Theorem 7.4. *Let K be a subfield of $\mathbf{C}$, finitely generated over $\mathbf{Q}$. Let m, n be positive integers. There exists a constant $B = B(m, n, K)$ depending only on m, n, K having the following property. Let $f(X, Y)$ be a* *polynomial with coefficients in K, of degree m in X and n in Y, mn ≠ 0. If f(x, y) = 0 for elements x, y ∈ C* satisfying*

$$M_K(x) \geqq B \qquad and \qquad M_K(y) > 0,$$

then f(X, Y) has an irreducible factor of the same form as in Theorem 7.3, that is an irreducible component of f(X, Y) = 0 is the translation of a subtorus.

Without loss of generality we may assume that $f(X, Y)$ is irreducible.

We shall first show how Theorem 7.4 implies Theorem 7.3. First by Lemma 7.1 we may assume that Γ_0 is its own division group in K^*. For each finite extension L of K we let

$$\Gamma_L = \Gamma \cap L$$

be the division group of Γ_0 in L. Suppose first that $x, y \in \Gamma$ range over elements such that $f(x, y) = 0$ and such that $M_K(x)$ and $M_K(y)$ are bounded, say by M. Then such elements x, y lie in $\Gamma_K^{1/M}$ which is finitely generated, so we can apply Theorem 3.2 to conclude the proof of Theorem 7.3 in this case.

Next suppose that $M_K(y)$ is bounded. If $f(X, y) = 0$ then we view $f(X, Y)$ as a polynomial in X with coefficients in $K[Y]$, and these co-efficients vanish at y so $f(X, Y)$ has a factor which is a polynomial in Y of the form $Y^s - b$ with $b \in \Gamma_K$. Since $f(X, Y)$ is irreducible, it has the desired form. Then without loss of generality, we may now assume that $f(X, y) \neq 0$. Then after replacing K by a finite extension L, we may assume that $y \in \Gamma_L$. Then $[L(x):L]$ is bounded by n since $f(x, y) = 0$. In that case, Kummer theory shows that $M_K(x)$ is also bounded, thus reducing the situation to one already settled. To apply Kummer theory in its usual form, one may first adjoin n' roots of unity to L for $n' \leqq n$, and then apply Kummer theory over this bigger field.

In the above manner, Theorem 7.3 is reduced to the case when both $M_K(x)$ and $M_K(y)$ tend to infinity, and therefore is implied by Theorem 7.4 which we now prove. In fact, the argument which we have just given shows that also in Theorem 7.4 we may assume without loss of generality that both $M_K(x)$ and $M_K(y)$ are sufficiently large. Thus the hypotheses are symmetric in x and y.

We let

$$a = M_K(x), \qquad b = M_K(y), \qquad x_0 = x^a, \qquad y_0 = y^b,$$

so that a, b are the smallest positive integers such that

$$x^a = x_0 \in K \qquad and \qquad y^b = y_0 \in K.$$

Let $L = K(\mu_a, \mu_b)$ and let

$$m_0 = [L(x) : L(x) \cap L(y)], \qquad n_0 = [L(y) : L(x) \cap L(y)].$$

As before, we may assume that $f(x, Y)$ and $f(X, y) \neq 0$. The former follows from $M_K(x) \geq B$. For the latter, if $f(X, y) = 0$ then $f(X, Y)$ is divisible by $\prod (Y - y^\sigma)$, where the product is over $\sigma \in \mathrm{Gal}(K(y)/K)$. Since

$$b = M_K(y) > 0,$$

this product is the irreducible polynomial $Y^b - y_0 \in K[X, Y]$, which has the desired form. $m_0 \leq m$, $n_0 \leq n$ and since the appropriate roots of unity are in L (so a fortiori in $L(x) \cap L(y)$) we know by abelian Kummer theory that $m_0 | a$, $n_0 | b$. We consider the following diagram of fields.

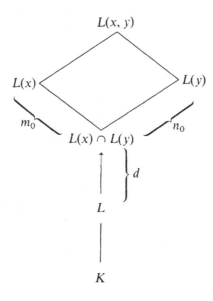

We let

$$d = [L(x) \cap L(y) : L].$$

The extension $L(x)$ over L is cyclic of degree dm_0 and $L(x) \cap L(y)$ is cyclic of degree d. Hence $L(x) \cap L(y)$ over L is a sub-Kummer extension of $L(y)$ over L, so $L(x) \cap L(y) = L(y^{n_0})$. But it is also a sub-Kummer extension of $L(x)$, so the intersection is also equal to $L(x^{m_0})$. Let $\{y^{n_0}\}$ be the cyclic group generated by y^{n_0}, and similarly for $\{x^{m_0}\}$.

We apply Lemma 6.5 to the group

$$G = \{y^{n_0}\}L^*/L^* = \{x^{m_0}\}L^*/L^*$$

of order d. We take

$$u = x^{m_0}L^*/L^* \qquad \text{and} \qquad v = y^{n_0}L^*/L^*.$$

There exists an element $z \in L^*$ such that

$$x^{m_0 s} = z y^{n_0 r}$$

and r, s satisfy the inequalities in the lemma. Then

$f(X, Y)$ has projective degree $m + n$,

$X^{m_0 r} - z Y^{n_0 s}$ has projective degree $\leq 2(m + n)\sqrt{d}$.

Furthermore, all the conjugates of (x, y) over L are common zeros. There are at least d such conjugates, so the polynomials have a common factor as soon as

$$d > 2(m + n)^2 \sqrt{d}, \quad \text{that is } \sqrt{d} > 2(m + n)^2,$$

and the proof is therefore complete if this inequality is satisfied.

There remains to analyze what happens when $\sqrt{d} \leq 2(m + n)^2$, or equivalently

$$d \leq 4(m + n)^4,$$

which we now assume. We shall reduce this case to the case when x, y are roots of unity as in Theorem 6.4. Before giving the formal proof, we indicate what happens in a typical case. Namely, it could happen that there exists an element $z \in K$ such that $z^a = x_0 \zeta$ with some root of unity ζ. Then x/z is a root of unity, say $x = zx_1$. Let $X = zX_1$. Then

$$f(X, Y) = f_1(X_1, Y),$$

and $f_1(x_1, y) = 0$. A similar phenomenon occurs with respect to y, so we end up with a polynomial to which we can apply Theorem 6.4 to conclude the proof. We shall see that the general situation essentially amounts to the one we have just described.

Let $\Gamma = \{x_0\}$ be the cyclic group generated by x_0, and suppose x_0 is not a root of unity. Let Γ' be the division group of Γ in K. We shall apply Theorem 8.2 of the next section, with $a = M$ and $N = ab$. Let e be the exponent

$$e = e_a(\Gamma) = \text{g.c.d.}(e(\Gamma'/\Gamma), a)$$

and $c = c(S)$ where S is the exceptional set of primes among the set of all primes, relative to K, as explained in §8. Since Γ is cyclic, we get

$$(\Gamma_\varphi : \Gamma^a) \leqq ce.$$

Since μ_a is cyclic of order a, from Theorem 8.2 we get the inequalities

$$a/ce \leqq |H_\Gamma(a, ab)| = m_0 d \leqq 4(m + n)^4 m_0.$$

Therefore e is large, in the sense that

$$e \geqq a/4c(m + n)^4 m_0.$$

Note that $e|a$ since e is a greatest common divisor of two integers, including a.

Let $w(K) = |\mu(K)|$ be the order of the group of roots of unity in K. Note that $\Gamma' \supset \mu(K)$, and $\Gamma'/\Gamma\mu(K)$ is cyclic. We have

$$e(\Gamma'/\Gamma\mu(K)) \geqq e(\Gamma'/\Gamma)/w(K).$$

Hence there exists an element $z \in \Gamma'$, a positive integer $e'|a$, and a root of unity $\zeta \in \mu(K)$ such that

$$z^{e'} = x_0 \zeta, \quad \text{and} \quad e' \geqq e_a(\Gamma)/w(K) \geqq a/\gamma,$$

where $\gamma = 4c(m + n)^4 m_0 w(K)$. Thus $a/e' \leqq \gamma$, and a/e' is bounded only in terms of m, n, K. Let

$$x_1 = x^{a/e'}.$$

Since a is the smallest power of x lying in K, it follows that e' is the smallest power of x_1 lying in K, and $x_1^e = x_0$. Let $x_2 = x_1/z$. Since $z \in K$ it follows that e' is the smallest power of x_2 lying in K, and we have

$$x_2^{e'} = \zeta^{-1}.$$

Therefore x_2 is a root of unity, primitive of order $\geqq e'$.

We must now find a polynomial closely related to $f(X, Y)$ but containing only powers of $X^{a/e'}$. It is natural to define

$$f_0(X, Y) = \prod_\omega f(\omega X, Y),$$

where the product is taken over $\omega \in \mu_{a/e'}$. Since $f_0(\omega X, Y) = f_0(X, Y)$, it follows that there is a polynomial f_1 such that

$$f_0(X, Y) = f_1(X^{a/e'}, Y) = f_1(X_1, Y)$$

where $X_1 = X^{a/e'}$. Let $X_2 = z^{-1}X_1$. Then

$$f_0(X, Y) = f_2(X_2, Y),$$

where f_2 has degree ma/e' in X_2 and n in Y. But now,

$$f_2(x_2, y) = 0,$$

and x_2 is a root of unity of high order, at least $e' \geqq a/\gamma$.

We may now perform the same construction with respect to Y, and end up with a polynomial $F(X_2, Y_2)$ which vanishes at roots of unity x_2, y_2 of high order. By Theorem 6.4, this means that $F(X_2, Y_2)$ has a factor in standard form

$$\alpha X_2^r + \beta Y_2^s \quad \text{or} \quad \alpha X_2^r Y_2^s + \beta.$$

If we substitute back the variables X, Y we again get a factor in standard form with respect to X, Y, dividing the polynomial

$$\prod_{\omega, \eta} f(\omega X, \eta Y),$$

where the product is taken over the appropriate roots of unity ω relative to X and η relative to Y. Let $f^*(X, Y)$ be an absolutely irreducible factor of $f(X, Y)$. Then $f^*(\omega X, \eta Y)$ is absolutely irreducible, and for some ω, η is therefore a polynomial in standard form. This implies that $f^*(X, Y)$ itself is in standard form, whence f has a K-irreducible factor in standard form. This concludes the proof of Theorem 7.4.

§8. Non-abelian Kummer Theory

This section is essentially an appendix which axiomatizes non-abelian Kummer theory, for applications at least to the multiplicative group and abelian varieties.

Lemma 8.1 (Sah). *Let G be a group and let E be a G-module. Let α be in the center of G. Then $H^1(G, E)$ is annihilated by the map $x \mapsto \alpha x - x$ on E. In particular, if this map is an automorphism of E, then $H^1(G, E) = 0$.*

Proof. The lemma is valid for all cohomology groups $H^r(G, E)$, as one sees at once from the standard chain complex defining the cohomology. W. Ellis pointed out to me that one can give a direct proof for H^1 as follows. Let f be a 1-cocycle of G in E. Then

$$f(\sigma) = f(\alpha\sigma\alpha^{-1}) = f(\alpha) + \alpha f(\sigma\alpha^{-1})$$
$$= f(\alpha) + \alpha[f(\sigma) + \sigma f(\alpha^{-1})].$$

Therefore

$$\alpha f(\sigma) - f(\sigma) = -\sigma \alpha f(\alpha^{-1}) - f(\alpha).$$

But $f(1) = f(1) + f(1)$ implies $f(1) = 0$, and

$$0 = f(1) = f(\alpha \alpha^{-1}) = f(\alpha) + \alpha f(\alpha^{-1}).$$

This shows that $(\alpha - 1)f(\sigma) = (\sigma - 1)f(\alpha)$ so f is a coboundary. This concludes the proof of the lemma.

In what follows, we let P be a set of prime numbers, and assume that all positive integers M, N, etc. which occur are divisible only by primes of P. We let G be a group, and let:

$A = G$-module such that the isotropy group of any element of A is of finite index in G.

A is assumed divisible by primes in P, that is $pA = A$ for $p \in P$.

Γ = finitely generated subgroup of A such that $\Gamma^G = \Gamma$ (that is, Γ is pointwise fixed by G).

For every N, the subgroup A_N of elements of order N in A is assumed to be finite. Then for every N, $\frac{1}{N}\Gamma$ is also finitely generated. Note that

$$\frac{1}{N}\Gamma \supset A_N.$$

Γ' = subgroup of A^G consisting of elements x such that for some N we have $Nx \in \Gamma$. We assume that Γ' is also finitely generated and that the index $(\Gamma' : \Gamma)$ is finite.

For convenience, a representation of G on an abelian group A as above will be called a **Kummer representation**. Once Γ is fixed, all the representations we are interested in factor through the **division group of** Γ, that is the union

$$\bigcup \frac{1}{N}\Gamma$$

taken over all positive integers N (divisible only by primes of P according to our convention).

The above data and conditions summarize some of the properties which hold in either of the following cases:

A = abelian variety over K and $G = G_K$;

$A = \mathbf{C}^*$ and $G = G_K$ the group of automorphisms over a finitely generated field over the rational numbers.

Next we define the appropriate groups analogous to the Galois groups of Kummer theory, as follows. For any G-submodule B of A, we let:

$G(B)$ = image of G in $\mathrm{Aut}(B)$.

$G(N) = G(A_N)$ = image of G in $\mathrm{Aut}(A_N)$.

$H(N)$ = subgroup of G leaving A_N pointwise fixed.

$H_\Gamma(M, N)$ (for $M \mid N$) = image of $H(N)$ in $\mathrm{Aut}\!\left(\dfrac{1}{M}\,\Gamma\right)$.

We have an exact sequence

$$0 \to H_\Gamma(M, N) \to G\!\left(\frac{1}{M}\,\Gamma + A_N\right) \to G(N) \to 0.$$

Remark. In the concrete cases mentioned above, the reader will easily recognize these various groups as Galois groups. For instance, letting A denote either an abelian variety or the multiplicative group, we could draw the following lattice of field extensions with corresponding groups:

$$G\!\left(\frac{1}{M}\,\Gamma + A_N\right)\left\{\begin{array}{c} K\!\left(A_N, \dfrac{1}{M}\,\Gamma\right) \\[4pt] \mid \\[4pt] K(A_N) \\[4pt] \mid \\[4pt] K \end{array}\right\}\begin{array}{l} \left.\vphantom{\begin{array}{c}a\\a\\a\end{array}}\right\} H_\Gamma(M, N) \\[18pt] \left.\vphantom{\begin{array}{c}a\\a\\a\end{array}}\right\} G(N). \end{array}$$

Note that in applications, we do want to know how much degeneracy there is when we translate $K\!\left(A_M, \dfrac{1}{M}\,\Gamma\right)$ over $K(A_N)$, with $M \mid N$. When $A = \boldsymbol{\mu}$ then $G(N)$ can be identified with a subgroup of $\mathbf{Z}(N)^*$. When A is an abelian variety, then $G(N)$ is a complicated group, which has a natural representation in $\mathrm{GL}_{2d}(N)$, where $d = \dim A$.

For our purposes, we are interested especially in that part of $\mathbf{Z}(N)^*$ contained in $G(N)$, namely the group of integers n (mod N) such that there is an element $[n] \in G(N)$ such that

$$[n]a = na \quad \text{for all } a \in A_N.$$

Such elements are always contained in the center of $G(N)$, and are called **homotheties**.

More generally, let p be one of the primes in P and let $A^{(p)}$ be the p-primary torsion subgroup of A (subgroup of elements of order equal to a power of p). By the general definition,

$$G(A^{(p)}) = \text{image of } G \text{ in } \text{Aut}(A^{(p)}).$$

Non-degeneracy assumptions. We assume that there is a finite subset S of P, called **special**, having the following properties.

ND 1. If $p \notin S$, then $G(A^{(p)}) \supset \mathbf{Z}_p^*$, and $2 \notin P - S$.

ND 2. If $p \in S$, then there is an integer $c(p) = p^{f(p)}$ with $f(p) \geq 1$ such that

$$G(A^{(p)}) \supset U_{c(p)},$$

where $U_{c(p)}$ is the subgroup of $\mathbf{Z}_p^*$ consisting of those elements $\equiv 1 \bmod c(p)$.

ND 3. Let $A_{P\text{-tor}}$ be the subgroup of A consisting of the torsion elements of order divisible by the primes in P. Then

$$G(A_{P\text{-tor}}) \supset \prod_{p \in S} U_{c(p)} \times \prod_{p \notin S} \mathbf{Z}_p^*.$$

The product decomposition on the right is relative to the direct sum decomposition

$$A_{P\text{-tor}} = \bigoplus_{p \in P} A^{(p)}.$$

In practice, the non-degeneracy assumptions are satisfied for the set P of all primes when $A = \mu$ is the group of roots of unity and $G = \text{Gal}(K^a/K)$ where K is finitely generated over $\mathbf{Q}$. By a fundamental theorem of Serre [Ser 3], these assumptions are also satisfied if A is an elliptic curve over K. For an abelian variety A, it is a conjecture that they are satisfied, see the comments at the end of the chapter.

In the above examples, we could take P to be the set of all prime numbers, while the set S is finite. On the other hand, let K be a p-adic field (finite extension of $\mathbf{Q}_p$) and let $P = S$ consist of the single prime p. Then the non-degeneracy assumptions are also satisfied. One of the reasons for the formulation of the assumptions was to give them in sufficient generality to include this case.

Under the non-degeneracy assumptions, we observe that

$$[2] \in G(A_{(P-S)\text{-tor}}) \quad \text{and} \quad [1 + c] \in G(A_{S\text{-tor}}),$$

where

$$c(S) = c = \prod_{p \in S} c(p).$$

Therefore $[2] - [1] = [1]$ and $[1 + c] - [1] = [c]$ can (and will) be used in the context of Lemma 8.1. For any positive integer M we let

$$c(M) = \prod_{\substack{p \mid M \\ p \in S}} c(p).$$

In the next theorem, we shall also need the **exponent**:

$e(\Gamma'/\Gamma) = $ smallest positive integer e such that $e\Gamma' \subset \Gamma$.

The exponent of any finite abelian group is defined in the same way.

It is clear that degeneracy in the Galois group $H_\Gamma(M, N)$ defined above can arise from lots of roots of unity in the ground field, or at least degeneracy in the Galois group of roots of unity; and also if we look at an equation

$$X^M - x_0 = 0,$$

from the fact that x_0 is already highly divisible in K. This second degeneracy would arise from the exponent $e(\Gamma'/\Gamma)$, if we look at the Galois group of the divisions of Γ. The next theorem shows that these are the only sources of degeneracy.

We have the abelian **Kummer pairing** (provided $M \mid N$)

$$H_\Gamma(M, N) \times \Gamma/M\Gamma \to A_M \quad \text{given by } (\tau, x) \mapsto \tau y - y,$$

where y is any element such that $My = x$. The value of the pairing is independent of the choice of y. Thus we have a homomorphism

$$\varphi: \Gamma \to \text{Hom}(H_\Gamma(M, N), A_M),$$

such that for $x \in \Gamma$,

$$\varphi_x(\tau) = \tau y - y \quad \text{where } My = x.$$

Theorem 8.2. *Let $M|N$. Let φ be the homomorphism*

$$\varphi: \Gamma \to \operatorname{Hom}(H_\Gamma(M, N), A_M),$$

and let Γ_φ be its kernel. Let $e_M(\Gamma) = \mathrm{g.c.d.}(e(\Gamma'/\Gamma), M)$. Under the non-degeneracy assumptions, we have

$$c(M)e_M(\Gamma)\Gamma_\varphi \subset M\Gamma.$$

Proof. Let $x \in \Gamma$ and suppose $\varphi_x = 0$. Let $My = x$. For $\sigma \in G$ let $y_\sigma = \sigma y - y$. Then $\{y_\sigma\}$ is a 1-cocycle of G in A_M, and by the hypothesis that $\varphi_x = 0$, this cocycle depends only on the class of σ modulo the subgroup of G leaving the elements of A_N fixed. In other words, we may view $\{y_\sigma\}$ as a cocycle of $G(N)$ in A_M. Let $c = c(N)$. By Lemma 8.1, it follows that $\{cy_\sigma\}$ splits as a cocycle of $G(N)$ in A_M. In other words, there exists $t_0 \in A_M$ such that

$$cy_\sigma = \sigma t_0 - t_0,$$

and this equation in fact holds for $\sigma \in G$. Let t be such that $ct = t_0$. Then

$$c\sigma y - cy = \sigma ct - ct,$$

whence $c(y - t)$ is fixed by all $\sigma \in G$, and therefore lies in Γ' by the definition of Γ'. Therefore

$$e(\Gamma'/\Gamma)c(y - t) \in \Gamma.$$

We multiply both sides by M, and observe that $cM(y - t) = cMy = cx$. This shows that

$$c(N)e(\Gamma'/\Gamma)\Gamma_\varphi \subset M\Gamma.$$

Since $\Gamma/M\Gamma$ has exponent M, we may replace $e(\Gamma'/\Gamma)$ by the greatest common divisor as stated in the theorem, and we can replace $c(N)$ by $c(M)$ to conclude the proof.

Corollary 8.3. *Assume that M is prime to $2(\Gamma' : \Gamma)$ and is not divisible by any primes of the special set S. Then we have an injection*

$$\varphi: \Gamma/M\Gamma \to \operatorname{Hom}(H_\Gamma(M, N), A_M).$$

If in addition Γ is free with basis $\{a_1, \dots, a_r\}$, and we let $\varphi_i = \varphi_{a_i}$, then the map

$$H_\Gamma(M, N) \to A_M^{(r)} \qquad \text{given by} \qquad \tau \mapsto (\varphi_1(\tau), \dots, \varphi_r(\tau))$$

is injective. If A_M is cyclic of order M, this map is an isomorphism.

Proof. Under the hypotheses of the corollary, we have $c(M) = 1$ and $e_M(\Gamma) = 1$ in the theorem.

Remark 1. In Theorem 8.2, the factors $c(M)$, $e_M(\Gamma)$ are uniformly bounded by $c(S)$, which depends only on the degeneracy of the Galois group of roots of unity; and by $e(\Gamma'/\Gamma)$, which depends only on Γ and K. Thus Theorem 8.2 gives uniform bounds for the degeneracy of $H_\Gamma(M, N)$.

Remark 2. Since $H_\Gamma(M, N)$ is a normal subgroup of $G\left(\dfrac{1}{M}\Gamma + A_N\right)$, the group $G\left(\dfrac{1}{M}\Gamma + A_N\right)$, and actually $G(N)$, operates by conjugation

$$\tau \mapsto \sigma\tau\sigma^{-1}, \quad \text{for } \sigma \in G\left(\frac{1}{M}\Gamma + A_N\right).$$

It is immediately verified that φ_x is a

$$G\left(\frac{1}{M}\Gamma + A_N\right)\bigg/ H_\Gamma(M, N) = G(N)\text{-homomorphism,}$$

that is

$$\varphi_x(\sigma\tau\sigma^{-1}) = \sigma\varphi_x(\tau).$$

Hence

$$\varphi : \Gamma/\Gamma_\varphi \to \mathrm{Hom}_G(H_\Gamma(M, N), A_M)$$

has its image actually in the G-homomorphisms. If $A = \mathbf{G}_m$ is the multiplicative group, then A_M is cyclic of order M, and $\mathrm{Hom}(H_\Gamma(M, N), A_M)$ is just the dual group.

If A is an elliptic curve or an abelian variety, then the situation is more complicated. Especially in these cases, the G-module structure becomes important. For a discussion and applications in such a context, cf. [L 7], Chapter 5, which includes results of Bashmakov and Ribet in this direction.

Notes and Comments

Integral Points

Siegel [Sie 1] showed how one can put the Mordell–Weil theorem together with diophantine approximations (the Thue–Siegel theorem) in order to prove that a curve of genus ≥ 1 has only a finite number of integral points. Since at that time he did not have the strong form of Roth's theorem he had to surmount considerable difficulties in the diophantine approximations because of the lack of uniformity.

He actually worked over the ring of integers of a number field. Mahler [Mah 1] noted that by using his p-adic version of the diophantine approximations, one could allow a finite number of primes in the denominator, at any rate for curves of genus 1 over the rationals. He conjectured that the same result would hold for curves of genus ≥ 1 over number fields.

Once the theory of the Jacobian was sufficiently developed and once Roth's theorem was obtained, it was then natural to reconsider this question. The general version used here was presented in [L 4] following Siegel's (and Mahler's) method. The Jacobian replaces the theta functions, as usual, and the mechanism of the covering already used by Siegel appears here in its full formal clarity. It is striking to observe that in [L 1] I used the Jacobian in a formally analogous way to deal with the class field theory in function fields. In that case, Artin's reciprocity law was reduced to a formal computation in the isogeny $u \mapsto u^{(q)} - u$ of the Jacobian. In the present case, the heart of the proof is reduced to a formal computation of heights in the isogeny $u \mapsto mu + a$.

I have transported Siegel's proofs to the more general case of finitely generated rings over $\mathbf{Z}$. The arguments go through because of the finite generation of the units and the divisor class group.

In *Diophantine Geometry*, I conjectured that Siegel's theorem in its absolute and relative form should be valid for abelian varieties, namely:

Let A be an abelian variety in characteristic 0, defined over a field of finite type K. Let U be an affine open subset of A, and R a finitely generated subring of K over Z. Then U has only a finite number of points in R.

In the relative case when K is a function field over a constant field k of characteristic 0, and R is a finitely generated subalgebra over k, the conjecture states that

integral points of U in R have bounded height, and thus lie in a finite number of cosets of $\tau B(k)$.

These conjectures are still unproved. In [L 4], I pointed out that Siegel's theorems for curves of genus 0 or ≥ 1 have a relative formulation. An analysis of the classical proofs shows among other things that they must be separated into two parts: one proving that certain sets of points have bounded height; and second showing what sets of bounded height look like.

Intersection with Certain Subgroups

After reducing certain diophantine problems to the equation

$$au + bu' = 1$$

in units, Siegel applies the coset argument to raise back this equation to a curve of genus ≥ 1, or even arbitrarily high genus, and then applies the finiteness of integral points on such curves. Hence the structural role of this equation was not independently clear. I gave a quite different approach to this same equation, reducing it to linear combinations of logarithms in [L 12]. Cf. [L 7], Chapter VI. Simultaneously I also showed how one can argue directly on the elliptic logarithms to get the appropriate inequality coming from diophantine approximations to bound the integral points. It remains a problem to show how to reduce the finiteness of integral points on abelian varieties to such an inequality with abelian logarithms. For a bibliography concerning such inequalities, referring especially to work of Coates-Lang and Masser following Baker, see for instance Waldschmidt [Wa].

Recently Vojta has obtained some interesting results proving finiteness of integral points lying outside proper algebraic subvarieties. He considers the linear system $L(D)$ where D is a positive divisor with sufficiently many distinct irreducible components and the rational map into affine space associated with such a system. He reduces the finiteness property to generalizations of the unit equation, namely $a_1 u_1 + \cdots + a_n u_n = 0$ in several variables. Cf. [Vo].

As Siegel already observed, the coset argument is formally the same when dealing with curves of genus ≥ 1 or subgroups of multiplicative groups. The analogy between toruses and abelian varieties was again observed by Chabauty, who in two papers [Cha 1] and [Cha 2] considers the infinite intersections of a subvariety of a torus or an abelian variety with particular subgroups of finite type, namely subgroups of units and groups of rational points in a number field, respectively. Thus one is led to generalize and reformulate a conjecture of Chabauty [Cha 1] in the following manner.

Let G be a torus (resp. an abelian variety) in characteristic 0. Let V be a subvariety of G having an infinite intersection with a subgroup of finite type Γ of G. Then V contains a finite number of translations of group subvarieties of G which contain all but a finite number of points of $V \cap \Gamma$.

This statement was proved in §3 for the case when V is of dimension 1 and G is a torus. One may ask whether it would not be valid also for a commutative group extension of an abelian variety by a torus.

Let us consider the statement when G is an abelian variety A and V is a curve. It then generalizes a classical conjecture of Mordell [Mo 1] to the effect that a curve of genus ≥ 2 over the rationals has only a finite number of rational points.

Let us mention Chabauty's partial result. He proves the following statement in [Cha 2].

Let C be a complete non-singular curve of genus ≥ 2 defined over a p-adic field K. Let C be embedded in its Jacobian over K, and let Γ be a finitely *generated subgroup of* J_K. *If the rank of* Γ *is smaller than (strictly) the genus of* C, *then* $C \cap \Gamma$ *is finite.*

We sketch his argument. An infinite number would have a point of accumulation, say P_0. In a neighborhood of P_0, there will be infinitely many points of $\Gamma \cap C$, and thus without loss of generality (by the non-archimedean behaviour) we may assume that Γ itself lies in such a neighborhood. The assumption on the rank would then imply a linear relation on the differentials of the first kind on C or on J (those of C being induced by those of J), which is impossible.

The general problem is to investigate the intersection of a subvariety of a group variety with certain subgroups. In the context of the Mordell conjecture, one considers the intersection of a curve with a finitely generated subgroup of its Jacobian. Manin's investigations of the Picard-Fuchs equations [Man 1] led him to ask whether the intersection of a curve with the torsion group of the Jacobian is finite. The same question was raised by Mumford at about the same time. In [L 9] I formulated a conjecture which covers both situations as follows:

Let A *be an abelian variety. Let* Γ_0 *be a finitely generated subgroup, and let* Γ *be its division group, that is the group of all points* $x \in A$ *such that* $nx \in \Gamma_0$ *for some positive integer n. Let* V *be a subvariety of* A *having an infinite intersection with* Γ. *Then* V *contains a finite number of translations of abelian subvarieties of* A *which contain all but a finite number of points of* $V \cap \Gamma$.

Of course, one can replace A by a linear torus, or an extension of A by a linear torus in the above statement. In other words, from a diophantine point of view, when one considered previously a finitely generated group, one may as well consider its division group. When dim $V > 1$, the statement is still not known when A is a linear torus (this would be the higher dimensional generalization of Liardet's theorem), but should be accessible in light of Raynaud's results, see below.

Taking Γ_0 to be the unit element of A yields the special case when Γ consists of all torsion points on A. In this case, I reduced the problem when V is a curve to a Galois theoretic statement as follows [L 9]. Assume that V has an infinite intersection with A_{tor}. Let K be a field of definition for A and V finitely generated over the rationals. Let m be an integer ≥ 1. Let $[m] = [m]_A$ be multiplication by m on A. As a cycle, $[m](V) = s \cdot V^{(m)}$, where s is some positive integer, and $V^{(m)}$ consists of all points mx with $x \in V$. Then $s \cdot V^{(m)}$ is algebraically equivalent to $m^2 \cdot V$. If $V \neq V^{(m)}$, then $V \cap V^{(m)}$ has at most $m^2(\deg V)^2$ points by the generalized Bezout's theorem, Chapter 3, Lemma 3.5. We can view V and $V^{(m)}$ as divisors on their sum in A.

If $V = V^{(m)}$ then $[m]$ gives an unramified covering of V over itself, of degree m^2, and hence V is of genus 1, so is an abelian subvariety.

The proof of the conjecture in this case is then reduced to the following hypothesis, which is the analogue of the irreducibility of the cyclotomic equation.

(∗) *Let A be an abelian variety defined over K. There exists an integer $c \geq 1$ with the following property. Let x be a point of period n on A. Let G_n be the multiplicative group of integers prime to n, mod n. Let G be the subgroup of G_n consisting of those integers d such that dx is conjugate to x over K. Then*

$$(G_n : G) \leq c.$$

To apply (∗), suppose that there exist points x_n of period n, $n \to \infty$, lying on V. Let d be a positive integer prime to n. By (∗), there exists an automorphism σ of $K(x_n)$ over K such that $\sigma x_n = d^r x_n$, where r is a positive integer bounded by c. Then

$$\sigma x_n = d^r x_n \in V \cap V^{(d^r)}.$$

Furthermore, if τ is in the group of automorphisms of $K(x_n)$ over K, then

$$\tau d^r x_n \in V \cap V^{(d^r)}.$$

If $V \neq V^{(d^r)}$ we obtain the inequalities, using (∗):

$$\frac{\phi(n)}{c} \leq \text{number of points on } V \cap V^{(d^r)} \leq d^{2r}(\deg V)^2.$$

We note that $\phi(n) \geq n^{1/2}$ for sufficiently large n. By Lemma 6.2, taking d to be a sufficiently small prime number not dividing n, we get a contradiction as soon as n is sufficiently large, as desired.

Conjecture (∗) is still open in general at the time this book is written, although Shimura pointed out that it was a corollary of known results in the theory of complex multiplication in that case, cf. [L 18]. For elliptic curves without complex multiplication, it is contained in Serre's results [Ser 3].

On the other hand, Bogomolov [Bo] has proved substantial parts of the conjecture (∗), and its application to the intersection of V with A_{tor}, essentially in those cases when Lie theory can be applied, that is for prime power torsion involving a finite number of primes.

Raynaud [Ra] proved the general conjecture concerning the intersection of V with A_{tor} by means of entirely different techniques, although still using Galois theory heavily. He also proved the conjecture when $\dim V > 1$. For general division groups of finitely generated groups, he reduces it to the Mordell conjecture.

Algebraic Families

In [L 4] and in *Diophantine Geometry*, I transposed Mordell's conjecture on curves of genus ≥ 2 to a geometrical context as follows. Let k be a field of characteristic 0 (algebraically closed, or equal to $\mathbf{C}$ if you want). Let $K = k(t)$ be a function field over k, where t is the generic point of a variety T, and let $C = C_t$ be a curve of genus ≥ 2 defined over K. Then C_t can be viewed as a generic member of an algebraic family. The conjecture asserts:

If C_t has infinitely many rational points in $k(t)$ (cross sections of the parameter variety T in the graph of the family), then C_t is birationally equivalent over $k(t)$ to a curve C_0 defined over k, and all but a finite number of these points come from points of C_0 in k.

In particular, the graph is birationally equivalent to a product.

This conjecture was proved by Manin [Ma 1]. My report of Manin's proof at the Arbeitstagung in Bonn in 1963 caught the interest of Grauert, who later gave another proof [Gra]. The special case when $C_t = C_0$ is already defined over k amounts to a classical theorem of de Franchis:

Let V be a variety and C a curve of genus ≥ 2 (in characteristic 0). There exists only a finite number of generically surjective rational maps of V on C.

I give a quick proof of this theorem. Taking a generic hyperplane section of V and inducing the rational map on it, one reduces the theorem to the case when V is itself a curve. Indeed, two distinct generically surjective rational maps $f, f': V \to C$ induce distinct generically surjective maps on the hyperplane section, as one sees by taking the induced homomorphisms on the Albanese varieties, using AV, Theorem 4 of Chapter VIII, §2.

Assuming now that V is a curve, we have the formula for the genus

$$2g(V) - 2 = d[2g(C) - 2] + \lambda,$$

where $\lambda \geq 0$. Thus the degree of V over C is bounded. Taking suitable projective embeddings, we see that the degree of the graph of our rational maps f must be bounded. Hence these graphs Γ_f lie in finitely many algebraic families on $V \times C$. On the other hand, a generic element of such families is likewise a generically surjective rational map of V onto C (as one sees by projecting on both factors). Taking the induced homomorphisms on the Jacobians, and using the fact that an abelian variety has no algebraic family of abelian subvarieties, we see that all induced maps coming from the same family differ by translations. We now use the fact that C is not equal to a non-zero translation of itself in its Jacobian. We conclude that a graph Γ_f actually must constitute by itself a maximal algebraic family on $V \times C$, and finally that there is only a finite number of such graphs, or maps f. This concludes the proof. (When V is a curve, we do not need characteristic 0, only the assumption that the map $f: V \to C$ is separable, to be able to use the genus formula above.)

The Mordell conjecture thus gives rise to diophantine criteria for lowering fields of definition, and we actually proved such a criterion in the context of integral points, Proposition 2.1, which after Manin's theorem gives something only in the case of genus 1. As mentioned above, higher dimensional cases would arise from integral points on abelian varieties.

Parsin [Par 1] reduced the Mordell conjecture to the Shafarevich conjecture that over a number field, there is only a finite number of isomorphism classes of curves of given genus with good reduction outside a given finite set of primes. For the Shafarevich conjecture in the function field case, see Arakelov [Ar 1] and [Par 1].

Higher Dimensional Complex Analytic Analogues

In [L 13] and [L 14] I conjectured higher dimensional analogues of the Mordell conjecture. Typically, if a projective non-singular variety V is "hyperbolic" or the quotient of a bounded domain by a discrete group of automorphisms operating freely, then it should have only a finite number of points rational over a field K finitely generated over $\mathbf{Q}$. For an algebraic family of such varieties, if there are infinitely many sections then the family should split birationally, and all but a finite number of these sections should come from constant sections. In particular, there should be only a finite number of generically surjective rational maps of a variety on V. The finiteness of generically surjective meromorphic maps on varieties "of general types" was proved by Kobayashi and Ochai [K–O]. This generalizes the theorem of de Franchis. See also Noguchi [No] and [No–S] as well as [MD–LM]. The theorem of Manin and Grauert's techniques were generalized to the higher dimensional case by Noguchi [No].

Also in connection with subvarieties of an abelian variety, in order to link the conjectures already expressed with the above, I suggested that a subvariety of an abelian variety which does not contain the translation of an abelian subvariety (of dimension ≥ 1) should be hyperbolic. This was proved by Mark Green [Gr].

Vojta [Vo] has recently described an extraordinary analogy between Nevanlinna defect theory and Roth's theorem, together with possible higher dimensional cases. Vojta's conjecture concerning such situations includes as special cases Mordell's conjecture, my conjecture on integral points, and the above conjectures in the higher dimensional cases, concerning rational points. It is one of the most beautiful insights into mathematics that I have ever seen.

CHAPTER 9

Hilbert's Irreducibility Theorem

In its simplest form, Hilbert's theorem asserts: let $f(t, X)$ be a polynomial in $\mathbf{Q}[t, X]$ (so in two variables), and assume that $f(t, X)$ is irreducible. Then there exist infinitely many rational numbers t_0 such that $f(t_0, X)$ is irreducible over $\mathbf{Q}$.

More generally, let k be a field, and $f(t_1, \ldots, t_r, X_1, \ldots, X_s)$ a polynomial with coefficients in k which is irreducible as a polynomial in $r + s$ variables. We ask whether there exist values (t') of (t) in k such that the polynomial $f(t', X)$ is irreducible as a polynomial in (X) with coefficients in k. We shall exhibit a large variety of fields for which this is true.

Suppose $f(t, X)$ has coefficients in $k(t)$, i.e. is in $k(t)[X]$, and assume it is irreducible as a polynomial in (X) over $k(t)$. Multiplying f by a polynomial $\phi(t)$, to make the coefficients lie in $k[t]$ does not change this irreducibility property. If we then divide the resulting polynomial by the greatest common divisor of its coefficients, we obtain a polynomial $f_1(t, X) \in k[t, X]$ which is irreducible over k. Conversely, if $f(t, X)$ is in $k[t, X]$ and irreducible over k, then it is irreducible in $k(t)[X]$.

Given $f(t, X) \in k(t)[X]$, we denote by $U_{f,k}$ (or simply U_f if the reference to k is clear) the subset of the affine space S_k^r consisting of those points $(t'_1, \ldots, t'_r)$, with $t'_i \in k$, at which the coefficients of f are defined and such that $f(t', X)$ is irreducible in $k[X]$ over k. Such $U_{f,k}$ will be called **basic Hilbert sets**. The intersection of a finite number of basic Hilbert sets with a finite number of nonempty Zariski open subsets of S_k^r will be called a **Hilbert subset** of S_k^r. If $r = 1$, then $S_k^1 = k$ and the non-empty Zariski open sets are the complements of finite subsets of k.

A field k is called **Hilbertian** if the Hilbert subsets of S_k^r are not empty (and thus are infinite). If R is a subset of k, we shall say that R is **Hilbertian** in k if every Hilbert subset of S_k^r contains some element $(t'_1, \ldots, t'_r)$ with $t'_i \in R$. A ring will be said to be **Hilbertian** if it is Hilbertian in its quotient field.

Among other results, we shall prove in this chapter that all rings of finite type over the prime ring, and of transcendence degree ≥ 1 in characteristic > 0 are Hilbertian. In particular, every number field is Hilbertian. The proof of this fact will depend only on elementary calculus.

It will be shown also that the essential case occurs when $r = 1$ and $s = 1$, i.e. when there is one parameter and one variable. The general case can be reduced to this one by simple formal arguments. The essential case will be seen to be equivalent to the lifting of rational points in coverings of curves. From this point of view one is thus led to diophantine questions.

In §6 we shall apply the irreducibility property to abelian varieties, and show for instance that if an abelian variety or a curve of genus ≥ 1 is defined over a field of finite type over $\mathbf{Q}$, then there exist infinitely many non-degenerate specializations of it into a number field which induce an injection on the sets of rational points, a result due to Néron. However, a much stronger result due to Silverman will be proved in Chapter 12.

We make one final observation, as Hilbert does [Hi], concerning applications to Galois theory. Suppose that we can construct a purely transcendental extension $k(t_1, \ldots, t_r)$ of k and a finite Galois extension E of $k(t_1, \ldots, t_r)$ with group G. If k is Hilbertian then we can specialize the (t) on (t') in k in such a way that the decomposition group remains of the same order as that of G, and thus obtain a Galois extension of k with group G. This can be done in particular with the symmetric groups over any field k, thereby allowing either the possibility of constructing a Galois extension with the symmetric group if k is Hilbertian or of proving that a field is not Hilbertian (for instance, a p-adic field). However, it does not seem to be known whether the power series ring in several variables over a field (or a unique factorization domain with infinitely many primes) is Hilbertian.

§1. Irreducibility and Integral Points

We consider the case of one t and one X. Let us write

$$f(t, X) = a_n(t)X^n + \cdots + a_0(t)$$

with $a_i(t) \in k[t]$. In studying special values of t in k, we always *exclude* the finite number which make a_n vanish.

Suppose that $f(t, X)$ is irreducible over $k(t)$. If f has a factorization

$$f(X) = a_n(t) \prod_{i=1}^{n} (X - \alpha_i)$$

in the algebraic closure of $k(t)$, every value t_0 of t in k determines a homomorphism

$$k[t] \to k[t_0] = k$$

which can be extended to the ring generated by the roots $\alpha_1, \ldots, \alpha_n$ of f under our assumption that $a_n(t_0) \neq 0$, because these roots are integral over

$k[t, a_n(t)^{-1}]$. The image of these roots is determined up to a permutation. Let α_i' ($i = 1, \ldots, n$) be these images. If we have a factorization

$$f(t_0, X) = g_0(X)h_0(X)$$

in $k[X]$, the coefficients of g_0 and h_0 are polynomial functions of the α_i'. Let g and h be the polynomials corresponding to these functions, giving a factorization

$$f(t, X) = g(X)h(X)$$

in the algebraic closure of $k(t)$. Since f is assumed to be irreducible, one of the coefficients of g or h does not lie in $k(t)$, say y. Then the ring $k[t, y]$ is the affine ring of a curve C (which may be reducible) over k. The factorization $f(t_0, X) = g_0 h_0$ gives rise to a rational point (t_0, y_0) on our curve C which has projection t_0 on the first factor.

If one writes $f(t, X) = g(X)h(X)$ in all possible manners in the algebraic closure of $k(t)$, with g and h of degree ≥ 1, there will be each time a coefficient which does not lie in $k(t)$, whence we get a finite number of curves $C_1, \ldots, C_m$ as above. Each value t_0 in k which cannot be lifted to a rational point of any of these curves will be such that $f(t_0, X)$ is irreducible in $k[X]$.

If k is the quotient field of a ring R, and if we replace y above by $\varphi(t)y$ for a suitable polynomial $\varphi(t) \in R[t]$, we may suppose that y is integral over $R[t]$. Excluding the finite number of points t_0 for which $\varphi(t_0) = 0$, one sees that y_0 and $\varphi(t_0)y_0$ are rational or irrational over k simultaneously. If, however, they are in k, then $\varphi(t_0)y_0$ will be integral over R, and hence in R if R is integrally closed (as is usual in practice).

By an **affine plane curve** C over a field k, we shall mean an affine curve whose ring is of type $k[t, y]$ over k, with t transcendental over k. We do not assume that C is absolutely irreducible, so that $k(t, y)$ need not be regular over k. If R is a subset of k, we denote by $U_{t,R}(C)$ or simply $U_{t,R}$ the set of elements $t_0 \in R$ such that there is no point $P \in C(k)$ for which $t(P) = t_0$, viewing t as a function on C. From our preceding discussion, we have:

Proposition 1.1. *Let k be a field, and R a subset of k. Every Hilbert subset of k contains a finite intersection of sets of type $U_{t,R}(C)$ for a finite number number of affine plane curves C over k.*

The Hibert subsets of k mentioned here are those determined by irreducible polynomials $f(t, X)$ with one t and one X. We shall see in §3 that the restriction to one X can be dispensed with.

Let us consider an affine plane curve over k, as above, with y not in $k(t)$. If y is not separable over $k(t)$, then a power y^{p^m} will be for m large and $p = $ characteristic. If y^{p^m} actually lies in $k(t)$, then to find elements of $U_{t,R}$ one analyzes k^*/k^{*p} directly. Let us exclude this case. We are therefore reduced to studying the case where y is separable over $k(t)$ but not in $k(t)$.

Let k_1 be the algebraic closure of k in $k(t, y)$. Then (t, y) determines an absolutely irreducible curve C_1 defined over k_1 having a conjugate C_1^σ over k not equal to C_1 if $k \neq k_1$. Every rational point of C in k must lie on both C_1 and C_1^σ, and hence there can be only a finite number of such points if C is not absolutely irreducible under the hypothesis that y is separable over $k(t)$. In the sequel, in trying to find points in $U_{t,R}$, we can therefore assume that C is absolutely irreducible.

Suppose that we have a variety C of dimension 1 over k and a non-constant function t in $k(C)$. If for instance k is of finite type over $\mathbf{Q}$, R of finite type over $\mathbf{Z}$, and C of genus ≥ 1 then we know that there is only a finite number of points P in $C(k)$ such that $t(P) \in R$. The worst case would therefore occur when C has genus 0, i.e., is a pure curve.

In that case, following Siegel, one can apply the analysis of curves of genus 0 to estimate the number of bad points, but using an idea of Néron, it is just as well to reduce the general case to that of curves of genus ≥ 1, by means of the following trick.

We start with our plane curve C, with generic point (t, y) over k. If we write $t = f(u)$ for a suitable polynomial $f(u)$ in $k[u]$, then we get a new curve C' with generic point (t, y, u) over k, which will have genus ≥ 1 (and in fact arbitrarily high for suitable f). We now look for special values u_0 which cannot be lifted to rational points (t_0, y_0, u_0) of C' in k. Of course, $t_0 = f(u_0)$ is determined and if f is taken to have coefficients in a subring R of k, and u_0 is selected in R, then t_0 will also lie in R, so that our integrality conditions are preserved.

In order to find polynomials f we simply use the genus formula for a covering. We have two cases:

Case 1.   *The function t has at least 3 singularities (zeros or poles) on the complete non-singular curve determined by C.*

We then take m large, relatively prime to the characteristic of k, to the degree $[k(C):k(t)]$, and to the orders of zeros and poles of t. Then the curve C' whose function field is $k(C)(u)$ over k is absolutely irreducible, and of large genus, because the formula

$$2g' - 2 = -2m + \sum (e_j - 1)$$

shows that the term $\sum (e_j - 1)$ will grow at least like $3(m - 1)$, each e_j being equal to m.

Case 2.   *The function t has exactly 2 singularities.*

Then we can find a function x in $k(C)$ such that $k(C) = k(x)$, and $t = x^r$ with an integer $r > 0$, prime to the characteristic p of k, or $t = x^{p^n} - a$

for some $a \in k$, and an integer $n \geq 0$. If $k(C) \neq k(t)$ then $r > 1$ or $n > 0$. Leaving aside the characteristic difficulties, and looking only at $t = x^r$, we take for $f(u)$ any polynomial of degree m prime to p with m distinct roots to see that the genus of C' again becomes large if m becomes large.

We have thus shown that any subring R of a number field is Hilbertian. In the next section, we shall given an elementary proof of this fact.

In characteristic $p > 0$, we do not have the finiteness of integral points because of the Frobenius automorphism. However it suffices to use the Mordell–Weil theorem, which gives us a sufficiently strong way of counting points, to see that they are Hilbertian. In fact, having obtained our curve of genus ≥ 1, we embed it in its Jacobian over k. If $J(k)$ is finitely generated, then we found an upper bound for the number of points of height $\leq B$ in Chapter 5, §7. In practice, if R is a subring of k, the number of points of R of height $\leq B$ is very much larger than a power of $\log B$, being in fact of the order of magnitude of B itself (for instance for the ordinary integers $\mathbf{Z}$ and, as the reader himself will verify immediately, for a polynomial ring in one variable over a finite constant field). Thus if we have a theory of heights on k, if $f: C' \to \mathbf{P}^1$ is a generically surjective rational map of C' on $\mathbf{P}^1$ defined over k, and if C' has genus ≥ 1, then we know that for a point $P' \in C'(k)$ we have

$$h(f(P')) \approx h(P')^d$$

where d is the degree of f (Proposition 3.3 of Chapter 4, which we are applying in a case where it could be verified more trivially).

Let us look at polynomials over a finite field. Let $R = k_0[w]$ be a polynomial ring in one variable over the finite field k_0 with q elements. The number of elements of R of degree $< r$ is q^r and one also sees that we could replace the degree by the logarithmic height, computed with respect to the obvious model. Let C be an affine plane curve with generic point (t, y) over the quotient field k of R. From the above discussion, we see that if y is separable over $k(t)$, then there exists a number α, with $0 < \alpha < 1$ such that the number of elements of $U_{t,R}(C)$ of degree $< r$ is

$$q^r + O(q^{\alpha r}) \quad \text{for } r \to \infty.$$

If y is not separable, then one must examine this case more closely. We shall deal with it in connection with arbitrary ground fields k in §4.

§2. Irreducibility Over the Rational Numbers

We consider the concrete case of the rational numbers, or rather the ring of integers $\mathbf{Z}$. Let (t, y) be a generic point of an affine plane curve C over $\mathbf{Q}$, and suppose $y \notin \mathbf{Q}(t)$. After multiplying y by a suitable polynomial in $\mathbf{Z}[t]$, we may assume that y is integral over $\mathbf{Z}[t]$. Thus if any value t_0 (excluding a finite number) in $\mathbf{Q}$ extends to a point (t_0, y_0) on C with $y_0 \in \mathbf{Q}$, then y_0 lies in $\mathbf{Z}$ if t_0 lies in $\mathbf{Z}$.

We may view y as an algebraic function of t over the reals, and as such it has an expansion at infinity:

$$y = \varphi(t) = at^{n/e} + \cdots + b + c\,\frac{1}{t^{1/e}} + \cdots$$

with coefficients $a, b, c, \ldots$ that are complex numbers, and $t^{1/e}$ one of the e-th roots of t taken as parameter. We choose in fact $t^{1/e}$ to be real. In that case, if there exist infinitely many values of t tending to infinity in $\mathbf{R}$ such that $\varphi(t)$ is real, then the coefficients are in fact real. Indeed, if one of them were complex, let us consider the one most to the left, say ξ. The angle of the term having ξ for its coefficient is $\neq 0, \pi$. As t tends to infinity, the term with this coefficient dominates the series to the right of it. Hence there cannot be any cancellations, and for $t \to \infty$ the values $\varphi(t)$ would be complex, contradicting the hypothesis.

We now study the integral points on the analytic curve $(t, \varphi(t))$ and prove the following theorem, concerning their distribution.

Theorem 2.1. *Let $\varphi(t)$ be a function of a real variable having a power series expansion*

$$\varphi(t) = at^{n/e} + \cdots + b + c\frac{1}{t^{1/e}} + \cdots$$

with coefficients $a, b, c, \ldots$ in $\mathbf{R}$, in terms of the real parameter $t^{1/e}$, and converging for all sufficiently large values of t. Assume that $\varphi(t)$ does not lie in $\mathbf{R}[t]$, i.e. is not a polynomial in t. Suppose that there is an infinity of positive integers

$$t_0 < t_1 < t_2 < \cdots$$

such that $\varphi(t_i)$ is in $\mathbf{Z}$. Then there exists an integer i_0, an integer $m > 0$, and a real number $\lambda > 0$, such that for all $i > i_0$ we have

$$t_{i+m} - t_i > t_i^\lambda.$$

In other words, the t_i become more widely spaced. As a corollary, we get an estimate which we can handle better in practice.

Corollary 2.2. *Let $\varphi(t)$ be as in the theorem. There exists a real number α with $0 < \alpha < 1$ such that the number $N(B)$ of $t_i \leq B$ for which $\varphi(t_i)$ is in $\mathbf{Z}$ satisfies the inequality*

$$N(B) \leq B^\alpha$$

for all B sufficiently large.

Proof. Choose $0 < \beta < 1$. Let N_1 be the number of integers t_i such that $t_i \leq B^\beta + 1$, and N_2 the number of t_i such that $B^\beta < t_i \leq B$. Write $N_2 = sm + m_0$ with $s \geq 0$ and $0 \leq m_0 < m$. By the theorem one verifies immediately, using a crude estimate, that

$$B^\beta + sB^{\beta\lambda} \leq B,$$

whence $s \leq B^{1-\beta\lambda}$ and thus

$$N(B) \leq N_1 + N_2 \leq B^\beta + 1 + m_0 + B^{1-\beta\lambda}.$$

From this one immediately deduces the existence of our number α.

Corollary 2.3. *Let U be a Hilbert subset of $\mathbf{Q}$. Then there exists a number α, with $0 < \alpha < 1$, such that the number of positive integers $\leq B$ lying in U is at least equal to*

$$B - B^\alpha$$

for all B sufficiently large.

Proof. Apply Proposition 1.1 and Corollary 2.2.

Corollary 2.4. *A Hilbert subset of $\mathbf{Q}$ contains infinitely many prime numbers.*

Proof. Apply the prime number theorem.

Corollary 2.5. *A Hilbert subset of $\mathbf{Q}$ is dense for the ordinary topology and every p-adic topology on $\mathbf{Q}$.*

Proof. If $f(t, X)$ is irreducible over $\mathbf{Q}(t)$, then so is $f(a + 1/t, X)$. Integral values of t tending to infinity and making the specialized polynomial irreducible will be such that $a + 1/t$ is close to a for the ordinary topology. For the p-adic topology, one takes $f(a + tp^v, X)$ with v large.

In order to prove Theorem 2.1, we need a mean value theorem due to H. A. Schwarz.

Lemma 2.6. *Let $\varphi(t)$ be a function m times continuously differentiable in the interval $t_i \leqq t \leqq t_{i+m}$. We suppose that $t_i < t_{i+1} < \cdots < t_{i+m}$ are real numbers (not necessarily in $\mathbf{Z}$). Then there exists a number τ with $t_i < \tau < t_{i+m}$ such that*

$$\frac{\varphi^{(m)}(\tau)}{m!} = \frac{\begin{vmatrix} 1 & t_i & t_i^2 & \cdots & t_i^{m-1} & \varphi(t_i) \\ & \cdots & & \cdots & & \cdots \\ 1 & t_{i+m} & t_{i+m}^2 & \cdots & t_{i+m}^{m-1} & \varphi(t_{i+m}) \end{vmatrix}}{V_m},$$

where V_m is the Vandermonde determinant.

(We see that the numerator differs from V_m only by the last column).

Proof. I owe this proof to Walter Strodt. Let $F(t)$ be the function

$$F(t) = \begin{vmatrix} 1 & t_i & t_i^2 & & t_i^{m-1} & \varphi(t_i) \\ \cdots & \cdots & & \cdots & & \cdots \\ 1 & t_{i+m-1} & t_{i+m-1}^2 & \cdots & & \varphi(t_{i+m-1}) \\ 1 & t & t^2 & & t^{m-1} & \varphi(t) \end{vmatrix}.$$

Then $F(t)$ is 0 when t is equal to the m points $t_i,\ldots,t_{i+m-1}$. For some constant C the function

$$G(t) = F(t) - C(t - t_i)(t - t_{i+1})\cdots(t - t_{i+m-1})$$

will also vanish at t_{i+m}, that is to say at $m + 1$ points. Hence $G^{(m)}(t)$ vanishes at least at one point τ between t_i and t_{i+m}. But $G^{(m)}(t) = F^{(m)}(t) - m!C$ (all the other terms will cancel). From this we conclude that

$$F^{(m)}(\tau) = m!C.$$

One sees immediately that $F^{(m)}(t)$ will have a 0 everywhere in the bottom line of its matrix, except for the last term, which will be $\varphi^{(m)}(t)$. Hence

$$F^{(m)}(\tau) = \varphi^{(m)}(\tau)V_{m-1},$$

where V_{m-1} is the small Vandermonde determinant. Since

$$C = \frac{F(t_{i+m})}{(t_{i+m} - t_i)\cdots(t_{i+m} - t_{i+m-1})}$$

and since $F(t_{i+m})$ is none other than the numerator of the expression in the statement of the lemma, we see that our lemma is proved, taking into account the fact that the product of V_{m-1} and the denominator of C is precisely equal to V_m.

Let us apply our lemma, taking for m an integer such that $\varphi^{(m)}(t)$ has no negative powers of $1/t^{1/e}$, hence its development is of the type

$$\varphi^{(m)}(t) = \xi \frac{1}{t^\lambda} + \cdots$$

with ξ real. Since φ is not a polynomial in t, its derivative $\varphi^{(m)}$ is not identically 0, and so we may assume $\xi \neq 0$ and $\lambda > 0$. This term dominates the series for t large. Hence for sufficiently large i, we have $\varphi^{(m)}(\tau) \neq 0$. Observe that the determinant in the numerator of the expression in the statement of the lemma is an integer. We have just seen that $\varphi^{(m)}(\tau)$, and a fortiori $\varphi^{(m)}(\tau)/m!$ is small, of order of magnitude $\tau^{-\lambda}$. This implies that V_m is large. But V_m is a product of differences of the t_j, and we can replace each such difference by $t_{i+m} - t_i$. This makes V_m only larger. There will be

$$m(m + 1)/2$$

such terms. We find therefore

$$t_{i+m} - t_i > t_i^{\lambda'}$$

with a suitable λ', taking an $m(m + 1)/2$-root.

§3. Reduction Steps

We study systematically the following reductions of the Hilbert polynomials to the case of one t, several X, several t and several X, and finite separable extensions.

a. One t, Several X

Let k be a field, and $f(X_1,\ldots,X_s)$ a polynomial with coefficients in k, of degree $< d$ in each X_i. We denote the set of such polynomials by $P(s, d, k)$. **Kronecker's specialization** S_d transforms f into a polynomial in 1 variable, namely

$$S_d f(Y) = f(Y, Y^d, \ldots, Y^{d^{s-1}}).$$

If we have a monomial $X_1^{i_1} \cdots X_s^{i_s}$, then S_d applied to this monomial yields

$$Y^{i_1 + i_2 d + \cdots + i_{s-1} d^{s-1}}.$$

Taking into account the d-adic expansion of an integer ≥ 0, we see S_d is a bijection of $P(s, d, k)$ on the polynomials in $k[Y]$ of degree $\leq d^s - 1$. Furthermore, if f, g and fg are in $P(s, d, k)$ then

$$S_d(fg) = S_d(f)S_d(g).$$

Let $f \in P(s, d, k)$ and suppose that $S_d f$ is reducible:

$$S_d f(Y) = G(Y)H(Y).$$

Then the degree of G and H is $\leq d^s - 1$ and hence $G = S_d g$ and $H = S_d h$ for some g, h uniquely determined by G, H respectively. From this we deduce **Kronecker's criterion:**

f is irreducible in k if and only if for every factorization $S_d f = GH$ with $G = S_d g$ and $H = S_d h$, the polynomial gh contains a monomial

$$\varphi X_1^{i_1} \cdots X_s^{i_s}$$

with $\varphi \in k$, $\varphi \neq 0$, of degree $\geq d$ for one of the X_i.

Proposition 3.1. Let k be a field, $f(t, X_1, \ldots, X_s)$ a polynomial with co-efficients in k, of degree $< d$ in each X_i, and let S_d be Kronecker's substitution. Let

$$S_d f = \prod g_j(t, Y)$$

be the factorization of $S_d f$ in $k(t)[Y]$ into irreducible factors in Y. Then, except for a finite number of values of t_0 in k, every t_0 in k such that each $g_j(t_0, Y)$ is irreducible in $k[Y]$ is also such that $f(t_0, X_1, \ldots, X_n) = f_0$ is irreducible in $k[X]$.

Proof. Except for a finite number of t_0, we have

$$S_d f_0 = \prod g_j(t_0, Y)$$

a factorization of $S_d f_0$ into irreducible factors in $k[Y]$. Excluding still another finite set of t_0 (those for which some coefficient $\varphi(t)$ in a monomial in Kronecker's criterion will vanish), it is clear that f_0 will be irreducible.

In particular, we see that every Hilbert subset of k determined by a polynomial $f(t, X_1, \ldots, X_s)$ contains a Hilbert set determined by a finite number of polynomials $g_j(t, Y)$ with a single Y.

b. Several t, Several X

We consider a polynomial $f(t_1, \ldots, t_r, X_1, \ldots, X_s)$ as a polynomial in t_1 and $r + s - 1$ variables $t_2, \ldots, t_r, X_1, \ldots, X_s$. We specialize t_1 to preserve irreducibility, and continue inductively. Thus on the Hilbert set $U_{f, k}$ in S_k^r we get a fibering, and we can determine its points by successive consideration of the preceding case.

c. Finite Separable Extensions

We shall prove that if k is Hilbertian, then any finite separable extension of k is also Hilbertian. We prove in fact a stronger result.

Lemma 3.2. *Let K be a finite Galois extension of a field k. Let*

$$f(t, X) \in K(t)[X]$$

be irreducible over $K(t)$. Assume that the leading coefficient of f in X is 1, and that if σ ranges over the distinct isomorphisms of K over k, then the conjugates f^σ of f are distinct. Put

$$F(t, X) = \prod_\sigma f^\sigma(t, X).$$

Then $F(t, X)$ is in $k(t)[X]$ and is irreducible. Excluding a finite number of $t_0 \in k$, $F(t_0, X)$ is irreducible over k if and only if $f(t_0, X)$ is irreducible over K.

Proof. By Galois theory, $F(t, X)$ is in $k(t)[X]$. If we had

$$F(t, X) = G(X)H(X),$$

a factorization over $k(t)$, then we could assume G and H have leading coefficient 1. We conclude that $f(t, X)$ divides G or H over $K(t)[X]$, say G. But then f^σ also divides G for each σ. As the f^σ are supposed to be distinct, their product divides G, a contradiction unless $F = G$. We now choose t_0 in k such that $F(t_0, X)$ is irreducible over k. Then it is clear that $f(t_0, X)$ will be irreducible over E, taking into account unique factorization in $K[X]$.

To apply the lemma even to the case where the conjugates f^σ might not be distinct, we make a simple transformation on f. Indeed, there exists an element $\varphi = \varphi(t) \in K(t)$ such that if we put

$$g(t, X) = f(t, X + \varphi)$$

all the g^σ are distinct. For instance, the constant term of g is $f(t, \varphi)$ and the existence of the desired φ is obvious. We can apply Lemma 3.2 to g, whose irreducibility over $K(t)$ is equivalent to that of $f(t, X)$.

Proposition 3.3. *Let E be a finite separable extension of a field k. Then a Hilbert subset of E contains a Hilbert subset of k.*

Proof. Lemma 3.2 and the above remarks prove the proposition when E is Galois over k. In general, let K be the Galois closure of E over k and let $f(t, X)$ be irreducible over $E(t)[X]$. Factor

$$f(t, X) = g_1(t, X) \cdots g_r(t, X) = \prod_\tau g^\tau(t, X),$$

where $g_i(t, X)$ is irreducible in $K(t)[X]$ for $i = 1, \dots, r$; and has leading coefficient 1, degree ≥ 1. The product over τ is taken over only some of the automorphisms of K over E. By the Galois case, there exists a Hilbert subset U of k such that for $t_0 \in U$ each polynomial $g_i(t_0, X)$ is irreducible over K, and omitting a finite number of elements of U we can achieve that $g_1(t_0, X), \dots, g_r(t_0, X)$ are distinct. In other words, the polynomials $g^\tau(t_0, X)$ are distinct, their product lies in $E(t)[X]$, and is equal to $f(t_0, X)$. Hence $f(t_0, X)$ is irreducible in E, thereby proving the proposition.

For a treatment of the purely inseparable case the reader may consult [In], where the analogous result is proved.

§4. Function Fields

In dealing with function fields, one is led by induction to the cases of purely transcendental extensions and finite extensions. Concerning the first, we show that Hilbert sets contain an abundance of elements.

Proposition 4.1. *Let k be an infinite field and w transcendental over k. Let t be transcendental over $k(w)$, and y separable over $k(w, t)$ but not in $k(w, t)$. Let U be the Hilbert subset of $k(w)$ determined by (t, y) as in §1. Let m be an integer ≥ 1. In the (w, t) plane, there exists a non-empty Zariski open set V such that for any $\lambda_0 \in k$ (except possibly a finite number depending on U, m, V) all the values t_0 of t of type*

$$t_0 = \tau_0 + \lambda_0(w - w_0)^m$$

with $\tau_0, w_0 \in k$ and $(\tau_0, w_0) \in V$ are in U.

Proof. After multiplying y by a polynomial in $k[w, t]$ we may assume that y is integral over $k[w, t]$. We consider the surface with generic point (w, t, y) over k, which can be viewed as a covering of the plane (w, t).

As we assume y separable over $k(w, t)$, there exists a relation

$$\varphi(w, t) = P(w, t, Y)f(w, t, Y) - Q(w, t, Y)f'(w, t, Y),$$

where $f(w, t, Y)$ is the irreducible polynomial of y over $k[w, t]$, f' its derivative with respect to Y, and P, Q are in $k[w, t, Y]$ while $\varphi \in k[w, t]$, $\varphi \neq 0$. Consequently, if (w_0, τ_0, y_0) is a point of the surface in k such that $\varphi(w_0, \tau_0) \neq 0$, then y has an expansion in formal power series:

$$y = \sum c_{ij}(t - \tau_0)^i(w - w_0)^j$$

with $c_{ij} \in k$. (Geometrically speaking, the covering is unramified.)

The zeros of $\varphi(w, t)$ form a closed Zariski set, whose complement is the open set V. From now on, we assume that (w_0, τ_0) is in V and $w_0, \tau_0 \in k$.

If there does not exist a point (w_0, τ_0, y_0) on the surface in k above (w_0, τ_0), then for all $\lambda_0 \in k$, there is no point on the (t, y)-curve in $k(w)$ above

$$\tau_0 + \lambda_0(w - w_0)^m.$$

Indeed, such a point would have polynomials in $k[w]$ as coordinates, and the homomorphism of this ring mapping w on w_0 would then give a point of the surface in k above (w_0, τ_0).

Suppose, on the other hand, that (w_0, t_0, y_0) is a point of the surface in k. If we substitute

$$\lambda(w - w_0)^m$$

for $(t - \tau_0)$ in the power series, we obtain a homomorphism of $k[w, t, y]$ into the power series ring $k[[w - w_0]]$. We have to choose λ in such a way that y does not map into $k[w - w_0] = k[w]$ (it will in any case be integral over $k[w]$). In other words, the series must not stop.

Since y does not lie in $k[w, t]$, infinitely many c_{ij} are not equal to 0. We can write

$$y_{\lambda, m} = \sum_{i, j} c_{ij}\lambda^i(w - w_0)^{mi + j}$$
$$= \sum_\nu c_\nu(\lambda)(w - w_0)^\nu.$$

Each $c_\nu(\lambda)$ is a polynomial in λ.

If, for some value λ_0 of λ in k, all but a finite number of $c_\nu(\lambda_0)$ are 0, then $y_{\lambda_0, m}$ considered as a rational function of w has a pole at infinity ($w = \infty$) of order equal to the largest ν such that $c_\nu(\lambda_0) \neq 0$. But we have

$$f(w, \tau_0 + \lambda_0(w - w_0)^m, y_{\lambda_0, m}) = 0.$$

Since m is fixed, still excluding a finite number of λ_0, we see that the order of a pole of $y_{\lambda_0, m}$ at infinity is bounded (as a function of the coefficients of f, which is fixed throughout). If we choose v_1 larger than this critical value and such that $c_{v_1}(\lambda_0) \neq 0$ (which is possible since infinitely many $c_{ij} \neq 0$) we get a $y_{\lambda_0, m}$ whose series cannot stop, because if it did, we would get a bad pole of $y_{\lambda_0, m}$. This concludes the proof.

In case y is not separable over $k(w, t)$, then either y^{p^r} is not in $k(w, t)$ for any r, in which case it becomes separable for r large enough and we can apply our proposition to it; or y^{p^r} does lie in $k(w, t)$ for some r, so that we may write

$$y^{p^r} = \varphi(w, t)$$

with $\varphi(w, t) \in k[w, t]$ and we may assume that $\varphi(w, t)$ is not a p-th power in $k[w, t]$. We distinguish two cases: the first, where there occurs some power of t or w in φ which is not divisible by p, and the second, where p divides the exponent of t and w in each term of φ. In the second case, some coefficient of φ in k is not a p-th power.

In the first case, take m greater than the degree of φ. Then setting $t = \lambda_0 w^m$ for any $\lambda_0 \in k$, $\lambda_0 \neq 0$, the polynomial

$$\varphi(w, \lambda_0 w^m)$$

will have the same set of non-zero coefficients as φ (this is the Kronecker substitution). If

$$\varphi(w, t) = \sum c_{ij} w^i t^j,$$

then

$$\varphi(w, \lambda_0 w^m) = \sum c_{ij} \lambda_0^j w^{i + mj}.$$

If one of the $e_{ij} \neq 0$, and p does not divide i or j, then we simply pick m so that p does not divide $i + mj$.

The same argument could be applied after making an arbitrary translation (w_0, τ_0) on (w, t).

In the second case, when p divides each i and j, then we can choose any m greater than the degree of φ and again any $\lambda_0 \in k$. If c_{ij} is not a p-th power in k, then $c_{ij} \lambda_0^j$ will also not be a p-th power and hence $\varphi(w, \lambda_0 w^m)$ will not be a p-th power in $k[w]$.

This takes care of the remaining case in our treatment of polynomial rings over finite fields. Thus, up to a purely inseparable extension (left open in our third reduction step) we have proved the following theorem.

Theorem 4.2. *Let R_0 be a ring, and R a finitely generated ring over R_0. Let k_0 and k be their quotient fields respectively. Then if $R_0 = \mathbf{Z}$, or if the transcendence degree of k over k_0 is $\geqq 1$, R is Hilbertian.*

Actually if R_0 is a finite field, k is separable over k_0, and hence our proof in this case is complete.

§5. Abstract Definition of Hilbert Sets

It will be convenient to reformulate the definition of Hilbert sets in the context of general rings. As usual, rings are without zero-divisors.

Let R be a ring. As we know, $\operatorname{spec}(R)$ has the Zariski topology, a basis for the open sets being the subsets $\operatorname{spec}(R_\xi)$ with $\xi \in R$, $\xi \neq 0$, and $\operatorname{spec}(R_\xi)$ is identified with the subset of prime ideals $\mathfrak{p}$ of $\operatorname{spec}(R)$ which do not contain ξ. These $\operatorname{spec}(R_\xi)$ will be called **basic Zariski sets**

Let S be a finitely generated ring over R. We shall say that it is **algebraic** over R if all its elements are algebraic over the quotient field $K(R)$ of R. Its quotient field $K(S)$ is then a finite extension of $K(R)$. Assume that S is such a ring. The **basic Hilbert set associated with** S is the subset U_S, or $U_{S/R}$, of $\operatorname{spec}(R)$, consisting of those $\mathfrak{p}$ having the following two properties:

(i)   There is exactly one point $\mathfrak{P}$ in $\operatorname{spec}(S)$ lying above $\mathfrak{p}$.
(ii) If we denote by $k(\mathfrak{p})$ and $k(\mathfrak{P})$ the residue class fields of these points, then

$$[k(\mathfrak{P}) : k(\mathfrak{p})] = [K(S) : K(R)].$$

By a **Hilbert subset** of $\operatorname{spec}(R)$ we shall mean the intersection of a finite number of basic Zariski sets with a finite number of basic Hilbert sets.

If S is of the above type, then for some $\xi \in R$, $\xi \neq 0$, S is integral over $R_\xi = R[1/\xi]$, and thus above each $\mathfrak{p} \in \operatorname{spec}(R_\xi)$ there exists at least one $\mathfrak{P}$ in $\operatorname{spec}(S)$. As we are interested only in intersections of subsets of $\operatorname{spec}(R)$ with basic Zariski sets, we may in practice assume that S is *finite* over R, i.e. integral and finitely generated over R.

Let us recall some facts from commutative algebra. If R is integrally closed, and S integral over R, and if $\mathfrak{p} \in \operatorname{spec}(R)$ and $\mathfrak{P} \in \operatorname{spec}(S)$ lies above $\mathfrak{p}$, then the separable degree

$$[k(\mathfrak{P}) : k(\mathfrak{p})]_s$$

is at most equal to $[K(S) : K(R)]$. There are cases where this does not hold true for the degree. However, if R is a *regular* ring (i.e. one for which each local ring $R_\mathfrak{p}$ is regular) then we do have

$$[k(\mathfrak{P}) : k(\mathfrak{p})] \leqq [K(S) : K(R)].$$

(This could be proved for instance by "blowing up" the given point p, i.e. taking a suitable discrete valuation ring $\mathfrak{o}$ of K, whose maximal ideal $\mathfrak{m}$ restricts to p, and whose residue class field $\mathfrak{o}/\mathfrak{m}$ is purely transcendental over $k(p)$. The point $\mathfrak{P}$ in $\mathrm{spec}(S)$ would lift to a point in the integral closure of $\mathfrak{o}$ in $K(S)$, and we could apply the known theorem for valuation rings.)

In practice, when we deal with rings R, they will be parameter rings, that is, we deal with schemes over R, and there will always exist an element $\xi \in R$, $\xi \neq 0$, such that R_ξ is regular (the simple points on a variety form a Zariski open set).

In order to ensure the validity of the conclusion of the next proposition, we see from our preceding discussion that we must make some restrictions on our rings. However, with such assumptions, guaranteeing the degree inequality, the proof is obvious.

Proposition 5.1. *Let $S' \supset S \supset R$ be three rings, with S, S' finite over R, and R, S integrally closed. Let $p \in U_{S'/R}$, and let $\mathfrak{P}$, $\mathfrak{P}'$ be the points lying above it in $\mathrm{spec}(S)$ and $\mathrm{spec}(S')$ respectively. Assume that S_p and R_p are regular, or that $k(\mathfrak{P}')$ is a separable extension of $k(p)$. Then $\mathfrak{P}$ is in $U_{S'/S}$, and p is in $U_{S/R}$.*

In our applications, $R = k[t_1, \ldots, t_r]$ is a finitely generated ring over a field k, with $t_1, \ldots, t_r$ algebraically independent over k. If, as in the proposition, we are concerned with a finite extension S' of a ring S, then using the Noether normalization theorem, one can find a polynomial ring R so that the search of points in $U_{S'/S}$ is reduced to that of points in $U_{S'/R}$.

Over the ring $k[t_1, \ldots, t_r]$ we can relate the sets U_f of §1 to our sets $U_{S/R}$ of the previous discussion:

Proposition 5.2. *Let k be a field, $k[t_1, \ldots, t_r]$ a purely transcendental ring extension R of k. Let S be finite over R. Then there exists an irreducible polynomial $f(t_1, \ldots, t_r, X)$ such that $U_{S/R}$ contains $U_{f,k} \cap Z$ where Z is a basic Zariski set of $\mathrm{spec}(R)$.*

(We have identified the point in $\mathrm{spec}(R)$ determined by values (t') of (t) in k.)

Proof. Without loss of generality, we may assume that k is infinite. Say $S = R[x_1, \ldots, x_n]$. Put

$$y = c_1 x_1 + \cdots + c_n x_n,$$

where the c_i are elements of k such that if $K = K(R)$ then

$$K(x_1, \ldots, x_n) = K(y).$$

Such elements can be found by the usual arguments used to prove the primitive element theorem of Galois theory. Then spec(S) and spec($R[y]$) contain a common Zariski basic set, and from this, taking for f an irreducible equation of y over R, our assertion is obvious.

Our rings S over R may be viewed as defining algebraic systems of 0-cycles, whose generic members are relatively irreducible over the quotient field of R. The Hilbert sets then give points of the parameter scheme over which special members of the family remain relatively irreducible. It is an easy matter (taking generic projections, as with Chow forms) to extend the definition to algebraic systems of arbitrary dimension, i.e. to assume merely that we are dealing with a scheme S over spec(R) which is of finite type, without divisors of zero, and whose generic member is irreducible as a scheme. One would see that the set of $\mathfrak{p} \in$ spec(R) for which the localized fiber $S_{\mathfrak{p}}$ is without divisors of zero and is irreducible over $k(\mathfrak{p})$ contains a Hilbert subset of spec(R).

One should not confuse this relative irreducibility with absolute irreducibility. For the convenience of the reader, let us recall here the standard result concerning the preservation of absolute irreducibility.

Proposition 5.3. *Let R be a ring, K its quotient field, and $f(X_1, \ldots, X_s) \neq 0$ a polynomial in $R[X]$ which is irreducible in $K^a[X]$. Then there exists $\xi \in R$, $\xi \neq 0$ such that for any $\mathfrak{p} \notin$ spec(R_ξ) the reduced polynomial $f_{\mathfrak{p}}(X)$ is irreducible in $k(\mathfrak{p})^a[X]$.*

Proof. Let f be of degree d. Consider two integers $m, n \geq 1$ such that $m + n = d$. Let M_α (resp. M_β) range over all monomials of degree $\leq m$ (resp. n) and put

$$g_u = \sum u_\alpha M_\alpha(X),$$
$$h_v = \sum v_\beta M_\beta(X),$$

where the u_α, v_β are a set of algebraically independent quantities over K. Then

$$g_u h_v = \sum \varphi_\lambda(u, v) M_\lambda(X),$$

where the $\varphi(u, v)$ are polynomials with integer coefficients and are "universal." We can write

$$f(X) = \sum c_\lambda M_\lambda(X), \qquad c_\lambda \in R.$$

By assumption, and Hilbert's nullstellensatz, the ideal generated by the $\varphi_\lambda(u, v) - c_\lambda$ in $K[u, v]$ is the unit ideal, and 1 has a finite expression in terms of these generators. Multiplying the coefficients by some element of R, we get

$$\xi = \sum \psi_\lambda [\varphi_\lambda(u, v) - c_\lambda]$$

with $\xi \in R$, $\xi \neq 0$, and $\psi_\lambda \in R[u, v]$.

For each m, n as above, we get such a relation, and we should actually index ξ by m, n, i.e. write $\xi_{m,n}$. Taking the product of these $\xi_{m,n}$ for each pair m, n together with the non-zero coefficients of f, we get an element ξ of R which will obviously satisfy our requirements.

§6. Applications to Commutative Group Varieties

Let R be an integrally closed ring, and A a commutative group variety scheme over spec(R), *or over R as we shall say. We assume that it is simple over R* (this amounts to the classical terminology of non-degeneracy). The only properties of such schemes which we shall use are the following:

If $S_1 \subset S_2$ are two rings containing R, and if $A(S_1)$, $A(S_2)$ denote the points of A in S_1 and S_2, respectively, then the natural map

$$A(S_1) \to A(S_2)$$

is an injection. We identify $A(S_1)$ in $A(S_2)$.

For each $\mathfrak{p} \in$ spec(R), the fiber $A_\mathfrak{p}$ over $\mathfrak{p}$ is a commutative group variety scheme over the residue class field $k(\mathfrak{p})$. If $x \in A(R)$ and $x(\mathfrak{p})$ denotes the corresponding point of $A_\mathfrak{p}$ in $k(\mathfrak{p})$, then the map

$$x \mapsto x(\mathfrak{p})$$

is a homomorphism of $A(R)$ into $A_\mathfrak{p}(k(\mathfrak{p}))$.

If $S \supset R$, $\mathfrak{P} \in$ spec(S), and m an integer > 1, prime to the characteristic of $k(\mathfrak{P})$, then the map $x \mapsto x(\mathfrak{P})$ defines an injection on the subgroup of points of $A(S)$ consisting of the points of period m.

If K is the quotient field of R, E a finite extension of K, and $a \in A(E)$, then there exists a subring S of E, finitely generated over R, such that $a \in A(S)$.

Given a point c' of period m in $A_\mathfrak{p}(k(\mathfrak{p}))$, there exists a point c of period m in some $A(S)$ for S finitely generated and algebraic over R such that $c' = c(\mathfrak{P})$ for some $\mathfrak{P} \in$ spec(S) lying above $\mathfrak{p}$.

If $b \in A(E)$ for some extension E of K, then there exists a smallest field $K(b)$ of rationality for b containing K. If $b \in A(S)$, where $R \subset S \subset K(b)$ and $\mathfrak{P} \in$ spec(S) lies above $\mathfrak{p}$, then $k(\mathfrak{P}) = k(\mathfrak{p})(b(\mathfrak{P}))$.

Lemma 6.1. *Let R be integrally closed and A a commutative group variety scheme over R. Let $a \in A(R)$ and let m be an integer > 0 prime to the* *characteristic of K. Suppose that $a \notin mA(K)$. Then there exists a ring S, finitely generated and algebraic over R, such that for all $\mathfrak{p} \in U_{S/R}$, the point $a(\mathfrak{p})$ does not lie in $mA_\mathfrak{p}(k(\mathfrak{p}))$.*

Proof. Since $a \notin mA(K)$, the field $K(b)$, for any point $b \in A(K^a)$ such that $mb = a$, has degree > 1 over K. We may assume that all these points lie in $A(S)$ where S is finitely generated and algebraic over R. Replacing R by $R[1/\xi]$ for a suitable $\xi \neq 0$ in R, we may assume that S is finite over R. We also adjoin $1/m$ to R. We see then that for $\mathfrak{p} \in U_{S/R}$ and $\mathfrak{P} \in \operatorname{spec}(S)$ lying above $\mathfrak{p}$, the extension

$$k(\mathfrak{p})(b(\mathfrak{P})) \quad \text{of} \quad k(\mathfrak{p})$$

has degree > 1. By hypothesis, the map $b \mapsto b(\mathfrak{P})$ induces a bijection of the m-th roots of a onto the m-th roots of $a(\mathfrak{p})$. Hence $a(\mathfrak{p})$ cannot lie in $mA_\mathfrak{p}(k(\mathfrak{p}))$.

If Γ is an abelian group, we denote by Γ_m the subgroup of points of period m.

Theorem 6.2. *Let R be an integrally closed ring, and A a commutative group variety scheme, simple over R. Let m be an integer > 1, prime to the characteristic of the quotient field K of R. Let Γ be a finitely generated subgroup of $A(R)$. Assume:*

(i) *The period of each torsion element of Γ divides m.*
(ii) *The canonical map $\Gamma'/m\Gamma' \to A(K)/mA(K)$ is an injection (denoting by Γ' the group $\Gamma + A(K)_m$).*

Then there exists a Hilbert subset U of $\operatorname{spec}(R)$ such that for $\mathfrak{p} \in U$, the map

$$x \mapsto x(\mathfrak{p})$$

induces an injection on Γ.

Proof. Extending R by a finite number of elements of type $1/\xi$ with $\xi \neq 0$ in R, we may assume that $1/m \in R$ and that all points in $A(K)$ of period m lie in $A(R)$. We may then assume, without loss of generality, that these points are contained in Γ' (considering Γ' instead of Γ).

We can write Γ as a direct sum of a free abelian group B and a finite torsion group C. Let $x_1, \ldots, x_n$ be a basis for B over $\mathbf{Z}$. Let $\{a_j\}$ be a set of representatives in B of the factor group B/mB, whose coordinates in $\mathbf{Z}$ are the integers α such that $0 \leqq \alpha \leqq m - 1$ or $0 \leqq -\alpha \leqq m - 1$. We consider the finite set of points $a_j + c$, $c \in C$. By assumption, we can determine for each such point a Hilbert set as in the lemma. Further, we can find a ring S finitely generated, algebraic over R, such that every point $z \in A(K^a)$ with $mz = 0$ is in $A(S)$. This condition implies in particular that if $\mathfrak{p} \in U_{S/R}$ and $K(z) \neq K$ then $z(\mathfrak{p})$ is not rational over $k(\mathfrak{p})$. We let U be the intersection of the finite number of Hilbert sets just mentioned.

Let $\mathfrak{p} \in U$ and let $x \in \Gamma$ be such that $x(\mathfrak{p}) = 0$. To show that $x = 0$, we may assume that $x \in B$ and that $x(\mathfrak{p})$ has period m and derive a contradiction. By construction, $x(\mathfrak{p})$ cannot be equal to $z(\mathfrak{p})$, where z is a point of period m in $A(K^a)$ and not in $A(K)$. Hence it must be of type $c(\mathfrak{p})$ with $c \in C$. Consequently,

$$x(\mathfrak{p}) - c(\mathfrak{p}) = 0.$$

For some a_j and some $y \in B$, we have

$$x = a_j + my$$

and hence

$$0 = a_j(\mathfrak{p}) - c(\mathfrak{p}) + my(\mathfrak{p}).$$

By construction, $a_j - c = 0$, and hence $y(\mathfrak{p})$ has period m.

If we write $x = \lambda_1 x_1 + \cdots + \lambda_n x_n$ with $\lambda_i \in \mathbf{Z}$, then we can express y in terms of our basis $x_1, \ldots, x_n$ in the form

$$y = \mu_1 x_1 + \cdots + \mu_n x_n,$$

where $m\mu_i = \lambda_i - \alpha_i$, the α_i being the coordinates of a_i with respect to our basis. If $x \neq 0$, then repeating the procedure a finite number of times, we get $a_j(\mathfrak{p}) + c(\mathfrak{p}) = 0$ for some a_j and some $c \in C$, which gives us our contradiction.

Corollary 6.3. *Let A_0 be an abelian variety defined over a field K, of finite type over $\mathbf{Q}$. Then there exists an integrally closed ring R of finite type over $\mathbf{Z}$ having K as quotient field, and an abelian variety scheme A over R such that A_0 is the extension of A to the ground field K, and*

$$A(R) = A(K).$$

There exist infinitely many prime ideals $\mathfrak{p}$ of R such that the residue class field $k(\mathfrak{p})$ is finite over $\mathbf{Q}$, and that the map

$$x \mapsto x(\mathfrak{p})$$

of $A(R)$ into $A_\mathfrak{p}(k(\mathfrak{p}))$ is an injection on $A(R)$. The above assertions remain true if A_0 is a complete non-singular curve of genus ≥ 1, and A is a scheme of such curves.

Proof. The first assertion is standard, and we can also make R integral over a ring $I[t_1, \ldots, t_r]$ where I is a ring generated over $\mathbf{Z}$ by $1/p_1, \ldots, 1/p_m$ for a finite number of primes p_i, and $t_1, \ldots, t_r$ are algebraically independent over **Q** (using the Noether normalization theorem or Proposition 5.1 of Chapter 2). We can now use Propositions 5.1 and 5.2, the latter to apply Hilbert's irreducibility theorem. The generalized Mordell–Weil theorem (Theorem 1 of Chapter 6) gives us the finite generation of $A(K)$, and consequently we can also apply Theorem 6.2, in the case of abelian varieties.

In the case of curves of genus ≥ 1, we select R large enough so that a canonical map of the curve into its Jacobian is defined over R, and that the family of such maps is non-degenerate. Then the points of the curve may be viewed as a subset of the points of its Jacobian, and our assertion is reduced to the case of abelian varieties.

By means of this corollary, one can reduce certain qualitative assertions concerning points of curves over finitely generated rings and fields over **Q** to assertions about points on curves in number fields. For instance, the finiteness of integral points proved in Chapter 8, for fields of finite type, is thus seen to be implied by the finiteness of integral points over finitely generated subrings of number fields over **Z**, *provided* one has Theorem 1 of Chapter 6, that is the Mordell–Weil Theorem for fields of finite type. In a like manner, the Mordell conjecture could be so reduced, if one could prove it in number fields.

Notes and Comments

The proof of Hilbert's theorem over the rationals is due to Dörge [Do], from which all of §2 has been more or less copied. The reduction steps occur in Hilbert's original paper [Hi].

Siegel [Sie 1] had already observed the effect of his theorem on Hilbert's. The idea of counting rational points on curves of genus ≥ 1 and finding that their number is much smaller than the number of points on curves of genus 0 is due to Néron [Ne 1].

The proof of the function field case is due to Franz [Fr].

The theorem of §6 concerning the possibility of specializing a finitely generated subgroup of an abelian variety injectively is due to Néron [Ne 1]. By this method, he gives examples of elliptic curves having non-trivial groups of rational points over certain number fields. It is unknown whether one can construct elliptic curves over the rationals such that the group of rational points has arbitrarily high rank.

The question for elliptic curves over a rational function field was already raised in *Diophantine Geometry*. Lapin [Lap] over an algebraically closed constant field, and Shafarevitch–Tate [Sh–T] over a finite field, showed that the rank can be arbitrary high in that case, and thereby changed the attitude toward the problem over the rational numbers, leading to the expectation that the rank can also be arbitrarily high for elliptic curves over **Q**.

Silverman's Theorem 6.3 in Chapter 11 (generalizing Demjanenko–Manin) will show that the injectivity of the specialization map actually occurs much more often than the Hilbert irreducibility theorem predicts.

In *Diophantine Geometry*, I raised the question whether Néron's specialization theorem would apply to non-commutative groups, say subgroups of GL_n. As Feit once remarked to me, the group generated by the matrices

$$\begin{pmatrix} 1 & 1 \\ 0 & 1 \end{pmatrix} \quad \text{and} \quad \begin{pmatrix} 1 & t \\ 0 & 1 \end{pmatrix}$$

immediately provides a counterexample. The question should therefore apply only when the group has only semisimple elements, and is finitely presented.

CHAPTER 10

Weil Functions and Néron Divisors

Weil in his thesis [We 2] gave a decomposition theorem which showed how the decomposition of the divisor of a rational function into irreducible components is reflected in the ideal decomposition of the values of this function at points in number fields. He extended his decomposition theorem to the case of arbitrary absolute values, including archimedean ones, in [We 1].

On the other hand, Néron [Ne 3] reformulated Weil's results in the framework of his "quasi-functions" which we have called here Néron divisors, since there is a natural homomorphism from the group of Néron divisors to the group of Cartier divisors. On elliptic curves, Tate made Néron's results more explicit, and formulated them in terms of a Néron function. Cf. [L 7]. I have kept the two notions separate although there is a natural bijection between Néron divisors and Weil functions. The exposition of Néron divisors and Weil functions is based on that of Néron [Ne 3]. The substance of the results of this chapter is due to Weil [We 1].

§1. Bounded Sets and Functions

Throughout this chapter we suppose that K is a field with a proper set of absolute values M_K. For an extension of an absolute value v to the algebraic closure K^a and $a \in K^a$ we define

$$v(a) = -\log|a|_v.$$

If $x = (x_1, \ldots, x_n)$ is a finite family of elements of K^a we denote

$$|x|_v = \max_i |x_i|_v \qquad \text{and} \qquad v(x) = -\log|x|_v = \inf_i v(x_i).$$

We agree that $v(0) = \infty$. If v is a discrete valuation on K, then up to a normalizing factor, $v(a)$ is the order of a at the valuation.

Remark. In first reading, it may be valuable to deal with only one absolute value v on an algebraically closed field, and to omit all references to the set M_K.

For our present purposes, we define an $\mathbf{M}_K$-constant γ to be a real valued function

$$\gamma: M_K \to \mathbf{R}$$

such that $\gamma(v) = 0$ for almost all v in M_K (all but a finite number of v in M_K). If v is an extension of an element v_0 in M_K to the algebraic closure K^a, then we define as before

$$\gamma(v) = \gamma(v_0).$$

Thus γ is extended to a function of $M(K^a)$ into $\mathbf{R}$. We view γ as a family $\{\gamma_v\}$ of constants, parametrized by M_K. The terminology is also adapted for later use, in connection with Néron functions, see below.

In the following we let V be a variety defined over K. All points will be assumed to be in $V(K^a)$ unless otherwise specified. Thus we omit the reference to K^a for simplicity of notation. Similarly, we shall write

$$M = M(K^a)$$

to denote the set of absolute values on K^a extending those in M_K.

Let E be a subset of $V \times M$ (so more accurately, $V(K^a) \times M(K^a)$ if we need to refer to K^a). We shall say that E is **affine M_K-bounded** or more simply **affine bounded**, if there exists a coordinated affine open subset U of V with coordinates $(x_1, \ldots, x_n)$ and an M_K-constant γ such that for all $(x, v) \in E$ we have

$$\max_i |x_i|_v = |x|_v \leqq e^{\gamma(v)}.$$

In additive notation, this is equivalent to

$$v(x) \geqq -\gamma(v)$$

for all (x, v) in E. If there is only one absolute value and K is algebraically closed, this notion coincides with the notion of a bounded set of points on an affine variety. It is useful to build in from the start the uniformity for the bounds, which are such that

$$e^{\gamma(v)} = 1 \quad \text{for almost all } v \in M_K,$$

or equivalently

$$\gamma(v) = 0 \quad \text{for almost all } v \in M_K.$$

A subset E of $V \times M$ is called M_K-**bounded** if it is contained in the finite union of affine M_K-bounded subsets. Unless otherwise specified, we shall always refer to the set M_K, and thus we shall omit reference to M_K in such expressions, referring only to bounded subsets of $V \times M$.

Remark on terminology. In situations like the above, we use M_K-constants γ to give uniform bounds for other functions

$$\lambda: V(K^a) \times M(K^a) \to \mathbf{R}.$$

If $\lambda(x, v)$ is independent of x and if we have $\lambda(x, v) = \gamma(v)$ for an M_K-constant γ, then we call λ a **constant Néron function**. For each v, it is constant on $V(K^a)$. If we had only one absolute value v in $M(K^a)$ this would obviously be the right terminology, and I have found it sufficiently suggestive to preserve it in the general case. The notion of M_K-constant differs from that of M_K-divisor in Chapter 2, §5 in that we make no restriction that $\gamma(v)$ has to be an element of the value group (or its logarithm, in the additive formulation which we are now using).

Let U be a Zariski open subset of V. A subset E of $V \times M$ will be said to be **subordinated** to U if $E \subset U \times M$. We say that E is **subordinated to an open covering** of V if it is subordinated to some element of the covering.

Lemma 1.1. *Let V be a projective variety defined over K. Given an open covering of V, there exists a projective embedding of V over K coordinatized by projective coordinates $(z_0, \dots, z_n)$ such that for each i the complement of the hyperplane $z_i = 0$ is contained in one of the open sets of the covering.*

Proof. One can find divisors $X_0, \dots, X_m$ in a suitably high multiple of the linear system corresponding to the given projective embedding such that the divisors $X_0, \dots, X_m$ have no point in common, and such that the complement of X_i is subordinated to the given covering for each i. Let $Y_0, \dots, Y_s$ be the basis divisors of any projective embedding. Then the divisors $X_i + Y_j$ define a projective embedding by means of homogeneous coordinates $(z_0, \dots, z_n)$ which satisfy our requirement, where

$$n = (m + 1)(s + 1) - 1.$$

Proposition 1.2. *Let V be a projective variety defined over K. Given an open covering of V over K, there exists a finite covering of $V \times M$ by bounded affine subsets which are subordinated to the given covering. In particular, $V \times M$ is bounded.*

Proof. Take a projective embedding as in the lemma. Given a point (P, v) in $V \times M$ we find some index j such that $|z_j(P)|_v$ is maximal. Then the affine coordinates (z_i/z_j) are such that

$$|z_i/z_j|_v \leqq 1.$$

If U_i is the complement of $z_i = 0$ then a point (P, v) with $|z_i/z_j(P)|_v \leqq 1$ lies in a bounded subset of U_i. The union of such bounded subsets covers $V \times M$, as desired.

Remark. We note that we could take 1 for a multiplicative bound.

A function

$$\alpha: V \times M \to \mathbf{R}$$

is called M_K-**bounded from above** if there exists an M_K-constant γ such that for all $(P, v) \in V \times M$ we have

$$\alpha(P, v) \leqq \gamma(v).$$

As before, we omit mention of M_K *and say simply bounded from above.* We define similarly **bounded from below** and **bounded**. We say that α is **locally bounded** if it is bounded on every bounded subset of $V \times M$; and define **locally bounded from above** or **below** similarly. We view a function α as a family of functions

$$\alpha_v: V \to \mathbf{R}$$

for each $v \in M$. In particular, if α is bounded, then $\alpha_v = 0$ for all v not lying above a finite set of elements of M_K.

Proposition 1.3. *Let V be a variety defined over K and let f be a rational function in $K(V)$, morphic on V (that is, f is in the local ring of every point of V). Then the map*

$$(P, v) \mapsto v \cdot f(P)$$

is locally bounded from below. If f is invertible on V, that is f and f^{-1} are rational functions morphic on V, then this map is locally bounded.

Proof. We may assume that V is affine. It follows that f is a polynomial function in the coordinates of V. Then the map

$$(P, v) \mapsto \log |f(P)|_v$$

is bounded from above on every bounded subset of $V \times M$, so the first statement follows from the definitions. The second statement about an invertible f is then a special case of the first, applied to f and f^{-1}.

Let $v \in M$. Then $V(K^a)$ has a topology determined by v, called the v-**topology** if we need to distinguish it from the Zariski topology. It is the smallest topology such that, for any Zariski open set U and a rational function f morphic on U, the function

$$U(K^a) \to \mathbf{R}$$

defined by

$$P \mapsto |f(P)|_v$$

is continuous. We have written this formula with the absolute value sign, to include the value $f(P) = 0$. On the other hand, the function

$$P \mapsto v \cdot f(P) = -\log|f(P)|_v$$

is defined on the complement of the set of zeros of f, and is also continuous for the v-topology.

We note that if Z is a Zariski-closed subset of V, then Z is v-closed for every absolute value v.

Lemma 1.4. *Let Z be a proper Zariski closed subset of V (that is $V(K^a)$). Then the complement of Z is v-dense in V for every absolute value v.*

Proof. We may assume that V is affine, with coordinates $x = (x_1, \ldots, x_n)$. Let z be a point of Z. We have to show that given ε, there exists $x \in V$ not lying in any component of Z such that

$$|x - z|_v < \varepsilon.$$

We do this by an appropriate projection to affine space where the proposition is obvious. Let $r = \dim V$. Let

$$s_i = \sum_{j=1}^{n} a_{ij} x_j \quad \text{for } i = 1, \ldots, r,$$

$$t = \sum_{j=1}^{n} a_{ij} x_j \quad \text{for } i = r + 1,$$

with coefficients $a_{ij} \in K$. For almost all choices of a_{ij} in the sense of the Zariski topology, depending on the given point $z \in Z$, the correspondence

$$x \mapsto (s, t) \mapsto s$$

gives Zariski local isomorphisms in a neighborhood of z,

$$V \xrightarrow{f} T \xrightarrow{g} S$$

from V to a hypersurface T, to affine r-space S. Then $g \circ f(Z)$ is a proper Zariski closed subset of S. We select a point $s \in S$ which is v-close to $g \circ f(z)$ and not in $g \circ f(Z)$. One of the points in $g^{-1}(z)$ is v-close to $f(z)$ by the continuity of the roots of an algebraic equation as functions of the coefficients, because T is a hypersurface, with the coordinate t satisfying an integral polynomial equation over the coordinate ring of S. Since $f: V \to T$ is a Zariski-isomorphism in a neighborhood of z, it then follows that $f^{-1}(g^{-1}(s))$ is v-close to z and does not lie in Z. This proves the proposition.

Let

$$\lambda: V \times M \to \mathbf{R}$$

be a function. We shall say that λ is **M-continuous** (or simply **continuous** if the reference to M is clear) if for each $v \in M$ the function

$$P \mapsto \lambda(P, v) = \lambda_v(P)$$

is continuous on $V(K^a)$.

Proposition 1.5. *Let U be a non-empty open subset of V and let*

$$\lambda: U \times M \to \mathbf{R}$$

be a continuous function. If λ extends to a continuous function on $V \times M$ then this extension is unique.

Proof. Immediate by Lemma 1.4.

§2. Néron Divisors and Weil Functions

Let V be a variety. Consider pairs (U, f) consisting of an open subset U and a rational function f. We say that (U, f) is **compatible** with (U_1, f_1) if $f_1 f^{-1}$ is invertible on $U \cap U_1$, that is both $f_1 f^{-1}$ and $f f_1^{-1}$ are morphisms on $U \cap U_1$. A family $\{(U_i, f_i)\}$ of compatible pairs such that the open sets U_i cover V is said to represent a **Cartier divisor** on V. Two such families are called **compatible** if the pairs in one are compatible to the pairs in the other. A maximal family of compatible pairs covering V is called a **Cartier divisor**. By abuse of language, a covering family of compatible pairs will often be referred to as a Cartier divisor, denoted say by D. If (U, f) is a compatible pair, we also say that D is **represented** by (U, f) on U, or simply by (U, f).

Let D', D'' be Cartier divisors on V. Then there exists a unique Cartier divisor D such that, if D' is represented on an open set U by a pair (U, f) and D'' is represented by a pair (U, g) (with the same open set U) then D is represented on the open set U by (U, fg). This divisor D is denoted by $D' + D''$ and is called the **sum** of D' and D''. The proof of the existence and uniqueness is immediate. Cartier divisors then form a group.

In case V is a non-singular variety, the group of Cartier divisors is isomorphic in a natural way with the group of divisors in the sense that we have used (free abelian group generated by the subvarieties of codimension 1) as follows.

Let

$$\sum n(W)W$$

be a formal sum of such subvarieties. Let P be a point of V. Then the local ring of V at P is a unique factorization ring. Each W is defined locally at P by a principal ideal (f_W) with some element f_W in this local ring. We let

$$f = \prod_{n(W) \neq 0} f_W^{n(W)}.$$

There exists an open subset U of V containing P such that if $n(W) \neq 0$, the restriction of f_W to U has only one component in its divisor $(f_W|U)$, with multiplicity 1. We then obtain a pair (U, f). It is easily verified that the family of such pairs is compatible, and thus defines a Cartier divisor. Thus we have a homomorphism from the group of divisors to the group of Cartier divisors.

Conversely, let D be a Cartier divisor. Let W be a subvariety of codimension 1. Let U be an open subset of V intersecting W in some point P, and take U sufficiently small that D has a representative pair (U, f) with some rational function f. Let $n_W = \text{ord}_W f$. It is easily seen that n_W is independent of the choice of U and f, and that the association

$$D \mapsto \sum n_W W$$

defines a homomorphism from the group of Cartier divisors to the group of divisors. Thus we obtain the desired isomorphism.

We return to the case when V is not assumed to be without singularities.

Suppose that V is defined over a field k as in the last section. We consider triples (U, f, α) where U is open (that is k-open), f is a rational function in $k(V)$, and

$$\alpha: U \times M \to \mathbf{R}$$

is a locally bounded continuous function. We say that (U, f, α) is **compatible** with (U_1, f_1, α_1) if (U, f) is compatible with (U_1, f_1) in the previous sense, and if in addition for every (P, v) with $P \in U \cap U_1$ we have

$$v((ff_1^{-1})(P)) = \alpha_1(P, v) - \alpha(P, v).$$

We also write this

$$v \circ (ff_1^{-1})(P) = \alpha_{1, v}(P) - \alpha_v(P).$$

A family of compatible triples $\{(U_i, f_i, \alpha_i)\}$ such that the open sets U_i cover V is said to **represent a Néron divisor**. A maximal family of compatible triples covering V is called a **Néron divisor**. As with Cartier divisors, a covering family of compatible triples is often referred to as a Néron divisor itself.

To each Néron divisor D one can associate a Cartier divisor simply by eliminating reference to the locally bounded functions α occurring as the third part of the triples.

If D', D'' are Néron divisors, one can define their **sum** $D' + D''$ as the unique Néron divisor such that on an open set U where D' is represented by (U, f, α) and D'' is represented by (U, g, β) then $D' + D''$ is represented by $(U, fg, \alpha + \beta)$. Existence and uniqueness are obvious. The Néron divisors form a group. If D is a Néron divisor, we let (D) be its associated Cartier divisor.

Suppose we have a Cartier divisor D represented in the neighborhood of a point P by (U, f). We shall say that P is in the **support** of D if f is not invertible locally at P, or in other words f is not a unit in the local ring at P. The set of points in the support of D is denoted by $|D|$ or $\operatorname{supp}(D)$. If V is normal, then this definition of the support is compatible with the usual definition of the support of an ordinary divisor (the set of points lying in some component of the divisor). This follows from the fact that the local rings of points are integrally closed, and that if a rational function is not defined at a point, then it has a pole along a subvariety of codimension 1 passing through that point.

Let D be a Néron divisor, represented in the neighborhood of a point P by a triple (U, f, α). Suppose that P is not in the support of (D) (also called the **support of** D). Then we may define a function λ_D at P by the formula

$$\boxed{\lambda_D(P, v) = v \circ f(P) + \alpha(P, v).}$$

In light of the compatibility relation for triples representing a Néron divisor, this value of λ_D is independent of the choice of triple (U, f, α), and λ_D is a continuous real valued function

$$\lambda_D : (V - |D|) \times M \to \mathbf{R}.$$

Conversely, let D be a Cartier divisor. By a **Weil function associated with** D we mean a function

$$\lambda:(V - |D|) \times M \to \mathbf{R}$$

having the following property. Let (U, f) be a pair representing D. Then there exists a locally bounded continuous function

$$\alpha: U \times M \to \mathbf{R}$$

such that for any point P in $U - |D|$ we have

$$\lambda(P, v) = v \circ f(P) + \alpha(P, v).$$

The function α is then uniquely determined by λ and the pair (U, f). In this way, we obtain a triple (U, f, α). It follows immediately from the definitions that two such triples derived from the same Weil function λ are compatible, and hence that the family of all such triples defines a Néron divisor. Thus there is a natural map

$$\text{Weil functions} \to \text{Néron divisors,}$$

and in fact this map is surjective, so the map is a bijection. To see this, it suffices to prove the following statement:

> Let D be a Néron divisor. Let (U, f) represent its Cartier divisor on some open set U. Then there exists a locally bounded continuous function α such that (U, f, α) represents D on U.

Proof. We can cover U by a finite number of open sets U_i $(i = 1, \ldots, m)$ such that the Néron divisor D has representatives (U_i, f_i, α_i). Let $P \in U$. Then P lies in some U_i and for some such i we define

$$\alpha(P, v) = \alpha_i(P, v) - v \circ (ff_i^{-1})(P).$$

The definition of compatibility of triples shows that this definition is independent of the choice of i, and since only a finite number of indices are involved, it follows that α is locally bounded and continuous. The compatibility of the triples (U_i, f_i, α_i) again shows that (U, f, α) is a triple compatible with them, and thus represents the Néron divisor D, as was to be shown.

If λ is a Weil function, we denote by (λ) its Cartier divisor.

Of course, if D', D'' are Néron divisors, we can add the Weil functions

$$\lambda_{D'} + \lambda_{D''}$$

to obtain a real valued function a priori defined on the complement of

$$|D'| \cup |D''|.$$

However, if $\lambda_{D'+D''}$ is the Weil function associated with the sum of the Néron divisors, then it follows from the definitions that $\lambda_{D'+D''}$ restricts to $\lambda_{D'} + \lambda_{D''}$ on the complement of $|D'| \cup |D''|$. Furthermore, we have:

Proposition 2.1. *Let D', D'' be Néron divisors. The function $\lambda_{D'+D''}$ is the unique Weil function whose Cartier divisor is $(D') + (D'')$, and which restricts to $\lambda_{D'} + \lambda_{D''}$ on the complement of $|D'| \cup |D''|$. The association*

$$D \mapsto \lambda_D$$

is a group isomorphism between Néron divisors and Weil functions.

The proof is immediate, since a continuous function on an open set U is uniquely determined by its values outside a proper Zariski-closed subset by Proposition 1.5. Again, it is here important that we are dealing with points in an algebraically closed field, otherwise we would not have enough points to insure this unless other types of assumptions are made.

Proposition 2.2. *Assume that V is projective. Let λ be a Weil function whose Cartier divisor is 0. Then λ is a bounded continuous function. If λ, λ' are Weil functions with the same Cartier divisor, then $\lambda - \lambda'$ is bounded.*

Proof. The first statement is immediate since V can be covered by a finite number of open sets on which λ is bounded continuous in view of Proposition 1.2 and Proposition 1.3. Note that if $(\lambda) = 0$ then λ is defined on all of $V \times M$. The second statement follows by additivity, namely $(\lambda - \lambda') = 0$.

A Weil function with a given divisor is said to be **associated with this divisor**. Under the conditions of Proposition 2.2, it is uniquely determined up to a bounded function. If V is non-singular, then we make no distinction between Cartier divisors and ordinary divisors.

For the converse, we must assume that V is non-singular.

Proposition 2.3. *Assume that V is non-singular. Let λ be a Weil function. Suppose there exists a proper Zariski-closed subset Z of V such that λ is bounded on $(V - |Z|) \times M$. Then $(\lambda) = 0$ and λ is bounded.*

Proof. Let $(\lambda) = D$ and suppose D corresponds to the sum $\sum n_W W$. Suppose $D \neq 0$ and let W be a component of D with multiplicity n_W. Say f has a zero along W so $n_W > 0$. Let w be a point of W which does not lie in any other component of D. There exists a pair (U, f) representing D on an open set U containing w. Then f is defined at w and $f(w) = 0$. By

Proposition 1.4, there exists a point $x \in V$ such that x is v-close to w for a given $v \in M$, but $x \notin |Z| \cup |D|$. Hence $f(x)$ is defined and is v-close to $f(w) = 0$. But then $v \circ f(x)$ is very large, so $\lambda(x)$ is very large, thus proving $D = 0$. That λ is bounded follows by continuity.

Corollary 2.4. *Assume that V is non-singular. Let λ, λ' be Weil functions. If there exists a proper Zariski-closed subset Z of V such that $\lambda = \lambda'$ on $(V - |Z|) \times M$ then λ and λ' have the same Cartier divisor and are equal.*

Proof. The corollary follows from Proposition 2.3 and Proposition 1.5 giving the uniqueness of the extension of a continuous function, by continuity.

Principal divisors. Let f be a rational function on V. For any open set U, the pairs (U, f) are equivalent, and define a Cartier divisor which is called **principal**. We say that this divisor is **locally represented** by f everywhere. Similarly, the function

$$\lambda_f : (V - |(f)|) \to \mathbf{R}$$

defined by $\lambda_f(P, v) = v \circ f(P)$ is a Weil function whose Cartier divisor is represented by f everywhere. For any open set U, the triple

$$(U, f, 0)$$

represents the associated Néron divisor on U, also called **principal**.

Functoriality. Let

$$\varphi : V' \to V$$

be a morphism. Let D be a Cartier divisor on V such that $\varphi(V')$ is not contained in the support of D. Then $\varphi^*(D)$ is defined as a Cartier divisor on V'. Indeed, let (U, f) represent D on an open subset U of V. Then the composite rational function $f \circ \varphi$ is defined on V', and the pairs

$$(\varphi^{-1}(U), f \circ \varphi)$$

are equivalent on V'. Thus they define a Cartier divisor denoted $\varphi^*(D)$ or $\varphi^{-1}(D)$. If D is principal, represented by the function f, then $\varphi^{-1}(D)$ is also principal, represented by $f \circ \varphi$. We have a similar analysis for Néron divisors:

Proposition 2.5. *Let D be a Néron divisor on V such that $\varphi(V')$ is not contained in the support of (D). Then there exists a unique Néron divisor D' on V' such that if (U, f, α) is a triple representing D on V, then*

$$(\varphi^{-1}(U), f \circ \varphi, \alpha \circ \varphi)$$

represents D' on V'. Furthermore we have

$$(D') = \varphi^*((D)).$$

The proof is immediate from the definitions. Note that if V and D are defined over a field k, and $\varphi: V' \to V$ is defined over k, then D' is also defined over k. We call D' the **inverse image of** D by φ, and denote it also by φ^*D or $\varphi^{-1}D$.

Of course we have a similar statement for Weil functions.

Proposition 2.6. *Let $\varphi: V' \to V$ be a morphism. Let λ be a Weil function on V, with Cartier divisor D. Assume that $\varphi(V')$ is not contained in the support of D. Then $\lambda \quad \varphi$ is a Weil function on V', with Cartier divisor D' such that $D' = \varphi^*D$. In other words,*

$$\varphi^*((\lambda)) = (\lambda \quad \varphi).$$

The proof is again immediate from the definitions, or from the natural correspondence between Néron divisors and Weil functions.

§3. Positive Divisors

We say that a Cartier divisor X is **positive** if there exists a defining family of compatible pairs $\{(U, f)\}$ covering V and such that for each pair (U, f), the function f is a morphism on U. In particular, one can take the open sets U to be affine, and then f is an element of the affine coordinate ring of U. We write $X \geqq 0$ to indicate that X is positive. The set of positive Cartier divisors is closed under addition. If D is a Néron divisor whose Cartier divisor is positive, then the Weil function λ_D can be extended to all of V if we define $\lambda_D(x) = \infty$ at points in the support of D. We regard ∞ as larger than any real number. For any given absolute value v we can define neighborhoods of ∞ in the usual manner, and this extension of λ_D to all of V is then v-continuous on all of V. In other words, if P lies in the support of D then λ_D is continuous at P in the sense that for all points x close to P the value $\lambda_D(x, v)$ is large positive.

If V is non-singular, then a Cartier divisor is positive if and only if the corresponding ordinary divisor is positive in the usual sense that all components occur with positive multiplicity. This is immediate from the definitions, and the fact that if a function is not defined at a point, then it has a pole along a subvariety of codimension 1 passing through that point.

Proposition 3.1. *Assume that V is projective, or merely that $V \times M$ is bounded. Suppose that λ is a Weil function whose Cartier divisor is positive. Then there exists a constant γ such that*

$$\lambda(P, v) \geqq \gamma(v)$$

for all points of $V \times M$.

Proof. Let U be an affine open subset of V and let f be a morphism on U. Then the map

$$(P, v) \mapsto v \circ f(P)$$

is locally bounded from below. The proposition follows from the fact that a projective variety is bounded by Proposition 1.2.

For the next proposition, we do not need to assume that V is projective.

Proposition 3.2. *Let* λ_i $(i = 1, \ldots, m)$ *be Weil functions whose Cartier divisors have the form*

$$(\lambda_i) = Y + X_i$$

with $X_i \geqq 0$ *for all* i, *and such that the supports of* $X_1, \ldots, X_m$ *have no points in common. Then the function defined by*

$$\lambda(P, v) = \inf_i \lambda_i(P, v)$$

for P *outside the support of* Y *is a Weil function with Cartier divisor* Y.

Proof. First we remark that if P is outside the support of Y, then for some i and all v, $\lambda_i(P, v)$ is not ∞ and so the $\inf_i$ is defined as a real number, recalling our convention that ∞ is greater than any real number.

Given a point P, for each i let (U_i, f_i, α_i) be a local representative of the Néron divisor of λ_i on an open set U_i containing P. Let $i(P)$ be an index such that $P \notin |X_{i(P)}|$, and let

$$U_P = \bigcap U_i \cap \mathscr{C}|X_{i(P)}|,$$

where $\mathscr{C}|X_{i(P)}|$ is the complement of $|X_{i(P)}|$. Then U_P is an open set containing P. Let $f_P = f_{i(P)}$ and let

$$\alpha_P : U_P \times M \to \mathbf{R}$$

be defined by

$$\alpha_P(x, v) = \inf_i [v \circ (f_i f_{i(P)}^{-1})(x) + \alpha_i(x, v)].$$

We claim that the family of triples $\{(U_P, f_P, \alpha_P)\}$ is compatible, and that it defines a Néron divisor whose associated Néron function has the desired property.

First, the rational function f_P has Cartier divisor $Y + X_{i(P)}$ on $U_{i(P)}$, and hence it has the Cartier divisor $Y | U_P$ on U_P. Hence the family $\{(U_P, f_P)\}$ defines the Cartier divisor Y on V.

Next, by definition we know that $f_i f_{i(P)}^{-1}$ represents the Cartier divisor X_i on U_P and hence is morphic on U_P. Therefore the function

$$(x, v) \mapsto v \cdot (f_i f_{i(P)}^{-1})(x)$$

is locally bounded from below on $U_P \times M$. Since α_i is locally bounded, it follows that α_P is locally bounded from below on $U_P \times M$. Since

$$f_{i(P)} f_{i(P)}^{-1} = 1,$$

it follows that α_P is also locally bounded from above. Hence α_P is locally bounded, and its definition shows directly that it is continuous.

Next we check the compatibility condition. Let P' be a point of V (which may be P) and let $i'(P')$ be an index such that $P' \notin |X_{i'(P')}|$. Define

$$(U_{P'}, f_{P'}, \alpha_{P'})$$

as we did (U_P, f_P, α_P). Let $x \in U_P \cap U_{P'}$. Then using the fact that for any family of real numbers r_i and r we have

$$\inf(r_i + r) = \inf(r_i) + r,$$

we get

$$\alpha_P(x, v) + v \cdot (f_{i(P)} f_{i'(P')}^{-1}) = \alpha_{P'}(x, v),$$

thus proving the compatibility, and defining a Néron divisor D whose Cartier divisor is Y.

Finally, to compute $\lambda_D(P, v)$ at a point P of V which is not in $|Y|$, we may use the representative (U_P, f_P, α_P) as above, in which case

$$\lambda_D(P, v) = v \cdot f_P(P) + \alpha_P(P, v).$$

By definition,

$$\lambda_i(P, v) = v \cdot f_i(P) + \alpha_i(P, v).$$

Substituting $x = P$ in the definition of $\alpha_P(x, v)$ we find that

$$\lambda_D(P, v) = \inf \lambda_i(P, v),$$

thus proving the proposition.

Corollary 3.3. *Suppose that V is projective, and let $X_1, \ldots, X_m$ be positive Cartier divisors with no points in common. Let $\lambda_1, \ldots, \lambda_m$ be Weil functions* *satisfying* $(\lambda_i) = X_i$. *Then there are constants* γ_1, γ_2 *such that*

$$\gamma_1(v) \leqq \inf \lambda_i(P, v) \leqq \gamma_2(v)$$

for all points P *and all* v.

Proof. Immediate from Proposition 3.2 and Proposition 2.2.

Example. If V is a non-singular projective curve and X_1, X_2 are two distinct points, then they have no point in common, and therefore λ_1, λ_2 are "almost relatively prime", that is their inf is bounded.

We now repeat Lemma 3.1 of Chapter 4 in the context of Cartier divisors. Specifically, we need:

Lemma 3.4. *Assume* V *is projective. Let* Y *be a Cartier divisor. Then there exist positive Cartier divisors* X_i $(i = 1, \ldots, m)$ *and* Y_j $(j = 1, \ldots, n)$ *such that the* $|X_i|$ *have no common point, the* $|Y_j|$ *have no common point, and*

$$Y + X_i \sim Y_j.$$

We shall recall the proof rapidly below, and we now use the lemma to prove the existence of a Weil function associated with a given Cartier divisor on a projective variety. By additivity, the set of Cartier divisors which can be lifted to Néron divisors is a group, which contains the principal Cartier divisors.

Theorem 3.5. *Let* V *be a projective variety. Let* Y *be a Cartier divisor on* V. *Then there exists a Weil function having this divisor.*

Proof. Apply the lemma. There exist rational functions f_{ij} such that

$$Y - Y_j + X_i = (f_{ij}).$$

By Proposition 3.2, there exists a Weil function λ_j such that $(\lambda_j) = Y - Y_j$ and

$$\lambda_j = \inf_i \lambda_{ij},$$

where λ_{ij} is the Weil function associated with the rational function f_{ij}. Then

$$(-\lambda_j) = Y_j - Y,$$

and we apply Proposition 3.2 once more to get the desired function λ, which is defined by the formula

$$\lambda = \sup_j \inf_i \lambda_{ij}.$$

This proves the theorem.

Remark. If V is non-singular, then we can identify the Cartier divisor with an ordinary divisor and the theorem will be interpreted as such.

The sup-inf procedure was already used by Weil in the proof of his decomposition theorem.

As to the proof of Lemma 3.4, it comes as follows. For any Cartier divisor D we can define $L(D)$ just as we defined this space for ordinary divisors, namely as the space of rational functions f such that $(f) \geqq -D$. Any set of generators $(f_0, \ldots, f_n)$ of $L(D)$ gives a rational map of V into $\mathbf{P}^n$. We say that D is **very ample** if this rational map is a projective embedding. We say that D is **ample** if there exists a positive integer n such that nD is very ample. If V is projective, that is V is already embedded in a projective space $\mathbf{P}^n$, and X is the Cartier divisor of a hyperplane section not containing V, then the pull back of X to V is very ample. In fact, one has

Theorem 3.6. Let V be projective. If D is a Cartier divisor on V and X is ample, then there exists a positive integer n such that $D + nX$ is very ample.

This can be viewed as a special case of Serre's theorem that under suitable finiteness conditions, a suitably high twist of a sheaf can be generated by global sections, cf. Hartshorne [Ha], Theorem 5.17 of Chapter II. Indeed, as explained in [Ha], preceding Proposition 6.13, if D is a Cartier divisor represented by (U, f) on an open set U, we let $\mathcal{C}(D, U)$ be the $\mathcal{C}_U$-module generated by f^{-1}, where $\mathcal{C}_U$ is the structure sheaf of U. Then the family $\{\mathcal{C}(D, U)\}$ defines a sheaf $\mathcal{C}(D)$ on V which is locally free of rank 1, or as is also said, invertible. The sections of that sheaf are precisely the elements of the vector space $L(D)$. Applying Serre's theorem we see that for large positive n, the space of sections $L(D + nX)$ generates the sheaf $\mathcal{C}(D + nX)$. This means that the rational map of V into projective space arising from generators of $L(D + nX)$ is a morphism. If mX is very ample, then

$$D + nX + mX$$

is very ample, i.e. the rational map arising from $L(D + nX + mX)$ gives a projective embedding, as desired.

We give a reformulation of the existence of Weil functions in the words of Weil's **decomposition theorem**

Theorem 3.7. *Let* V *be a projective non-singular variety over* K. *Let* $\varphi \in K(V)$ *be a rational function, and let*

$$(\varphi) = \sum m_W W$$

be its divisor, where each W *is a prime rational divisor of* V *over* K. *Then there exist constants* γ_1, γ_2 *such that for any* $P \in V(K^a)$ *we have*

$$\gamma_1(v) + \sum m_W \lambda_W(P, v) \leq v \circ \varphi(P) \leq \sum m_W \lambda_W(P, v) + \gamma_2(v).$$

§4. The Associated Height Function

In this section we assume that $\mathbf{F}$ *is a field with a proper set of absolute values satisfying the product formula, and that* k *is a finite extension of* $\mathbf{F}$. *We let* V *be a variety defined over* k.

Let λ be a Weil function. For any point $P \in V(\mathbf{F}^a)$ not in the support of (λ), we define the associated **height**

$$h_\lambda(P) = \frac{1}{[K : \mathbf{F}]} \sum_{v \in M(K)} N_v \lambda(P, v),$$

for any finite extension K of $\mathbf{F}$ over which P is rational. The value is independent of the choice of K.

We define two Weil functions λ, λ' to be **linearly equivalent** if there exists a rational function f such that $\lambda = \lambda' + \lambda_f + \gamma$ where γ is a constant Weil function. If $P \notin \text{supp}(\lambda) \cup \text{supp}(\lambda')$, then for λ linearly equivalent to λ' we have

$$h_\lambda(P) = h_{\lambda'}(P) + \text{constant},$$

by the product formula. Given any Cartier divisor X and a point P, there exists a rational function f such that P does not lie in the support of $X - (f)$. Suppose $X = (\lambda)$. Put $\lambda' = \lambda - \lambda_f$. We then define

$$h_\lambda(P) = h_{\lambda'}(P).$$

This value is independent of the choice of f, and thus h_λ is defined for all $P \in V(\mathbf{F}^a)$, that is everywhere defined on V. The function h_λ depends only on the linear equivalence class of λ.

Functoriality. *Let*

$$\varphi: V' \to V$$

be a morphism. Let λ be a Weil function on V. Then for P in V' we have

$$h_{\lambda \circ \varphi}(P) = h_\lambda(\varphi(P)).$$

The formula is obvious.

Proposition 4.1. *Suppose that V is projective. If λ, λ' are Weil functions with the same Cartier divisor then $h_\lambda \sim h_{\lambda'}$. In other words $h_\lambda - h_{\lambda'}$ is bounded as a real valued function on $V(\mathbf{F}^a)$.*

Proof. This is immediate from Proposition 2.2. In fact, there exists a positive M_k-constant γ such that

$$|\lambda(P, v) - \lambda'(P, v)| \leqq \gamma(v)$$

for all $P \in V(\mathbf{F}^a)$ and all $v \in M(k^a)$. Then

$$|h_\lambda - h_{\lambda'}| \leqq \frac{1}{[k:\mathbf{F}]} \sum_{v \in M(k)} N_v \gamma(v).$$

Two Cartier divisors D, D' are called **linearly equivalent** if there exists a rational function f such that $D = D' + (f)$. The linear equivalence classes of Cartier divisors form a group denoted by **CaPic**(V). A class c is called **very ample** if it contains a very ample Cartier divisor, and similarly c is called **ample** if it contains an ample Cartier divisor.

On a projective variety, we see that to each Cartier divisor we can associate an equivalence class of height functions on $V(\mathbf{F}^a)$, which depends only on the linear equivalence class of the divisor. Furthermore, we have **additivity**

$$\boxed{h_{\lambda + \lambda'} = h_\lambda + h_{\lambda'}.}$$

In other words we have a natural homomorphism from the group of Weil functions into the group of heights.

Proposition 4.2. *Let X be a hyperplane section on $\mathbf{P}^n$, and let λ be a Weil function whose divisor is X. Let h be the height on $\mathbf{P}^n$ determined by the coordinate functions. Then*

$$h_\lambda \sim h,$$

that is $h_\lambda - h$ is bounded on $\mathbf{P}^n(\mathbf{F}^a)$.

Proof. Let $X_0, \ldots, X_n$ be the hyperplane sections corresponding to the coordinate functions. There exist rational functions f_i such that

$$(f_i) = X_i - X.$$

For any point $P \notin \mathrm{supp}(X)$, we have

$$h(P) = \frac{1}{[K : \mathbf{F}]} \sum_{v \in M(K)} N_v \sup_i \log |f_i(P)|_v$$

for any finite extension K of $\mathbf{F}$ such that $P \in \mathbf{P}^n(K)$. But

$$\sup_i \log |f_i(P)|_v = -\inf_i v \circ f_i(P).$$

We conclude the proof by applying Proposition 3.2 and Proposition 2.2.

Theorem 4.3. *Let V be a projective variety. To each Cartier class*

$$c \in \mathrm{CaPic}(V)$$

one can associate an equivalence class of functions h_c (so up to bounded functions on $V(\mathbf{F}^a)$) uniquely satisfying the following conditions:

4.3.1. *For two classes c, c' we have $h_{c+c'} = h_c + h_{c'} + O(1)$.*

4.3.2. *If c contains a very ample Cartier divisor D, and φ is a projective embedding derived from the space $L(D)$, then $h_c \sim h_\varphi$.*

Proof. Property 4.3.1 is clear from the definitions. Property 4.3.2 follows from Proposition 3.2 by pull-back to an arbitrary variety, using the functoriality properties of the height associated with Weil functions, and of Cartier divisors. The uniqueness comes from Theorem 3.6.

Remark. *This chapter can be substituted for all of Chapter* 4, since it derives ab ovo the properties of heights from the properties of Néron divisors. We have used a different set of foundational material to provide the same basic auxiliary statements like Theorem 3.6. Otherwise, formally, the development is entirely similar but more general (and also more elaborate) in the case of Weil functions and Néron divisors.

CHAPTER 11

Néron Functions on Abelian Varieties

On an arbitrary variety, a Weil function associated to a divisor is defined only up to a bounded function. On abelian varieties, Néron showed how to define a function more canonically, up to a constant function. This chapter develops Néron's results, but in §1 we shall prove existence by a method due to Tate, which is much simpler than Néron's original construction, and is the analogue of Tate's limit procedure for the height.

The first two sections of this chapter bear to Chapter 10 the same relation as Chapter 5 bears to Chapter 4. This entire chapter is due to Néron, and reproduces the results of [Ne 3], except for §6.

In §1 through §4 we let A be an abelian variety defined over a field K with a proper set of absolute values.

Some readers may be interested especially in the application of the next chapter, concerning algebraic families. Only §1 and §2 of this chapter are necessary for the understanding of Chapter 12, except for the definition of a Néron model in §5.

§1. Existence of Néron Functions

Let A be an abelian variety defined over K. If D is a divisor on A which is principal, $D = (f)$, then the rational function f is determined only up to a constant, and so the Weil functions λ_f for such f are determined only up to an additive constant. We let Γ be the group of M_K-**constant** functions.

Theorem 1.1. *Let A be an abelian variety defined over K. To each divisor D on A there exists a Weil function λ_D associated with D, uniquely determined mod Γ by the following properties.*

(1)   *If D, D' are divisors on A then*

$$\lambda_{D+D'} = \lambda_D + \lambda_{D'} \bmod \Gamma.$$

(2) If $D = (f)$ is principal, then $\lambda_D = \lambda_f$ mod Γ.

(3) We have $\lambda_{[2]^*D} = \lambda_D \circ [2]$ mod Γ.

Such functions λ_D also satisfy the property that if

$$\varphi: B \to A$$

is a homomorphism of abelian varieties defined over K, then

(4) $$\lambda_{\varphi^*D} = \lambda_D \circ \varphi \quad \text{mod } \Gamma.$$

Proof. Suppose first that D is even or odd, that is $D = D^-$ or $D = -D^-$. Let $m = 4$ if D is even, $m = 2$ if D is odd. Then

$$[2]^*D = mD + (f),$$

where f is a rational function. Let λ be any Weil function whose divisor is D. By functoriality and linearity, there exists a unique bounded continuous function γ on $A \times M_K$ such that

$$\lambda(2x, v) = m\lambda(x, v) + v \circ f(x) + \gamma(x, v)$$

for all x outside a suitable Zariski closed subset. However, by Proposition 2.3 and Corollary 2.4 of Chapter 10, γ is everywhere defined because it is a Weil function with divisor 0. We now need a lemma.

Lemma 1.2. Let $\mathscr{A}$ be a topological abelian group, and let

$$\gamma: \mathscr{A} \to \mathbf{R}$$

be a bounded continuous function. Let $m > 1$ be an integer. Then there exists a unique bounded continuous function α such that

$$\gamma(x) = \alpha(2x) - m\alpha(x).$$

If $\| \ \|$ is the sup norm, then $\|\alpha\| \leq \dfrac{3}{m-1} \|\gamma\|$.

Proof. The $\mathbf{R}$-vector space $BC(\mathscr{A}, \mathbf{R})$ of bounded continuous functions on $\mathscr{A}$ is a Banach space. We consider the mapping

$$S: BC(\mathscr{A}, \mathbf{R}) \to BC(\mathscr{A}, \mathbf{R})$$

given by

$$S\alpha(x) = \frac{1}{m} [\alpha(2x) - \gamma(x)].$$

It is immediately verified that S is a shrinking map, and in fact

$$\|S\alpha - S\beta\| \leq \frac{1}{m} \|\alpha - \beta\|,$$

where $\| \ \|$ is the sup norm. By the standard shrinking lemma of elementary analysis, the map S has a unique fixed point which satisfies the desired conditions. For the bound, recall that

$$\alpha = \lim_{n \to \infty} S^n \gamma \qquad \text{and so} \qquad \alpha - S\gamma = \sum_{n=1}^{\infty} (S^{n+1}\gamma - S^n\gamma).$$

We apply the lemma to the *finite number* of functions γ_v which are not 0, for $v \in M_K$. We define

$$\lambda_v' = \lambda_v - \alpha_v,$$

so λ' differs from λ only at those v extending a finite set in M_K. It is immediate that λ' satisfies (3), and indeed,

(*) $$\lambda_v'(2x) = m\lambda_v'(x) + v \circ f(x).$$

Suppose that λ' and λ'' are two Weil functions with the same divisor D and both satisfy (*) mod Γ. Let

$$\beta = \lambda' - \lambda''.$$

Then β is a Weil function whose divisor is 0, and is therefore bounded. Furthermore, β satisfies the hypotheses of the following lemma, which then shows that β is constant.

Lemma 1.3. *Let m be an integer > 1. Let $\beta \colon \mathscr{A} \to \mathbf{R}$ be a bounded function on an abelian group $\mathscr{A}$ satisfying*

(**) $$\beta(2x) = m\beta(x) + \gamma, \quad \text{for all } x \in \mathscr{A},$$

where γ is constant. Then β is constant, $\beta = -\gamma/(m-1)$, so $\beta = 0$ if $\gamma = 0$.

Proof. By iterating one gets

$$\beta(x) = -\sum_{i=1}^{n} \frac{1}{m^i}\gamma + \frac{1}{m^n}\beta(2^n x).$$

Since β is bounded, it follows that $\beta(x)$ is constant. In fact, we have explicitly

$$\beta = - \sum_{i=1}^{\infty} \frac{1}{m^i} \gamma$$

as desired.

Given an even or odd divisor D we have therefore determined a Weil function λ', which we now denote by λ_D, with divisor D, and satisfying (∗) relative to a given choice of f in the formula

$$[2]^*D = mD + (f).$$

This function λ_D is uniquely determined by the lemma. It follows at once that the association $D \mapsto \lambda_D$ also satisfies conditions (1) and (2) of the theorem.
    For any divisor D we write

$$2D = (D + D^-) + (D - D^-),$$

and a divisor which is both odd and even is equal to 0. Hence we have associated the desired kind of Weil function to divisors of type $2D$. We define

$$\lambda_D = \tfrac{1}{2}\lambda_{2D} \bmod \Gamma.$$

This definition is compatible with the previous one. It is immediate that this gives a Weil function satisfying (1), (2) and (3), thus proving existence and uniqueness of the functions satisfying these three conditions.
    Finally, we note that homomorphisms $\varphi: B \to A$ commute with multiplication by 2. From this (4) follows at once by applying φ^* to the formula

$$[2]^*D = mD + (f)$$

This proves the theorem.

Observe that instead of multiplication by 2, we could have used multiplication by any positive integer in condition (3).

A Weil function λ_D satisfying the conditions of Theorem 1.1 will be called a **Néron function associated with the divisor** D, **or having the divisor** D. It is well-defined up to a constant Néron function.

For convenience of reference, we summarize the process by which we constructed the Néron function in a proposition.

Proposition 1.4. *Let D be an even (resp. odd) divisor, and let λ be a Weil function with divisor D. Let $[2]^*D = mD + (f)$, and let γ be the unique bounded continuous function on $A \times M_k$ such that*

$$\lambda(2x, v) = m\lambda(x, v) + v \circ f(x) + \gamma(x, v)$$

for all x outside a suitable Zariski closed subset. Let S be the operator of Lemma 1.2, let $\alpha = \lim S^n \gamma$, and $\lambda' = \lambda - \alpha$. Then λ' is the unique Néron function with divisor D, satisfying

$$\lambda'(2x, v) = m\lambda'(x, v) + v \circ f(x).$$

Next we consider the characterization of Néron functions by the property of invariance under translations.

Theorem 1.5. *In Theorem 1.1, the existence and uniqueness assertions remain valid if instead of property (3) one assumes invariance under translation, that is:*

(3$_{tr}$) $\lambda_{D_a}(x) = \lambda_D(x - a) \bmod \Gamma.$

The unique function $\lambda_D \bmod \Gamma$ satisfying (1), (2), (3$_{tr}$) then also satisfies (3).

Proof. In the next section, it will be proved that the function of Theorem 1.1 satisfies (3$_{tr}$) and that invariance under translation characterizes the function, so the proof will follow from Theorems 2.1 and 2.2.

On an abelian variety, unless otherwise specified, we always let λ_D denote a Néron function having divisor D. In light of Theorem 1.1 or 1.5, it is the unique Weil function (up to a constant function) satisfying (1), (2), (3) or (1), (2), (3$_{tr}$).

In Chapter 10, §4 we saw how to recover the height from Weil functions (modulo bounded functions) in case the proper set of absolute values satisfies the product formula. This height was on arbitrary varieties. On the other hand, on abelian varieties, we have a canonical height as a unique quadratic function in its class modulo bounded functions as in Chapter 5. It is no surprise that the Néron function gives rise to this canonical height, as in the following theorem.

Theorem 1.6. *Assume the product formula. Let λ be a Néron function having divisor D. Let h_λ be the associated height. Let $c = \mathrm{Cl}(D)$ be the linear equivalence class of D, and let h_c be the quadratic height function associated with the class c. Then*

$$h_\lambda = h_c + constant.$$

Proof. There exists a rational function F_3 on $A \times A \times A$ such that

$$s_3^{-1}(D) - s_{12}^{-1}(D) - s_{13}^{-1}(D) - s_{23}^{-1}(D)$$
$$+ p_1^{-1}(D) + p_2^{-1}(D) + p_3^{-1}(D) = (F_3).$$

For (x, y, z) outside some Zariski closed set, the product formula gives

$$\sum_v N_v v \circ F_3(x, y, z) = 0.$$

By the additivity property of the Néron function, we conclude that h_λ is a quadratic function (up to an additive constant), first outside some Zariski closed set, and then everywhere after changing D by the divisor of a rational function. By Proposition 4.2 of Chapter 10, we conclude that $h_\lambda = h_c + \gamma$, where γ is constant, as was to be shown.

§2. Translation Properties of Néron Functions

Each relation of Chapter 5, §2 among divisor classes immediately yields a relation among Néron functions. Under suitable homomorphisms like the sum or projections, every divisor becomes principal and so one can recover the value of a Néron function from that of a rational function on a sufficiently large product (double product or triple product, for instance). We already used one of these properties at the end of the last section when analyzing the height. We now use the others in analyzing translations.

Let D be a divisor on A. For any $a \in A$ let

$$\beta_{a, D}(x) = \lambda_{D_a}(x) - \lambda_D(x - a).$$

A priori, $\beta_{a, D}$ is defined for those x such that $x - a \notin |D|$. However, $\beta_{a, D}$ is a Weil function with divisor 0, and so extends uniquely to a bounded continuous function everywhere defined. Since λ_D is defined only up to a constant function, the function $\beta_{a, D}$ likewise is defined only up to a constant function.

Theorem 2.1. *The function $\beta_{a, D}$ is constant. In other words, there exists a constant function $\gamma_{a, D}$ such that*

$$\lambda_{D_a} = \lambda_D \circ T_{-a} + \gamma_{a, D},$$

where T_{-a} is translation by $-a$.

Proof. It is certainly the case that $\beta_{a,D}$ is constant if D is the divisor of a function. Suppose first D is odd. There exists a rational function F_1 on $A \times A$ such that

$$s_2^{-1}(D) - p_1^{-1}(D) - p_2^{-1}(D) = (F_1).$$

We translate by a point (a, b) on $A \times A$, so for $Y = (F_1)$ we have

$$\lambda_Y(x - a, y - b) = \lambda_{Y_{(a,b)}}(x, y) + \gamma$$

for some constant γ. We use the trivial fact that for any homomorphism $\varphi: B \to A$ and $u \in B$ we have

$$\varphi^{-1}(D)_u = \varphi^{-1}(D_{\varphi(u)}).$$

By functoriality with respect to homomorphisms and additivity we find

$$\lambda_D(x - a, y - b) - \lambda_D(x - a) - \lambda_D(y - b)$$
$$= \lambda_{D_{a+b}}(x + y) - \lambda_{D_a}(x) - \lambda_{D_b}(y) \quad \mod \Gamma.$$

In terms of $\beta_{a,D} = \beta_a$ this can be rewritten in the form

$$\beta_{a+b}(x + y) = \beta_a(x) + \beta_b(y) \mod \Gamma.$$

Put $b = 0$. Then $\beta_a(x + y) = \beta_a(x) \mod \Gamma$, so β_a is invariant under translations and is therefore constant.

Next suppose that D is even. By Lemma 1.3 it suffices to prove that

(∗) $$\beta_{a,D}(2x) = 2\beta_{a,D}(x) + \gamma_a,$$

where γ_a is a constant depending on a but not on x. To prove (∗) we derive a somewhat more general formula. Let Y be the "parallelogram" on $A \times A$ formed with D, that is

$$Y = s_2^{-1}(D) + d_2^{-1}(D) - 2p_1^{-1}(D) - 2p_2^{-1}(D) = (F_2)$$

for some rational function F_2 on $A \times A$. Then the property we are trying to prove is true for Y since Y is principal. We again translate by a point (a, b) on $A \times A$ and find

$$\lambda_D(x - a + y - b) + \lambda_D(x - a - y + b) - 2\lambda_D(x - a) - 2\lambda_D(y - b)$$
$$= \lambda_{D_{a+b}}(x + y) + \lambda_{D_{a-b}}(x - y) - 2\lambda_{D_a}(x) - 2\lambda_{D_b}(y) \quad \mod \Gamma.$$

In terms of $\beta_{a,D}$ this can be rewritten in the form

$$\beta_{a+b}(x + y) + \beta_{a-b}(x - y) = 2\beta_a(x) + 2\beta_b(y) + \gamma_{a,b}.$$

Now put $b = 0$ and $x = y$ to get (∗) and prove the theorem.

The Néron function can be characterized by its property under translations rather than by its functoriality under homomorphisms. First we need some notation. Let

$$\mathfrak{a} = \sum n(x)(x)$$

be a 0-cycle on A of degree 0. If the support of $\mathfrak{a}$ is disjoint from the support $|D|$ of a divisor, and v is a fixed absolute value, we define the **Néron pairing** or **symbol**

$$\langle D, \mathfrak{a} \rangle_v = \sum n(x)\lambda_D(x, v).$$

Of course λ_D is defined only up to a constant function, but taking $\mathfrak{a}$ of degree 0 makes the value on the right hand side independent of the choice of λ_D.
Similarly, if $D = (f)$ is principal, we let

$$f(\mathfrak{a}) = \prod f(x)^{n(x)}.$$

This value depends only on D since the constant disappears when taking the product over the points of a cycle of degree 0. We let:

$\mathrm{Div}(A)_K$ = group of divisors on A rational over K.

$Z(A(K))$ = group of 0-cycles whose components are rational over K.

$Z_0(A(K))$ = subgroup of $Z(A(K))$ consisting of elements of degree 0.

Then the symbol satisfies the properties of the next theorem, which characterize it.

Theorem 2.2. *Let A be an abelian variety defined over a field K with a proper absolute value v, such that $A(K)$ is Zariski dense in A. To each pair $(D, \mathfrak{a})$ consisting of a divisor $D \in \mathrm{Div}(A)_K$ and $\mathfrak{a} \in Z_0(A(K))$ with disjoint supports, one can associate in one and only one way a real number $\langle D, \mathfrak{a} \rangle_v = \langle D, \mathfrak{a} \rangle$ satisfying the following properties.*

NS 1. *The symbol $\langle D, \mathfrak{a} \rangle$ is bilinear.*

NS 2. *If $D = (f)$ is principal, then $\langle D, \mathfrak{a} \rangle = v(f(\mathfrak{a}))$.*

NS 3_{tr}. *The symbol is invariant under translation, that is*

$$\langle D_u, \mathfrak{a}_u \rangle = \langle D, \mathfrak{a} \rangle \quad \text{for } u \in A(K).$$

NS 4.   *Let $x_0 \in A(K)$ but $x_0 \notin |D|$.   Then the map*

$$x \mapsto \langle D, (x) - (x_0) \rangle$$

is bounded on every bounded subset of $A(K) - |D|(K)$.

The symbol is then necessarily v-continuous in **NS 4.**

Proof. The properties of a Néron function and Theorem 2.1 prove the existence of a symbol as above. As to uniqueness, suppose we have two such symbols. Let $\tau(D, \mathfrak{a})$ be their difference. Then $\tau(D, \mathfrak{a})$ is bilinear. Furthermore $\tau(D, \mathfrak{a}) = 0$ if D is principal and the supports of $D, \mathfrak{a}$ are disjoint. We can then extend the definition of τ to an arbitrary pair, namely given any $D, \mathfrak{a}$ we find a rational function f in $K(A)$ such that the support of $D + (f)$ is disjoint from the support of $\mathfrak{a}$. The value $\tau(D + (f), \mathfrak{a})$ is independent of the choice of f, and is 0 if D is principal. The symbol $\tau(D, \mathfrak{a})$ is then everywhere defined, and satisfies properties **NS 1**, **NS 3$_{tr}$**, **NS 4**, while **NS 2** is replaced by the property that $\tau(D, \mathfrak{a}) = 0$ if D is principal. That $\tau(D, \mathfrak{a})$ is 0 in all cases follows from the following lemma, which axiomatizes the situation because we shall not need the full subgroup $\mathrm{Div}(A)_K$ or $Z_0(A(K))$. We can deal with more general subgroups of divisors and 0-cycles.

Lemma 2.3.   *Let A^* be a subgroup of A. Let $\mathscr{D}^*$ be a subgroup of divisors on A, stable under addition and translation by elements of A^*. Let Z^* be the group of 0-cycles of degree 0 with support in A^*. Let*

$$\tau : \mathscr{D}^* \times Z^* \to \mathbf{R}$$

be a map satisfying:

(i)   *τ is bilinear.*
(ii)   *If D is principal, then $\tau(D, \mathfrak{a}) = 0$.*
(iii)   *We have $\tau(D_u, \mathfrak{a}_u) = \tau(D, \mathfrak{a})$ for $u \in A^*$.*
(iv)   *Fix $x_0 \in A^*$ and $D \in \mathscr{D}^*$. The map $x \mapsto \tau(D, (x) - (x_0))$ is bounded.*

Then $\tau(D, \mathfrak{a}) = 0$ for all $D, \mathfrak{a}$.

Proof. We let $S : Z_0(A) \to A$ be the usual sum map which to a 0-cycle associates the sum of its points on A, so $S(\mathfrak{a}) \in A$. We write $\mathfrak{a} \sim 0$ to mean that $S(\mathfrak{a}) = 0$, or in other words, that $\mathfrak{a}$ is in the kernel of Albanese. We shall first prove that if $\mathfrak{a} \sim 0$ then $\tau(D, \mathfrak{a}) = 0$. It is easy to see by induction that if $\mathfrak{a} \sim 0$ then there exists $\mathfrak{b} \in Z^*$ such that

$$\mathfrak{a} = \mathfrak{b}_u - \mathfrak{b}$$

with some element $u \in A^*$. Then

$$\tau(D, \mathfrak{a}) = \tau(D, \mathfrak{b}_u - \mathfrak{b}) = \tau(D_{-u} - D, \mathfrak{b}).$$

The cycle

$$\mathfrak{c}_m = (-mu) - (0) - m((-u) - (0))$$

is in the kernel of Albanese, and hence

$$D_{\mathfrak{c}_m} \sim 0.$$

Therefore $\tau(D_{\mathfrak{c}_m}, \mathfrak{b}) = 0$, whence

$$\tau(D_{(-mu)-(0)}, \mathfrak{b}) = m\tau(D_{(-u)-(0)}, \mathfrak{b}),$$

and using the invariance under translations,

$$\tau(D, \mathfrak{b}_{mu} - \mathfrak{b}) = m\tau(D, \mathfrak{b}_u - \mathfrak{b}) = m\tau(D, \mathfrak{a}).$$

But $\mathfrak{a}$, $\mathfrak{b}$ are fixed, and property (iv) shows that $\tau(D, \mathfrak{b}_{mu} - \mathfrak{b})$ is bounded. Dividing by m and letting m tend to infinity proves that $\tau(D, \mathfrak{a}) = 0$, and concludes the proof that $\tau(D, \mathfrak{a}) = 0$ if $\mathfrak{a} \sim 0$.

We now conclude that $\tau(D, \mathfrak{a})$ depends only on the class of $\mathfrak{a}$ modulo the kernel of Albanese. Therefore

$$\tau(D, (mx) - (0)) = m\tau(D, (x) - (0))$$

for all $x \in A^*$. Again we divide by m, let m tend to infinity, and use property (iv) to see that $\tau(D, (x) - (0)) = 0$. But cycles of degree 0 are generated by cycles of the form $(x) - (0)$, so this concludes the proof of the lemma.

To characterize the symbol we did not need any continuity property in the fourth property. However, since the existence was via Néron functions which are continuous, it follows that the symbol is necessarily continuous in the last property **NS 4**.

This concludes the proof of Theorem 2.2.

The next section extends the definition of the Néron symbol and canonical Néron functions to arbitrary varieties, but only when the divisor is algebraically equivalent to 0. As an application we then give Néron's reciprocity laws for the symbol. These sections are independent of §5, which gives the interpretation of the Néron function in terms of local intersection multiplicities.

There is another variation of Theorem 2.2 which is also useful when one deals with 0-cycles $\mathfrak{a} \in Z_0(A)_K$ instead of $Z_0(A(K))$. For this purpose, write

$$\mathfrak{a} = \sum_{i=1}^{d} [(P_i) - (Q_i)],$$

where no point P_i is equal to a point Q_j. Then we call d the **positive degree** of $\mathfrak{a}$, and denote it by

$$d = \deg^+(\mathfrak{a}).$$

Theorem 2.4. *Let A be an abelian variety defined over a field K with a proper absolute value v. Assume that $A(K)$ is Zariski dense in A. Suppose that to each $D \in \mathrm{Div}(A)_K$ and $\mathfrak{a} \in Z_0(A)_K$ with disjoint support one can associate a symbol $[D, \mathfrak{a}]$ satisfying the following properties.*

(1)   *The symbol $[D, \mathfrak{a}]$ is bilinear.*
(2)   *If $D = (f)$ is principal, then $[D, \mathfrak{a}] = v(f(\mathfrak{a}))$.*
(3)   *The symbol is invariant under translation, that is*

$$[D_u, \mathfrak{a}_u] = [D, \mathfrak{a}] \quad \textit{for } u \in A(K).$$

(4)   *Let $\tau(D, \mathfrak{a}) = [D, \mathfrak{a}] - \langle D, \mathfrak{a} \rangle$. Then for D fixed, and $\deg^+(\mathfrak{a})$ bounded, the values $\tau(D, \mathfrak{a})$ are bounded.*

Then $[D, \mathfrak{a}] = \langle D, \mathfrak{a} \rangle$.

Proof. The proof is the same as that of Theorem 2.2., except for a minor change at the end corresponding to the alternate version of condition (4). We leave this to the reader.

§3. Néron Functions on Varieties

The next theorem gives Néron functions on arbitrary varieties (complete non-singular), associated with divisors which are obtained by pull back from abelian varieties. The divisors will be restricted, however, to those which are algebraically equivalent to 0. Actually, a trivial argument could be given a posteriori to show how to extend the Néron functions to divisors D such that mD is algebraically equivalent to 0 for some positive integer m, namely

$$\lambda_D = \frac{1}{m} \lambda_{mD}.$$

Theorem 3.1. *Let V be a projective non-singular variety defined over K. To each $D \in \mathrm{Div}_a(V)_K$ there is a Weil function λ_D, unique mod constant functions, satisfying the following conditions:*

(1) *If $D, D' \in \mathrm{Div}_a(V)_K$ then $\lambda_{D+D'} = \lambda_D + \lambda_{D'} \bmod \Gamma$.*
(2) *If $D = (f)$ is principal then $\lambda_D = \lambda_f \bmod \Gamma$.*
(3) *If $\varphi \colon V \to W$ is a morphism defined over K and $Y \in \mathrm{Div}_a(W)_K$ is such that $\varphi^{-1}(Y)$ is defined, then*

$$\lambda_{\varphi^{-1}(Y)} = \lambda_Y \circ \varphi \bmod \Gamma.$$

The function λ_D of Theorem 3.1 will be called a **Néron function** associated with D.

Before proving the theorem, we make some remarks.

Remark 1. The Néron function λ_D should really be indexed by K since we are considering only divisors rational over K. However, if E is any algebraic extension of K, then the uniqueness property shows that modulo constant functions we have $\lambda_{K,D} = \lambda_{E,D} \bmod \Gamma$, so we may omit the K from the subscript.

Remark 2. If M'_K is a subset of M_K then again the uniqueness shows that for those v on K^a extending absolute values in M'_K the values $\lambda_D(P, v)$ are the same, whether we refer to M'_K or M_K.

Remark 3. For an abelian variety, when the morphism φ is restricted to a morphism of abelian varieties, then the theorem has already been proved in §1 and §2.

We now come to the proof. First, uniqueness follows from the same type of result as Theorem 2.2, but on arbitrary varieties.

For uniqueness, we may assume that we deal with one absolute value v, which will therefore be left out of the notation.

Let λ_D, λ'_D be two Néron functions satisfying the conditions of the theorem. Let g_D be their difference. Then g_D is a Weil function whose divisor is 0, and is consequently everywhere defined and bounded. Define

$$\langle D, \mathfrak{a} \rangle = g_D(\mathfrak{a}).$$

Because $\mathfrak{a}$ has degree 0, this value is well defined, independent of the constant up to which λ_D or λ'_D are defined. Furthermore, we have $\langle D, \mathfrak{a} \rangle = 0$ if D is the divisor of a function. Since g is bounded, we have an inequality

$$|\langle D, \mathfrak{a} \rangle| \leqq \gamma_D \deg^+(\mathfrak{a}),$$

where γ_D is a positive number depending only on g_D. The following lemma then suffices to conclude the proof.

Lemma 3.2. *Suppose that for each projective variety V over K, we have a bilinear pairing*

$$(D, \mathfrak{a}) \mapsto \langle D, \mathfrak{a} \rangle$$

from $\mathrm{Div}_a(V)_K \times Z_0(V)_K \to \mathbf{R}$, *satisfying the following conditions:*

(i)   *If* $D \sim 0$ *then* $\langle D, \mathfrak{a} \rangle = 0$.

(ii) *For every morphism* $\varphi: V' \to V$ *over* K *and* $\mathfrak{b} \in Z_0(V')_K$ *we have*

$$\langle \varphi^{-1}(D), \mathfrak{b} \rangle = \langle D, \varphi(\mathfrak{b}) \rangle.$$

(iii) *For D fixed, and* $\deg^+(\mathfrak{a})$ *bounded, the values* $\langle D, \mathfrak{a} \rangle$ *are bounded.*

Then $\langle D, \mathfrak{a} \rangle = 0$ *for all D, $\mathfrak{a}$.*

Proof. First suppose $V = A$ is an abelian variety. Let $D \in \mathrm{Div}_a(A)_K$, so D is algebraically equivalent to 0. Then $[m]^{-1}D \sim mD$, whence

$$\langle D, [m](\mathfrak{a}) \rangle = \langle [m]^{-1}D, \mathfrak{a} \rangle = m \langle D, \mathfrak{a} \rangle.$$

Divide by m and let m tend to infinity. By (iii) we find that $\langle D, \mathfrak{a} \rangle = 0$.

Second let V be arbitrary. It may be that V does not have a rational point over K and that there is no canonical map of V into its Albanese variety A defined over K. However, something almost as good exists, namely a suitable positive integral multiple of such a map exists which is suitably universal for our purposes. We now recall how this is done and derive the needed lemma to apply to the present situation.

There exists a separable field extension K' of K such that V has a rational point P_0 over K'. There is a canonical map

$$\alpha: V \to A$$

defined over K'. The map

$$F: V \times V \to A$$

given by $F(x, y) = \alpha(x) - \alpha(y)$ is actually defined over K (cf. AV, Chapter 2, §3). Let

$$\varphi(x) = \sum_{\sigma} F(x, \sigma P_0) = nf(x) - x_0,$$

where the sum is taken over all conjugates of K' over K, and

$$x_0 = \sum \sigma \alpha(P_0) \in A(K)$$

is rational over K. Thus φ is defined over K, and is equal to $n\alpha$ up to a K-rational translation where $n = [K' : K]$.

Lemma 3.3. *Let $\varphi: V \to A$ be as above, n times a canonical map of V into its Albanese variety, defined over K. Let $\psi: V \to B$ be a map into an abelian variety defined over K. Then $n\psi$ factors through φ.*

Proof. Exercise left for the reader. Thus we have a commutative diagram, where η is a homomorphism composed with a translation, and is defined over K.

Lemma 3.4. *Let $D \in \mathrm{Div}_a(V)_K$. There exists a divisor $Y \in \mathrm{Div}_a(A)_K$ and a positive integer m such that*

$$mD \sim \varphi^{-1}(Y).$$

Proof. First observe that if $D \in \mathrm{Div}_a(V)_K$, then there exists a divisor $Y_0 \in \mathrm{Div}_a(A)_K$ such that

$$D \sim \alpha^{-1}(Y_0),$$

by the theory of the Picard variety. Let $Y = \sum \sigma Y_0$. Then

$$n^2 D = n \sum \sigma D \sim n \cdot \alpha^{-1}(Y) \sim (n\alpha)^{-1}(Y).$$

But Y is rational over K, and $n\alpha$ differs from φ by a K-rational translation, which does not change the linear equivalence of Y. This proves the lemma.

We may now complete the proof of Lemma 3.2. Let us write

$$mD \sim \varphi^{-1}(Y)$$

as in Lemma 3.4. Then

$$m\langle D, \mathfrak{a} \rangle = m\langle \varphi^{-1}(Y), \mathfrak{a} \rangle = m\langle Y, \varphi(\mathfrak{a}) \rangle = 0$$

by the triviality of the pairing on abelian varieties already proved. This concludes the proof of Lemma 3.2, and therefore of the uniqueness of the Néron functions in Theorem 3.1.

We now prove existence. Suppose first that we work over the algebraic closure K^a of K. Let $\alpha: V \to A$ be a canonical map of V into its Albanese variety, which may in fact be defined over a separable extension of K. By the theory of the Picard variety, given $D \in \mathrm{Div}_a(V)$ (always assumed rational over K^a) there exists $X \in \mathrm{Div}_a(A)$ such that

$$D = \alpha^{-1}(X) + (f)$$

for some rational function f on V. Given $P \notin |D|$, we can choose X in its linear equivalence class such that $\alpha(P) \notin |X|$, and hence $f(P)$ is defined and $\neq 0$. Define

$$\lambda_D(P, v) = \lambda_X(\alpha(P), v) + v \circ f(P).$$

It is immediately verified that λ_D is well-defined up to a constant function. If $D = (f)$ is a principal divisor, then λ_D as we have defined it above coincides with λ_f up to a constant. Thus one sees that λ_D is a Weil function with divisor D. If $\varphi: V \to W$ is a morphism of projective non-singular varieties, then one has a commutative diagram:

$$
\begin{array}{ccc}
V & \overset{\varphi}{\longrightarrow} & W \\
\Big\downarrow{\alpha_V} & & \Big\downarrow{\alpha_W} \\
A(V) & \underset{\varphi_A}{\longrightarrow} & A(W)
\end{array}
$$

up to a translation, and pulling back divisors from $A(W)$ around either side of the diagram shows that condition (3) in Theorem 3.1 is satisfied.

We now take the restriction of the association $D \mapsto \lambda_D$ to divisors in $\mathrm{Div}_a(V)_K$, that is divisors rational over K. Using Lemma 3.4 and the fact that the functions λ_D take values in a uniquely divisible group (namely $\mathbf{R}$), one sees immediately that the conditions of Theorem 3.1 are satisfied, thereby proving the theorem.

Theorem 3.5. *Let K be a field with a proper absolute value v. To each projective non-singular variety V over K such that $V(K)$ is Zariski dense in V, and to each couple $(D, \mathfrak{a})$ consisting of $D \in \mathrm{Div}_a(V)_K$ and $\mathfrak{a} \in Z_0(V(K))$ with disjoint supports one can associate uniquely a real number $\langle D, \mathfrak{a} \rangle_v$ satisfying the following conditions.*

NS 1. *The symbol is bilinear in D and $\mathfrak{a}$.*

NS 2. *If $D = (f)$ is principal then $\langle D, \mathfrak{a} \rangle_v = v \circ f(\mathfrak{a})$.*

NS 3. *For every morphism $\varphi: V' \to V$ defined over K, and $\mathfrak{a}' \in Z_0(V')_K$ we have*

$$\langle \varphi^{-1}(D), \mathfrak{a}' \rangle_v = \langle D, \varphi(\mathfrak{a}') \rangle_v.$$

NS 4. *Fix* $x_0 \in V(K) - |D|$. *Then the map* $V(K) - |D| \to \mathbf{R}$ *defined by*

$$x \mapsto \langle D, (x) - (x_0) \rangle_v$$

is locally v-bounded.

The symbol is then necessarily v-continuous in **NS 4.**

Proof. The uniqueness is proved by using the functoriality of **NS 3** and reducing the problem to the Albanese variety, in which case we can quote Theorem 2.2. The existence follows at once from Theorem 3.1 by putting

$$\langle D, \mathfrak{a} \rangle_v = \lambda_D(\mathfrak{a}, v).$$

The right-hand side is defined by linearity from points, that is,

$$\text{if} \quad \mathfrak{a} = \sum (P_i) \quad \text{then} \quad \lambda_D(\mathfrak{a}, v) = \sum \lambda_D(P_i, v).$$

The ambiguous constant disappears when we evaluate Néron functions at a cycle of degree 0.

If the variety is a curve, then one can characterize the Néron pairing entirely on the curve as follows. Note first that

$$\text{Div}_a(V) = Z_0(V),$$

in other words, divisors are cycles of dimension 0. If $\mathfrak{a} = (f)$ is the divisor of a rational function, and $\mathfrak{b} = \sum n_i(P_i)$ has support disjoint from $\mathfrak{a}$, then we define as usual

$$f(\mathfrak{b}) = \prod f(P_i)^{n_i}.$$

Theorem 3.6. *Let V be a complete non-singular curve defined over a field K with a proper absolute value v, and such that $V(K)$ is Zariski dense in V. To each pair of elements $\mathfrak{a}, \mathfrak{b} \in Z_0(V(K))$ with disjoint supports, one can define in one and only one way a real number $\langle \mathfrak{a}, \mathfrak{b} \rangle_v$ satisfying the following properties:*

(i) *The pairing is bilinear in each variable.*
(ii) *If $\mathfrak{a} = (f)$ is principal, then $\langle \mathfrak{a}, \mathfrak{b} \rangle_v = v \circ f(\mathfrak{b})$.*
(iii) *The pairing is symmetric, that is $\langle \mathfrak{a}, \mathfrak{b} \rangle_v = \langle \mathfrak{b}, \mathfrak{a} \rangle_v$.*
(iv) *Fix $x_0 \in V(K) - |\mathfrak{a}|$. Then the map $V(K) - |\mathfrak{a}| \to \mathbf{R}$ defined by*

$$x \mapsto \langle \mathfrak{a}, (x) - (x_0) \rangle_v$$

is continuous and locally bounded.

Proof. The existence of symmetry will follow from the reciprocity law to be proved in the next section, taking the divisor on the product to be the diagonal. Existence of the other properties has already been proved. For uniqueness, on curves, we do not need either functoriality or translations to carry out the usual argument, because the Riemann–Roch theorem is available as follows. Subtracting two pairings satisfying the first three conditions, we obtain a pairing $\tau(\mathfrak{a}, \mathfrak{b})$ which vanishes if $\mathfrak{a}$ is the divisor of a function, and similarly on the other side, by the symmetry of (iii). We have to prove that $\tau(\mathfrak{a}, \mathfrak{b}) = 0$ for all $\mathfrak{b}$. By the Riemann–Roch theorem, there exist positive divisors X, Y of degree g (the genus) such that $\mathfrak{b} + X \sim Y$. Therefore

$$\tau(\mathfrak{a}, \mathfrak{b}) = \tau(\mathfrak{a}, X - Y),$$

and the fourth property shows that $\tau(\mathfrak{a}, \mathfrak{b})$ is bounded. Replacing $\mathfrak{b}$ by $m\mathfrak{b}$ for large positive integers m shows that $\tau(\mathfrak{a}, \mathfrak{b}) = 0$, thus proving uniqueness. and concluding the proof of the theorem.

It is possible to define Néron functions for any divisor, uniquely eliminating the ambiguous constant which disappears when one deals with divisors of degree 0. To do so requires heavier machinery. It will be done in Chapter 13 over the complex numbers. At the non-archimedean primes, one uses intersection numbers in a manner similar to §5. Cf. Arakelov [Ar 1], [Ar 2], and recently Hriljac [Hri], Faltings [Fa 1], [Fa 2].

Remark. The same remark as for Theorem 3.1 applies to Theorems 3.5 and 3.6. Let L be a finite extension of K, and let w be an extension of v to L. Then the restriction of the symbol $\langle D, \mathfrak{a} \rangle_w$ to elements $D, \mathfrak{a}$ with

$$D \in \mathrm{Div}_a(V)_K$$

and $\mathfrak{a} \in Z_0(V(K))$ in Theorem 3.5 coincides with $\langle D, \mathfrak{a} \rangle_v$ because it satisfies the axioms. Similarly, the restriction of $\langle \mathfrak{a}, \mathfrak{b} \rangle_w$ to elements $\mathfrak{a}, \mathfrak{b} \in Z_0(V(K))$ coincides with $\langle \mathfrak{a}, \mathfrak{b} \rangle_v$.

Just as for abelian varieties, one can state and prove the alternate version when cycles are rational over K, rather than when their components are rational over K, as follows.

Theorem 3.7. *Let V be a complete non-singular curve defined over a field K with a proper absolute value. To each pair of elements $\mathfrak{a}, \mathfrak{b} \in Z_0(V)_K$ with disjoint supports, suppose that we have associated a real number $[\mathfrak{a}, \mathfrak{b}]$ satisfying the following properties:*

(1)   *The pairing is bilinear in each variable.*
(2)   *If $\mathfrak{a} = (f)$ is principal, then $[\mathfrak{a}, \mathfrak{b}] = v \circ f(\mathfrak{b})$.*

(3) *The pairing is symmetric, that is* $[a, b] = [b, a]$.
(4) *Let* $\tau(a, b) = [a, b] - \langle a, b \rangle$. *Then for* a *fixed, and* $\deg^+(b)$ *bounded, the values* $\tau(a, b)$ *are bounded.*

Then $[a, b] = \langle a, b \rangle$.

Proof. The proof is essentially the same as the proof of Theorem 3.6. However, since we are dealing with cycles rational over K rather than having their components rational over K, there is a slight variation in the use of the Riemann–Roch theorem as follows. Let d_0 be the least common multiple of the degrees of divisors on V rational over K. We fix some positive multiple $m_0 d_0 = d > g$, and let $X \in Z(V)_K$ have degree d. Then for any $b \in Z_0(V)_K$ there is a positive divisor Y of degree d such that $b + X \sim Y$. Therefore at this point of the proof we again get $\tau(a, b) = \tau(a, X - Y)$, and the fourth property shows that $\tau(a, b)$ is bounded. Replacing b by mb for large positive integers m shows that $\tau(a, b) = 0$, thus proving the theorem.

§4. Reciprocity Laws

Let V, W be projective non-singular varieties over K. Let E be a divisor on the product. If $a \in Z_0(V)$ we let

$$E(a) = \mathrm{pr}_2[E \cdot (a \times W)].$$

Similarly, if $b \in Z_0(W)$ we let

$$'E(b) = \mathrm{pr}_1[E \cdot (V \times b)].$$

Then $E(a)$ is a divisor on W and $'E(b)$ is a divisor on V. Of course, E and $a \times W$ or $V \times b$ must intersect properly.

We shall also use the following notation on an abelian variety A. Let $a, b \in Z(A)$ be 0-cycles on A, say

$$a = \sum (a_i) \quad \text{and} \quad b = \sum (b_j).$$

We denote

$$a \otimes b = \sum_{i, j} (a_i + b_j).$$

Also as usual,

$$a^- = \sum (-a_i).$$

Theorem 4.1. *Let A be an abelian variety defined over K, with a proper absolute value v. For $D \in \mathrm{Div}(A)_K$ and $\mathfrak{a}, \mathfrak{b} \in Z_0(A)_K$ such that $D_\mathfrak{a}$ and $\mathfrak{b}$ have disjoint supports, we have*

$$\langle D_\mathfrak{a}, \mathfrak{b} \rangle_v = \langle D^-{}_\mathfrak{b}, \mathfrak{a} \rangle_v = \langle D_{\mathfrak{b}^-}, \mathfrak{a}^- \rangle_v.$$

Proof. By Theorem 2.2, the symbol is invariant under translations, and therefore

$$\langle D_\mathfrak{a}, \mathfrak{b} \rangle_v \quad \text{and} \quad \langle D_{\mathfrak{b}^-}, \mathfrak{a}^- \rangle = \langle D, \mathfrak{a}^- \otimes \mathfrak{b} \rangle_v.$$

Since the symbol is also invariant by automorphisms of the variety, the invariance under $x \mapsto -x$ shows that

$$\langle D^-{}_\mathfrak{b}, \mathfrak{a} \rangle_v = \langle D_{\mathfrak{b}^-}, \mathfrak{a}^- \rangle_v,$$

thereby proving the theorem.

Theorem 4.2. *Let V, W be projective non-singular varieties defined over the field K with a proper absolute value v. Let E be a divisor on the product, rational over K. Let $\mathfrak{a} \in Z_0(V)_K$ and $\mathfrak{b} \in Z_0(W)_K$ be such that $\mathfrak{a} \times \mathfrak{b}$ has disjoint support from E. Then*

$$\langle E(\mathfrak{a}), \mathfrak{b} \rangle_v = \langle {}'E(\mathfrak{b}), \mathfrak{a} \rangle_v.$$

Proof. By the general theory of divisorial correspondences on varieties, a divisor E on $V \times W$ is congruent to the pull back of a divisor on the product of Albanese varieties $A(V) \times A(W)$, modulo trivial correspondences, that is divisors of the form

$$(F) + X \times W + V \times Y,$$

where F is a rational function on the product, X a divisor on V, and Y a divisor on W. Cf. AV, Theorem 2 of Chapter VI, §3, and the seesaw principle, especially Theorem 6 of Appendix §2. Hence a simple pull back argument shows that it is sufficient to prove the theorem when V, W are abelian varieties A, B respectively. In this case, the theorem is also immediately verified when the divisor E is a trivial correspondence as above, since for a rational function F the value $F(a, b)$ at a point $(a, b) \in A \times B$ is equal to ${}'F(b, a)$. In each correspondence class we shall now find a suitable divisor for which we can verify the truth of the theorem by reducing it to Theorem 4.1.

Let $\gamma_E: A \to B'$ be the homomorphism $x \mapsto c_E(x) - c_E(0)$, where c_E is the linear equivalence class of E. Let Y be an ample divisor on B and as usual let $\varphi_Y \colon B \to B'$ be the homomorphism $\varphi_Y(u) = \mathrm{Cl}(Y_u - Y)$. Replacing E by a suitable positive integral multiple if necessary, there exists a homomorphism

$$\beta \colon A \to B$$

making the following diagram commutative.

Let G_β be the graph of β and let $Z = \sigma^{-1}(Y)$ where

$$\sigma \colon B \times B \to B$$

is the twisted sum, that is $\sigma(u, w) = w - u$. We take the twist so that

$$\sigma^{-1}(Y)(u) = Y_u$$

instead of Y_{-u}. Let

$$D = Z \circ G_\beta = \mathrm{pr}_{13}[(A \times Z) \cdot (G \times B)]$$

be the composition of correspondences as in AV, Appendix §2, Theorem 7. Then D is a divisor on $A \times B$ and for any cycle $\mathfrak{a}$ of degree 0 on A such that $D(\mathfrak{a})$ is defined, we have

$$D(\mathfrak{a}) = Y_{\beta(\mathfrak{a})}.$$

Hence D lies in the same correspondence class as E, and it will suffice to prove the theorem for D.

The formula for $D(\mathfrak{a})$ yields $\langle D(\mathfrak{a}), \mathfrak{b} \rangle_v = \langle Y_{\beta(\mathfrak{a})}, \mathfrak{b} \rangle_v$. On the other hand,

$${}^t\!D(\mathfrak{b}) = \beta^{-1}(Y^-{}_\mathfrak{b}).$$

Then we find

$$\langle {}^t\!D(\mathfrak{b}), \mathfrak{a} \rangle_v = \langle \beta^{-1}(Y^-{}_\mathfrak{b}), \mathfrak{a} \rangle_v$$
$$= \langle Y^-{}_\mathfrak{b}, \beta(\mathfrak{a}) \rangle_v \qquad \text{by NS 3}$$
$$= \langle Y_{\beta(\mathfrak{a})}, \mathfrak{b} \rangle_v \qquad \text{by Theorem 4.1.}$$

This concludes the proof.

§5. Néron Functions as Intersection Multiplicities

Let $S = \text{spec}(\mathfrak{o})$ where $\mathfrak{o}$ is a Dedekind ring, with fraction field K. By a **Néron model** over S (or over $\mathfrak{o}$) for an abelian variety over K we mean a group scheme $\mathbf{A}/S$ ($\mathbf{A}$ over S), which to simplify the notation we write as $\dot{\mathbf{A}}_S$, or $\mathbf{A}$, satisfying the following properties.

NM 1. $\mathbf{A}$ is smooth over S.

NM 2. The general fiber $\mathbf{A}_K$ is the given abelian variety.

NM 3. For every smooth morphism $X \to S$, a morphism $X_K \to \mathbf{A}_K$ extends uniquely to a morphism $X \to \mathbf{A}$ over S. In other words, the natural map

$$\text{Mor}_S(X, \mathbf{A}) \to \text{Mor}_K(X_K, \mathbf{A}_K)$$

obtained by extending the base from S to K is a bijection, and hence an isomorphism of abelian groups.

We are now using strict scheme terminology. For the definition of smooth morphism, see Hartshorne [Har], Chapter 3, §10; and for the definition of group scheme, see Chapter 4, §4. Note that the smoothness assumption implies in particular that $\mathbf{A}$ is regular, that is all the local rings of points on $\mathbf{A}$ are regular; and therefore any divisor on $\mathbf{A}$ is locally principal.

Néron [Ne 4] has proved the existence of Néron models.

The definition is phrased in such a way as to make $\mathbf{A}_S$ represent the functor

$$X \mapsto \text{Mor}_K(X_K, \mathbf{A}_K),$$

for X smooth over S. In the special case that $X = S$, we see that

$$\text{Mor}_S(S, \mathbf{A}) = \text{Sec}(S, \mathbf{A}) = \text{Sec}(\text{spec } K, A_K).$$

In other words, if $X = S$ the morphisms are just the sections. The values of the sections lie in the Néron model.

It may be useful for the reader to summarize here some other properties of Néron models.

First, Néron proves in [Ne 4] that if $\mathfrak{o}$ is a discrete valuation ring and $\hat{\mathfrak{o}}$ is its completion, then the Néron model over $\hat{\mathfrak{o}}$ is simply obtained by base extension from $\mathbf{A}$ over $\mathfrak{o}$. Note that $\hat{\mathfrak{o}}$ is not smooth over $\mathfrak{o}$, so this property does not follow automatically from the fact that $\mathbf{A}$ represents the point functor for smooth X over $S = \text{spec}(\mathfrak{o})$.

A special fiber $\mathbf{A}_s$ is an algebraic group, but need not be connected. However, the components are disjoint.

Furthermore, for all but a finite number of primes, the special fiber is connected, and is an abelian variety.

Suppose now that o is a discrete valuation ring.

By the third property of the Néron model, every rational point P of the generic fiber A_K in K comes from a unique section

$$\mathbf{P}: S \to \mathbf{A}.$$

Let D be a divisor on the generic fiber A_K, rational over K, so $D \in \mathrm{Div}(A_K)_K$ according to our notations. Suppose first that $D = W$ is a prime divisor. Let $\mathbf{W}$ be the closure of W in $\mathbf{A}$. Then $\mathbf{W}$ is a prime divisor of $\mathbf{A}$. Assume that $\mathbf{W}$ does not contain the image of $\mathbf{P}$. Since the divisor group on S is cyclic, generated by the special point (s), we can write the inverse image

$$\mathbf{P}^*\mathbf{W} = i(\mathbf{W}, \mathbf{P})(s),$$

where $i(\mathbf{W}, \mathbf{P})$ is an integer sometimes called the **intersection multiplicity** of $\mathbf{W}$ and $\mathbf{P}$ on $\mathbf{A}$ at s. Note that since W uniquely determines its closure, we can write $i(W, P)$ instead of $i(\mathbf{W}, \mathbf{P})$ without fear of confusion.

We then extend the association $W \mapsto \mathbf{W}$ to divisors by linearity, so to each divisor $D \in \mathrm{Div}(A_K)_K$ we associate a divisor $\mathbf{D}$ on $\mathbf{A}$, which we call its **thickening** (closure here might be confusing since we are dealing with multiplicities). We then obtain an injection

$$\mathrm{Div}(A_K)_K \to \mathrm{Div}(\mathbf{A}) \quad \text{by} \quad D \mapsto \mathbf{D}.$$

Similarly, we extend the intersection symbol $i(D, P)$ by linearity. For $P \notin |D|$, we then have by definition

$$\mathbf{P}^*\mathbf{D} = i(D, P)(s).$$

Let $\mathbf{A}_s$ be the special fiber of the Néron model. Let $\mathbf{A}_s^{(j)}$ $(j = 0, \ldots, r)$ be the connected components. If $P \in A(K)$, then $\mathbf{P}(s)$ lies in one of these components.

Theorem 5.1. *Let A be an abelian variety defined over K. Let $\mathbf{A}$ be its Néron model over the discrete valuation ring o, and let v be the valuation of o. Let $D \in \mathrm{Div}(A)_K$ and let λ_D be a Néron function with divisor D. For each connected component there is a constant γ_j such that for all*

$$P \in A(K) - |D|$$

and $\mathbf{P}(s) \in \mathbf{A}_s^{(j)}$ we have

$$\lambda_D(P, v) = i(D, P) + \gamma_j.$$

Proof. Let **D** be the thickening of D on **A**. Then **D** is represented by pairs (U, f) where U is open on **A** and f is a rational function in $K(A) = K(\mathbf{A})$. By definition, for any pair (U, f) such that f is defined at P and $f(P) \neq 0, \infty$ we have

$$i(\mathbf{D}, P) = v \circ f(P),$$

and the value on the right-hand side is independent of the choice of pair. Furthermore, such a pair exists since $P \notin |D|$. Hence the function $P \mapsto i(D, P)$ on $U \cap A(K)$ is equal to the Weil function λ_f. If (V, g) is another pair representing D, then $\lambda_f = \lambda_g$ on $U \cap V \cap A$. Hence $i(D, P)$ for $P \in A(K)$ is equal to the value of a Weil function with divisor D.

Let $D = (f)_A$ be the divisor of a rational function $f \in K(A) = K(\mathbf{A})$. The function f is determined only up to a scalar multiple in K, i.e. a rational function on the base S. Thus we have

$$\mathbf{D} = (f)_\mathbf{A} + Z_f,$$

where Z_f is a fibral divisor, that is the support of Z_f is contained in the inverse image of s in **A**. By assumption, $f(P)$ is defined and $\neq 0, \infty$, and we have

$$\mathbf{P}^*(f)_\mathbf{A} = v \circ f(P) \cdot (s).$$

Let $Z_f = \sum m(j, f) \mathbf{A}_s^{(j)}$. Each component $\mathbf{A}_s^{(j)}$ (which is a prime divisor on **A**) is represented locally by a parameter in $\mathfrak{o}$, that is an element in $\mathfrak{o}$ of order 1 at the valuation. Therefore

$$i(D, P) = \lambda_f(P, v) + m(j, f)$$

for all P such that $P(s) \in \mathbf{A}_s^{(j)}$.

Next a translation τ_u for $u \in A(K)$ induces an automorphism (for the scheme structure) of **A** over S. This follows from property **NM 3** because

$$\tau_u \in \mathrm{Mor}_K(A, A) \approx \mathrm{Mor}_S(\mathbf{A}, \mathbf{A}).$$

The local intersection multiplicity is invariant under automorphisms, and so $i(D, P)$ is invariant under translations.

Now consider only points P such that $\mathbf{P}(s)$ lies in the connected component $\mathbf{A}_s^{(0)}$, and consider the group of 0-cycles $\mathfrak{a}$ of degree 0, whose support lies in those points. Then by linearity, we have defined a symbol $i(D, \mathfrak{a})$. By Lemma 2.3, we have $i(D, \mathfrak{a}) = \langle D, \mathfrak{a} \rangle$, where $\langle D, \mathfrak{a} \rangle = \lambda_D(\mathfrak{a})$ is the symbol constructed by means of the Néron function. This proves Theorem 5.1 for points P such that $\mathbf{P}(s) \in \mathbf{A}_s^{(0)}$.

Finally, fix some index j and let $u \in A(K)$ be a point such that $u(s) \in \mathbf{A}_s^{(j)}$. (If there is no such point, we have nothing to worry about.) For any P such that $\mathbf{P}(s) \in \mathbf{A}_s^{(j)}$ we then get:

$$i(D, P) = i(D_{-u}, P_{-u}) = \lambda_{D_{-u}}(P_{-u}, v) + \gamma_0$$
$$= \lambda_D(P, v) + \gamma(u) + \gamma_0,$$

with some constant $\gamma(u)$, according to the translation properties of Néron functions. This proves the theorem.

Theorem 5.2. *Assume that λ_D is chosen so that $\gamma_0 = 0$. Let n be the exponent of the factor group $\mathbf{A}_s/\mathbf{A}_s^0$. Then $\gamma_j \in (1/n^2)\mathbf{Z}$ if the class of D is even; and $\gamma_j \in (1/n)\mathbf{Z}$ if the class of D is odd.*

Proof. We need the existence of enough rational points (e.g. Zariski dense), so we have to extend the base. One way is to use the fact that the Néron model does not change if we pass to the completion of K at v, which we assume. Also we may assume that the class of D is even or odd, say even. Write

$$[n]^*D = n^2 D + (f).$$

Then for $P \in A(K)$ such that $P \notin |D|$ and $nP \notin |D|$ we have

$$\lambda_D(nP) = n^2 \lambda_D(P) + v \circ f(P) + \text{constant},$$
$$i(D, nP) = i([n]^*D, P) = n^2 i(D, P) + v \circ f(P) + \text{integer}.$$

The above two expressions are equal by the definition of n and the assumption that $\gamma_0 = 0$. Taking first P such that $\mathbf{P}(s) \in \mathbf{A}_s^0$, we see from the right-hand side that the constant is an integer. Then taking P arbitrary, we conclude that the constant lies in $(1/n^2)\mathbf{Z}$, thereby proving the theorem.

Remark. The intersection number allows us to choose a Néron function in its equivalence class modulo constants. This is due to the thickening operation, which to each divisor on the generic fiber associates a unique divisor on the Néron model.

In Chapter 12, §3 we shall reprove in the context of heights a theorem which follows trivially from the considerations of this section. But there we shall work in a geometric context, and I thought it would be useful for the reader to have available self-contained treatments in each chapter.

§6. The Néron Symbol and Group Extensions

We begin by recalling the correspondence between extensions of an abelian variety by the multiplicative group $\mathbf{G}_m$, and divisor classes algebraically equivalent to 0. This is due to Weil [We 7]. For a more complete discussion of generalizations, see Serre [Se 1], Chapter VII.

First consider an exact sequence

$$0 \to \mathbf{G}_m \to \mathbf{G} \to A \to 0,$$

where A is an abelian variety, and $\mathbf{G}$ is some commutative group extension by the multiplicative group. Then an automorphism $\alpha: \mathbf{G} \to \mathbf{G}$ which is the identity on $\mathbf{G}_m$ (viewed as subgroup of $\mathbf{G}$) and induces the identity on A must be the identity, so the group extension has no automorphisms. Indeed, the map $x \mapsto \alpha(x)x^{-1}$ is equal to 1 for $x \in \mathbf{G}_m$; and for all $x \in \mathbf{G}$, we have $\alpha(x)x^{-1} \in \mathbf{G}_m$. But $\alpha(x)x^{-1}$ is defined as a morphism of A into $\mathbf{G}_m$, so is constant, equal to 1.

By a **section** s on an open set U of A we always mean a section which is a morphism. A rational map $s: U \to \mathbf{G}$ which is a section on some smaller open set will be called a **rational section**. The fiber in $\mathbf{G}$ above a generic point of A is a principal homogeneous space for $\mathbf{G}_m$, and consequently has a rational section by Hilbert's Theorem 90. This gives rise to some section over a suitably small open set. By translations, we then see that there is a covering $\{U_i\}$ by open sets, such that a section $s_i: U_i \to \mathbf{G}$ exists over each U_i.

Let U be any open set, with a section $s: U \to \mathbf{G}$. Then for each i we get a section

$$ss_i^{-1}: U \cap U_i \to \mathbf{G}_m.$$

which is actually into $\mathbf{G}_m$. Then ss_i^{-1} may be viewed as a rational function f_i on A. It is immediately verified that the family $\{(U_i, f_i)\}$ defines a divisor, which is independent of the choices of covering family U_i and sections s_i, and which will be called the **divisor of the section** s, denoted by (s).

The section s, a priori defined only on an open set U, is in fact a rational map of A into $\mathbf{G}$. It follows immediately from the definitions, and the fact that A is non-singular that s is a morphism on the complement of the support of its divisor. Cf. Proposition 4.3 of Chapter 2, applied to ss_i^{-1} for each i.

If f is a rational function on A, then s viewed as a rational map of A into $\mathbf{G}$ can be multiplied by f, namely for all $x \in A$ outside some proper closed subset, we have

$$(fs)(x) = f(x)s(x).$$

Then fs is again a rational section, and it is immediately verified that the corresponding divisors satisfy the relation

$$(fs) = (f) + (s).$$

Conversely, if s, t are two rational sections (morphic on possibly different open sets), then st^{-1} is a rational function on A. If s, t have the same divisor then st^{-1} is everywhere defined, as a morphism of A into $\mathbf{G}_m$, and is therefore constant. Thus we see that the divisor class of divisors of sections is a well-defined element of $\mathrm{Pic}(A)$ associated with the extension, and denoted by c_G. If D is a divisor in this class, the unique section having divisor D (up to a constant) will be denoted by s_D.

For example, if $\mathbf{G} = A \times \mathbf{G}_m$ and $D = (f)$ is the divisor of a rational function, then in this case the section s_D can be chosen (up to the constant factor) to be

$$s_D(x) = (0, f(x)).$$

As we know, $\mathrm{Pic}(A)$ has a group structure. One can also define a group structure on isomorphism classes of extensions of A by $\mathbf{G}_m$ as follows. Let $\mathbf{G}_1$, $\mathbf{G}_2$ be two such extensions. Then $\mathbf{G}_1 \times \mathbf{G}_2$ is an extension of $A \times A$. Let $\mathbf{G}_3$ be the pull back of $\mathbf{G}_1 \times \mathbf{G}_2$ to the diagonal.

$$\begin{array}{ccc} \mathbf{G}_3 & \longrightarrow & \mathbf{G}_1 \times \mathbf{G}_2 \\ \downarrow & & \downarrow \\ A & \longrightarrow & A \times A \end{array}$$

Then the isomorphism class of $\mathbf{G}_3$ is the sum of $\mathbf{G}_1$ and $\mathbf{G}_2$.

The trivial extension $A \times \mathbf{G}_m$ is the neutral element in this group law.

Theorem of Weil–Barsotti. *The association* $\mathbf{G} \mapsto c_G$ *gives an isomorphism*:

$$\left\{ \begin{array}{c} \textit{isomorphism classes of} \\ \textit{group extensions of } A \textit{ by } \mathbf{G}_m \end{array} \right\} \xrightarrow{\approx} \mathrm{Pic}_0(A).$$

In other words, the divisors of sections (s) obtained above are algebraically equivalent to 0; conversely, each element of $\mathrm{Pic}_0(A)$ can be obtained from an extension; and the correspondence is an isomorphism for the group laws. For a proof in more general contexts, see [Se 1], Chapter VII, especially §15, Theorem 5 and §16, Theorem 6.

If $\mathbf{G}$ is the group extension corresponding to the divisor class c then we write $\mathbf{G} = \mathbf{G}_c$, or $\mathbf{G} = \mathbf{G}_D$ if $D \in \mathrm{Div}_a(A)$ is an element of that class. We have an exact sequence

$$0 \to \mathbf{G}_m \to \mathbf{G}_c \to A \to 0.$$

We call $\mathbf{G}$ the **group extension of A by $\mathbf{G}_m$ defined by the divisor D or the divisor class** c.

Now suppose that A is defined over a locally compact field K, so $K = \mathbf{R}, \mathbf{C}$, finite extension of $\mathbf{Q}_p$, or power series over a finite field. Taking points of all the group varieties rational over K, we obtain an exact sequence of locally compact groups:

$$0 \to K^* \to \mathbf{G}_c(K) \xrightarrow{\text{pr}} A(K) \to 0.$$

The homomorphism on the right is surjective, because the inverse image of a point in $A(K)$ is a principal homogeneous space over K for $\mathbf{G}_m$, and has a K-rational point by Hilbert's Theorem 90. Note that $A(K)$ is compact since A is projective.

Let v be the normalized absolute value on K, so the p-adic absolute value; or the ordinary absolute value of $\mathbf{R}$ or $\mathbf{C}$; or the discrete valuation of a power series field. Then v defines a continuous homomorphism

$$v \colon K^* \to \mathbf{R}, \qquad \text{by} \qquad v(z) = -\log|z|_v.$$

Lemma 6.1. *There exists a unique extension of v to a continuous homomorphism $v_c \colon \mathbf{G}_c(K) \to \mathbf{R}$.*

The lemma is a special case of a standard result in the theory of locally compact groups as follows.

Let G be a commutative locally compact group, H a closed subgroup such that G/H is compact. Let $\lambda \colon H \to \mathbf{R}$ be a continuous homomorphism. Then λ has a unique extension to a continuous homomorphism of G into $\mathbf{R}$.

See Bourbaki, *Intégration*, Chapter VII, §3, No. 2, Proposition 4. Note that the extension is unique, because two extensions differ by a homomorphism of a compact group into $\mathbf{R}$, which must therefore be 0.

For the convenience of the reader, I reproduce a proof due to Tate. First we reduce the problem to the case when $H = \mathbf{R}$. We form the push-out of the homomorphisms:

$$
\begin{array}{ccc}
H & \xrightarrow{\text{inc}} & G \\
{\scriptstyle \lambda}\downarrow & & \downarrow \\
\mathbf{R} & \longrightarrow & G \oplus_H \mathbf{R}.
\end{array}
$$

This push out is the direct sum $G \oplus \mathbf{R}$ modulo the subgroup of elements $(h, \lambda(h))$ for $h \in H$. It is easily verified that we obtain another exact sequence below the given one:

$$
\begin{array}{ccccccc}
0 \to & H & \to & G & \to & G/H & \to 0 \\
& {\scriptstyle \lambda}\downarrow & & \downarrow & & \downarrow & \\
0 \to & \mathbf{R} & \to & G \oplus_H \mathbf{R} & \to & G/H & \to 0.
\end{array}
$$

If we can find a splitting of the second sequence, then we immediately get the desired homomorphism of G into $\mathbf{R}$.

We are now reduced to proving that an exact sequence

$$0 \to H \to G \to G/H \to 0$$

of locally compact groups, with $H = \mathbf{R}$ and G/H compact, splits. The sequence admits local continuous sections, and since the kernel is $\mathbf{R}$, by a partition of unity, we can construct a continuous section $s: G/H \to G$, whose image is therefore compact, and such that $s(0) = 0$. Let

$$u: G \to \mathbf{R}$$

be defined by $u(x) = x - s(\mathrm{pr}(x))$. Then $u(x) = x$ if $x \in H$. Now for $x \in G$, one verifies easily that the sequence $\{u(2^n x)/2^n\}$ is Cauchy. Indeed, writing $2^n = 2^{m+k}$ with m large, and k an arbitrary positive integer, the Cauchy property depends on an estimate for

$$2^k u(a) - u(2^k a) = 2^k u(a) - 2^{k-1} u(2a) + 2^{k-1} u(2a) - 2^{k-2} u(2^2 a) + \cdots.$$

The compactness of the image of s then shows that

$$\frac{2^k s(a) - s(2^k a)}{2^k}$$

is bounded for $a \in G/H$, and the Cauchy property follows at once. We define the limit

$$V(x) = \lim u(2^n x)/2^n.$$

Then $V: G \to H = \mathbf{R}$ gives the desired splitting. This proves the lemma.

The lemma now yields the following diagram:

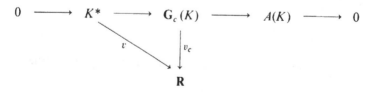

Theorem 6.2. *Assume that K is locally compact, A is defined over K. Let $D \in \mathrm{Div}_a(A)_K$ and $\mathfrak{a} \in Z_0(A(K))$ have disjoint supports. Then the Néron pairing is given by the expression*

$$\langle D, \mathfrak{a} \rangle_v = v_c \circ s_D(\mathfrak{a}),$$

where $c = \mathrm{Cl}(D)$, and $s_D(\mathfrak{a})$ is the extension of s_D to Z_0 by linearity.

Proof. It suffices to prove that $v_c \circ s_D(\mathfrak{a})$, as function of the pair $(D, \mathfrak{a})$, satisfies the characterizing properties of the Néron pairing.

The additivity in $\mathfrak{a}$ is obvious. For the additivity in D, let E be another divisor in $\mathrm{Div}_a(A)_K$. The construction of $\mathbf{G}_D$ is functorial, and one has a commutative diagram

$$
\begin{array}{ccc}
\mathbf{G}_{D+E} & \longrightarrow & \mathbf{G}_D \times \mathbf{G}_E \\
\downarrow & & \downarrow \\
A & \longrightarrow & A \times A
\end{array}
$$

where $\mathbf{G}_{D+E}$ is the pull-back of the product $\mathbf{G}_D \times \mathbf{G}_E$ by the diagonal map. Then the diagram of sections

$$
\begin{array}{ccc}
\mathbf{G}_{D+E} & \longrightarrow & \mathbf{G}_D \times \mathbf{G}_E \\
\uparrow s_{D+E} & & \uparrow s_D \times s_E \\
A & \longrightarrow & A \times A
\end{array}
$$

commutes up to a constant factor. The map $v_D \otimes v_E$ on $\mathbf{G}_D(K) \times \mathbf{G}_E(K)$ such that

$$(v_D \otimes v_E)(g, g') = v_D(g) + v_D(g')$$

gives a splitting of $\mathbf{G}_D \times \mathbf{G}_E$, so by uniqueness in Lemma 6.1, the additivity of $v_D \circ s_D(\mathfrak{a})$ with respect to D follows.

If $D = (f)$ is the divisor of a function, then $v_D \circ s_D(\mathfrak{a})$ is what it should be, since the section s_D is represented by f itself.

To see invariance under translation, let $u \in A(K)$. Let g be a point of $\mathbf{G}_D = \mathbf{G}_c$ above u. Let τ_u, τ_g be the translations by u and g respectively. Then we can take

$$s_{D_u} = \tau_g \circ s_D \circ \tau_{-u}.$$

Hence for $\mathfrak{a} \notin |D|$, we get $s_{D_u}(\mathfrak{a} + u) = g s_D(\mathfrak{a})$. If $\mathfrak{a} = \sum n(a)(a)$, then

$$v_c \circ s_{D_u}(\mathfrak{a}_u) = \sum n(a)[v_c(g) + v_c \circ s_D(\mathfrak{a})],$$

whence the invariance under translation follows since $\mathfrak{a}$ has degree 0.

Finally, the definition shows that the function

$$a \mapsto v_c \circ s_D((a) - (a_0))$$

is continuous on the complement of $|D|$. This concludes the proof.

Historical Note

Zarhin [Za] showed that the Néron symbols and height could be obtained from splitting Mumford's biextension [Mu 4]. At the time, most people did not realize its significance. Bloch [Bl] used the connection with simple extensions to reformulate the Birch–Swinnerton–Dyer conjecture in terms of Tamagawa numbers. To do biextensions here would have been too great an undertaking, so I limited myself to the single extensions, and I found the exposition of [Oe] useful. The group extension G_D associated with a divisor is one side of the biextension, and its splitting is one side of Zarhin's splitting of the biextension.

Mazur–Tate [Ma–T] pursue the biextension point of view. In the classical case of the Néron pairing, when the Néron functions are real valued, the splitting as in Lemma 6.1 is unique. However, if instead of the real valued v one takes representations into certain locally compact totally disconnected groups, then the splitting need not be unique. One of the main points of the Mazur–Tate paper is to obtain a canonical splitting, which is better than any other.

As Weil [We 7] originally remarked, group extensions of A by the multiplicative group represent differentials of third kind by pull back. In the case of curves over the complex numbers, the description of the Néron symbol in terms of the differentials of third kind will be given in Chapter 13, §7.

CHAPTER 12

Algebraic Families of Néron Functions

It is a general problem to estimate the difference between the Néron–Tate height h_c and the height h_φ coming from a projective embedding φ of A, where c is the class of the linear system containing the inverse image of a hyperplane by φ. Estimates for elliptic curves have been given by Manin [Man 4], Demjanenko [De 1], Zimmer [Zi]; and arising from local considerations by Lang [L 7], see also [L 15], Conjecture 5.

In the first edition of this book, I proposed the problem of investigating how the height depends on the parameter in an algebraic family, and what kind of uniformity one can obtain. Some results in this direction have been obtained by Colliot-Thélène [C–T], who remarks that base points in the linear system on elliptic curves cause lack of uniformity and severe variations in the height in special cases. A much deeper analysis is made by Manin–Zarhin [Ma–Z], who connect this problem on abelian varieties with that of estimating the difference between the Néron–Tate height and the projective height, and give an estimate in terms of a "canonical basis" for the linear system, using Mumford's fundamental papers on the equations satisfied by abelian varieties [Mum 3]. The estimate is very explicit, and for elliptic curves, can be viewed as having the same structure as that of [De 1], [Zi] or [L 7].

Silverman [Sil 1], [Sil 4] and Tate [Ta 5] have investigated the height of a point varying in an algebraic family and have also given a fundamental inequality for the difference between the canonical height and the height arising from a projective embedding. I am much indebted to Silverman for letting me have a copy of his paper [Sil 4]. The main result of §1 is a theorem of Silverman–Tate giving estimates for heights in algebraic families, and whose proof I have transcribed to apply to Néron functions. I am also indebted to Tate for his paper, where he refines and proves a conjecture of Silverman, reproduced in §5.

As already mentioned, only §1 and §2 of Chapter 11 are necessary for this chapter, except for the definition of a Néron model in Chapter 11, §5.

§1. Variation of Néron Functions in an Algebraic Family

Since we shall now deal with abelian varieties over function fields as in Chapter 3, we adopt the following notation.

We let k be a field with a proper set of absolute values. Let A, T be projective non-singular varieties defined over k, let y be a generic point of T, and let

$$\pi: A \to T$$

be a flat morphism defined over k, such that the generic fiber A_y of π is an abelian variety over $k(y) \approx k(T)$. We also assume that k has characteristic 0, because at some point we need the resolution of singularities. If this were known in arbitrary characteristic, the assumption would not be necessary.

For flat morphisms, see Hartshorne, Chapter III, §9. But as Hartshorne points out, p. 256, the condition of flatness is probably not fully needed here since we are dealing with non-singular varieties.

Let T^0 be a Zariski open subset of T so that for all points $t \in T^0(k^a)$ the fiber $A_t = \pi^{-1}(t)$ is an abelian variety, defined over k^a. Let

$$U = \pi^{-1}(T^0) = A \times_T T^0.$$

Then for each integer n, multiplication by n is a morphism

$$[n]: U/T \to U/T;$$

and this gives a rational map, also denoted by $[n]$,

$$[n]: A/T \to A/T.$$

Let $D \in \mathrm{Div}(A)_k$. Then we have a Weil function

$$\lambda_D = \lambda_{A,D}: [A(k^a) - |D|] \times M(k^a) \to \mathbf{R}$$

well-defined up to a bounded function. Furthermore, for each $t \in T^0(k^a)$ we let

$$D_t = D|A_t$$

be the restriction to any fiber A_t not contained in the support of D. We then have a Néron function λ_{D_t}, well defined up to a constant function. In fact, we shall see that there is a way of selecting the constants such that the functions λ_{D_t} differ only by suitable bounds from $\lambda_D | A_t$ on $A_t(k^a)$. We are interested in seeing how this bound varies for $t \in T^0$. We shall give such a bound in terms of Weil functions on T.

Theorem 1.1. *Let A, T be projective non-singular varieties, and let $\pi: A \to T$ be a flat morphism defined over k. Let T^0 be a Zariski open subset of T so that for all $t \in T^0(k^a)$ the fiber $A_t = \pi^{-1}(t)$ is an abelian variety defined over k^a. Let $D \in \mathrm{Div}(A)_k$, and λ_D a Weil function. There exists a positive divisor $Y \in \mathrm{Div}(T)_k$; a Weil function λ_Y; and for each $t \in T^0$ with $A_t \not\subset D$ there exists a Néron function λ_{D_t} such that*

$$|\lambda_{D_t} - \lambda_D | A_t| \leq \lambda_Y(t) \quad \text{for } t \notin |Y|.$$

The formula has been abbreviated as usual. Written in full, it means that for all $(P, v) \in [A_t(k^a) - |D_t|] \times M(k^a)$ we have

$$|\lambda_{D_t}(P, v) - \lambda_D(P, v)| \leq \lambda_Y(t, v).$$

Proof. The map $[2]: A/T \to A/T$ is merely rational. By [Har], Example 7.17.3 and the resolution of singularities, there exists a non-singular projective variety A'/T with a birational morphism

$$\varphi: A'/T \to A/T$$

satisfying:

(1) $\varphi: \varphi^{-1}(U) \to U$ is an isomorphism;
(2) $[2] \circ \varphi: A'/T \to A/T$ extends to a morphism.

We then define:

$U' = \varphi^{-1}(U)$;

$\psi: A'/T \to A/T$ the morphism extending $[2] \circ \varphi$;

$\pi' = \pi \circ \varphi: A' \to T$ the projection of A' onto T.

Let A_y be the generic fiber above a generic point of T over k, and let D_y be the restriction of D to A_y. Since $2D_y$ is the sum of an even and an odd divisor, the additivity of the height as function of D_y and the triangle inequality show that we may deal separately with the cases when D_y is even and when D_y is odd, that is when $D_y = D_y^-$ and $D_y = -D_y^-$. We take $D_y = D_y^-$ even for definiteness. For D_y odd, replace the usual 4 by 2 in the arguments, as in Chapter 11, §1, which we use freely. There exists a rational function f_y on A_y such that

$$\psi^*D_y = 4\varphi^*D_y + (f_y).$$

Consequently, there exists a divisor D_1 on A' and a rational function f_1 on A, in fact $f_1 \in k(A)$, such that

(1) $\psi^*D = 4\varphi^*D + (f_1 \circ \varphi) + D_1$;
(2) $\pi'(|D_1|)$ is a closed subset of T not equal to T.

We note the divisorial relation

$$(f_1 \circ \varphi)_{A'} = \varphi^*((f_1)_A).$$

For simplicity, we shall omit the subscripts A' and A on the divisors $(f_1 \circ \varphi)$ and (f_1) respectively.

Lemma 1.2. *There exists a positive divisor* $Y \in \mathrm{Div}(T)_k$, *a Weil function* λ_Y, *and for each* $t \in |T^0| - \pi'(|D_1|)$ *a Néron function* λ_{D_t} *such that*

$$|\lambda_{D_t} - \lambda_D|A_t| \leqq \lambda_Y(t) \quad \text{for } t \notin |Y|.$$

Proof. Take $P \in A_t$ such that

$$P \notin |D| \quad \text{and} \quad 2P \notin |D|.$$

By assumption on t, we also have $P \notin |D_1|$. Then

$$
\begin{aligned}
\lambda_D(2P, v) - 4\lambda_D(P, v) &= \lambda_D(\psi P', v) - 4\lambda_D(\varphi P', v) \\
&= (\lambda_D \circ \psi)(P', v) - 4(\lambda_D \circ \varphi)(P', v) \\
&= v \circ f_1 \circ \varphi(P') + \lambda_{D_1}(P', v) \\
&= v \circ f_1(P) + \lambda_{D_1}(P', v).
\end{aligned}
$$

Here we may choose λ_{D_1} in its class modulo bounded functions to make the equality true.

There exists a divisor Y on T, rational over k, such that

$$\pi'^*(Y) + D_1 > 0 \quad \text{and} \quad \pi'^*(Y) - D_1 > 0.$$

Indeed, we simply take Y to be a positive divisor with components whose union covers $\pi'(|D_1|)$, and with sufficiently high multiplicities (say the maximum of the multiplicities of the components of D_1). By Proposition

3.1 of Chapter 10, we can choose the Weil functions in their class modulo bounded functions such that

$$|\lambda_{D_1}(P', v)| \leq \lambda_{\pi'*Y}(P', v)$$
$$= \lambda_Y(\pi'P', v)$$
$$= \lambda_Y(\pi P, v) = \lambda_Y(t, v).$$

Hence we find

$$|\lambda_D(2P, v) - 4\lambda_D(P, v) - v \circ f_1(P)| \leq \lambda_Y(t, v).$$

We now replace λ_D by $\lambda_D | A_t$ and f_1 by $f_1 | A_t$ in the left-hand side to obtain precisely the expression defining $\gamma(P, v)$ in Chapter 11, Proposition 1.4, because of our assumption that $t \notin \pi'(|D_1|)$. By Chapter 11, Proposition 1.4, the bound in Lemma 1.2 of Chapter 11, and the definition $\lambda' = \lambda - \alpha$, there is a Néron function λ_{D_t} such that

$$|\lambda_{D_t}(P, v) - (\lambda_D | A_t)(P, v)| \leq 3\lambda_Y(t, v).$$

This proves Lemma 1.2.

Lemma 1.2 amounts to Theorem 1.1 with a supplementary restriction on t. We now move around by linear equivalence to get rid of this restriction. Write

$$D_1 = \sum_{i=1}^{r} m_i W_i,$$

where W_i are the absolutely irreducible components of D_1. For each i, $\pi'(W_i)$ is a subvariety of T, contained in some absolutely irreducible divisor X_i. If $X_i \cap T^0$ is not empty, then $\pi'^*(X_i)$ is an absolutely irreducible divisor on A' because for $t \in T^0$, $\pi'^{-1}(t)$ is an absolutely irreducible variety with multiplicity 1 (namely A_t) by assumption. Hence if $X_i \cap T^0$ is not empty, we have $W_i = \pi'^*(X_i)$. Therefore there exists a divisor X on T such that

$$D_1 = \pi'^*(X) + D_1' \quad \text{and} \quad \pi'(|D_1'|) \cap T^0 \text{ is empty.}$$

Let $t \in T^0$. There exists a rational function g on T such that

$$t \notin |X - (g)|.$$

Let

$$g_1 = g \circ \pi, \quad D_2 = D_1 - (g_1 \circ \varphi), \quad f_2 = f_1 g_1.$$

Then

$$(f_1 \circ \varphi) + D_1 = (f_2 \circ \varphi) + D_2.$$

Furthermore $t \notin \pi'(|D_2|)$.

Since T is compact, we can apply Lemma 1.2 a finite number of times and obtain positive divisors $Y_1, \ldots, Y_m$ such that for any given $t \in T^0$ and $A_t \not\subset |D|$ there exists some Y_j such that

$$|\lambda_{D_t} - \lambda_D|A_t| \leq \lambda_{Y_j}(t).$$

Finally we let $Y = \sup Y_j$ to conclude the proof of Theorem 1.1.

We can now formulate the corresponding result for heights. Assume that the product formula is satisfied. Let $c = \text{Cl}(D)$ and $c_t = \text{Cl}(D_t)$ so $c \in \text{Pic}(A)$ and $c_t \in \text{Pic}(A_t)$. For each $t \in T^0(k^a)$ we have the *canonical height*

$$h_{c_t} : A_t(k^a) \to \mathbf{R},$$

which is a quadratic function, differing by a bounded function from the restriction of h_c on $A_t(k^a)$. We are interested in seeing how this bound varies for $t \in T^0$. We shall give a bound in terms of the height on T, which is given with a projective embedding. Thus h_T refers to this given embedding.

Theorem 1.3 (Silverman–Tate). *Assume the product formula. Let*

$$c \in \text{Pic}(A).$$

There are numbers

$$\gamma_1 = \gamma_1(\pi, c) \qquad and \qquad \gamma_2 = \gamma_2(T, h_c)$$

such that for all $t \in T^0$ and $P \in \pi^{-1}(t)$ we have

$$|h_{c_t}(P) - h_c(P)| \leq \gamma_1 h_T(t) + \gamma_2.$$

Proof. This is an immediate consequence of Proposition 5.4 or Proposition 1.7 of Chapter 4, and of Theorem 1.1. In fact the proof of Theorem 1.1 was a transcription for Néron functions of the Silverman–Tate proof for heights.

Example 1. As Silverman remarks [Sil 4], the preceding theorem gives on abelian varieties estimates which can be expressed on elliptic curves as follows. Consider the 3-dimensional variety A defined on an affine open set in characteristic $\neq 2, 3$ by

$$y^2 = x^3 + ax + b,$$

viewing a, b as variables, and thus affine coordinates of $\mathbf{P}^2$. Let D be $3(O)$ where O is the zero section (an irreducible divisor on A). Then

$$h_D(P) = h((x(P), y(P), 1)).$$

Theorem 5.3 says that for all $t = (a, b, 1) \in \mathbf{P}^2(k^a)$ with $4a^3 + 27b^2 \neq 0$ and all $P \in A_t(k^a)$ with $P \neq O$ we have

$$|h_{D_t}(P) - h((x(P), y(P), 1))| \leq \gamma_1 h((a, b, 1)) + \gamma_2,$$

where γ_1, γ_2 are absolute constants. This is essentially the inequality mentioned at the beginning of the section for elliptic curves, except that one can determine these constants more explicitly for elliptic curves.

Of course, for abelian varieties, Manin–Zharin give an estimate of the same kind with respect to canonical coordinates [Ma–Z].

Example 2. Suppose that D is the polar divisor of a rational function $f \in k(A)$, and suppose again that we are in the elliptic curve case, that is $\dim A_t = 1$. We may select $h_D = h_f$, so that for all $P \in A_t(k^a)$ we have

$$h_D(P) = h(f(P)) = h(f_t(P)).$$

The Silverman–Tate theorem allows comparison between such families of heights h_{f_t} on A_t with the canonical height, and hence given two functions f, f' we can compare h_{f_t} and $h_{f'_t}$. This gives a theoretical background for the following example. The Fermat type curve

$$X^3 + Y^3 = dZ^3$$

can be mapped birationally on the Weierstrass curve

$$y^2 = x^3 + b \quad \text{with} \quad b = -2^4 3^3 d^2,$$

or in homogeneous form

$$w^2 z = u^3 - 2^4 3^3 d^2 z^3 = u^3 + bz^3,$$

by the transformation

$$u' = 2^2 3 dZ, \qquad w' = 2^2 3^2 d(Y - Z), \qquad z' = X + Y.$$

If one takes $d \in \mathbf{Z}$, and one wants to solve the equation for x, y in lowest form, with

$$x = u/z \quad \text{and} \quad y = w/z, \qquad u, w, z \in \mathbf{Z},$$

one must divide u', z' by their g.c.d. to get u, z. There are tables in the literature, for instance [Sel 1], [Sel 2] giving solutions of the Fermat form. If one puts these into Weierstrass form, one notices how the height h_x drops to about two-thirds of the height of $h_{X/Z}$. The Silverman–Tate theorem provides theoretical grounds for this, since X/Z has degree 3 and x has degree 2. Actually, the uniformity seems to be somewhat better than that arising from the Silverman–Tate theorem.

§2. Silverman's Height and Specialization Theorems

This section is due to Silverman. Throughout, we let k be a finite extension of the field F with a proper set of absolute values satisfying the product formula. We let T, A be projective non-singular varieties defined over k, and we let

$$\pi\colon A \to T$$

be a flat morphism defined over k, such that the generic fiber A_y is an abelian variety over the function field $k(y) \approx k(T)$.

As in the preceding section, there is a non-empty Zariski open set T^0 such that for all $t \in T^0(k^a)$ the fiber A_t is an abelian variety over k^a (in fact over $k(t)$).

In the main theorems, we shall also assume that the Néron–Severi group of T is cyclic. Recall the definition,

$$NS(T) = \mathrm{Div}(T)/\mathrm{Div}_a(T).$$

If it is cyclic, then the projective degree gives an embedding

$$\deg\colon NS(T) \to \mathbf{Z}.$$

This allows us to normalize heights on T a little more than previously. Let $X \in \mathrm{Div}(T)$ with $\deg X \neq 0$. We let

$$h_T = \frac{1}{\deg X} h_X.$$

If X' is another divisor with $\deg X' \neq 0$ and

$$h'_T = \frac{1}{\deg X'} h_{X'},$$

then by quasi-equivalence, Proposition 5.3 of Chapter 4, we may take the limit as any height $h(t)$ with respect to any positive divisor on T tends to infinity. We obtain

$$\lim_{h(t) \to \infty} h_T(t)/h'_T(t) = 1$$

because $\deg(X')X$ is algebraically equivalent to $\deg(X)X'$. This limit will be all that matters in the next theorem.

We shall follow the notation of the preceding section. In particular, if D is a divisor on A then we have the height h_D determined only up to a bounded function, and we have the canonical height h_{D_t} for $t \in T^0$, which is a quadratic function defined on $A_t(k^a)$. As these heights depend only on the linear equivalence classes, we shall denote them by h_c and h_{c_t} respectively, where $c = \mathrm{Cl}(D)$ and $c_t = \mathrm{Cl}(D_t)$.

Let $P: T \to A$ be a **section**, by which we here will mean a *morphism* such that $\pi \circ P = \mathrm{id}$. If $\dim T = 1$, so T is a curve, then any rational map $P: T \to A$ such that $\pi \circ P = \mathrm{id}$ generically is necessarily a section in the above sense, because any rational map of a non-singular complete curve is a morphism. In the higher dimensional case, this is a relatively rare occurrence, but when $A = A_0 \times T$ is a product, so the family splits, then again any rational map $P: T \to A_0$ is a morphism since T is assumed non-singular. (This is a theorem of Weil, cf. AV, Chapter II, §1, Theorem 2.)

To suggest families of points, we shall write P_t instead of $P(t)$, for $t \in T^0(k^a)$. On the other hand, we use y for a generic point. Then $P_y = P(y)$ is a rational point of A_y over $k(y)$. As in Proposition 3.2 of Chapter 3, we then have the height

$$h_{T,L}(P_y) = \deg P^*L,$$

where L is a hyperplane section of A in its given projective embedding, arising from the composition of mappings

$$T \overset{P}{\to} A \overset{\mathrm{inc}}{\to} \mathbf{P}^N.$$

On the other hand, we have the family of discrete valuations of $k(y)$ over k satisfying the product formula again as in §3 of Chapter 3. The restriction L_y of L to A_y is a hyperplane section of A_y in the embedding obtained by composition

$$A_y \overset{\mathrm{inc}}{\to} A \overset{\mathrm{inc}}{\to} \mathbf{P}^N,$$

and consequently the projective height of P_y as a point in $A_y(k(y))$ with respect to this embedding is precisely $\deg(P^*L)$. For the canonical height, this implies:

Lemma 2.1. *Under the basic assumptions, let $c \in \mathrm{Pic}(A)$, let c_y be its restriction to the generic fiber, and h_{c_y} the canonical height associated with the class. Let $P: T \to A$ be a section. Then*

$$h_{c_y}(P_y) = \deg(P^*c) + O(1).$$

Proof. The arguments before the lemma prove it when c is very ample. The general case follows by the additivity of both sides of the equation mod $O(1)$.

Theorem 2.2. *In addition to the basic assumptions, assume $NS(T)$ is cyclic. Let $P: T \to A$ be a section such that for arbitrarily large integers n, $[n]P$ is also a section (so a morphism). Let $c \in Pic(A)$, and let h_{c_y} be the canonical height on $A_y(k(y)^a)$ associated with the class c_y. Then*

$$\lim_{\substack{h(t) \to \infty \\ t \in T^0(k^a)}} h_{c_t}(P_t)/h_T(t) = h_{c_y}(P_y).$$

Proof. Constants and $O(1)$ in what follows will depend on A, c, π only. Dependence on P in addition will be stated explicitly in the notation when it occurs.

Since the height takes values in a uniquely divisible group, namely $\mathbf{R}$, without loss of generality it suffices to prove the desired relation with $2c$ replacing c. Again without loss of generality, we may prove the theorem separately in the cases when c_y is even and when c_y is odd.

For $t \in T^0(k^a)$ we find:

$$|h_{c_t}(P_t) - h_{c_y}(P_y)h_T(t)| \leq |h_{c_t}(P_t) - h_c(P_t)| + |h_c(P_t) - h_{c_y}(P_y)h_T(t)|$$

$$\leq \gamma_1 h_T(t) + \gamma_2$$

$$+ |h_{P*c}(t) + O_P(1) - [\deg(P*c) + O(1)]h_T(t)|$$

by Theorem 1.3, the functoriality of the height, and Lemma 2.1 respectively. We divide by $h_T(t)$ to get:

$$|h_{c_t}(P_t)/h_T(t) - h_{c_y}(P_y)| \leq \gamma_1 + \varepsilon_P(t)$$

$$+ |h_{P*c}(t)/h_T(t) - \deg(P*c) + O(1)|$$

with $\lim \varepsilon_P(t) = 0$ as $h(t)$ goes to infinity. By quasi-equivalence, Corollary 3.4 of Chapter 4, for $X \in Div(T)$,

$$\lim_{h(t) \to \infty} h_X(t)/h_T(t) = \deg X.$$

Letting $X = P*c$ and letting $h(t) \to \infty$ yields

$$|h_{c_t}(P_t)/h_T(t) - h_{c_y}(P_y)| \leq \gamma_3$$

with a constant γ_3 independent of P. We now replace P by nP for a large positive integer n such that $[n]P$ is a morphism. Since the heights of P_t and P_y on the left are canonical heights, they are either purely quadratic or homogeneous linear in n, so a factor of n^2 or n will come out. We divide by this factor and take the limit as $n \to \infty$. Then the right-hand side goes to 0, thereby proving the theorem.

Remark. The two conditions that $NS(T)$ is cyclic and that $[n]P$ is a section, are satisfied when T has dimension 1, which is one of the main cases of application for the theorem. When T has dimension 1, any rational map of T is a morphism.

We shall now give Silverman's application of his height theorem to his specialization theorem. Again we let y be a generic point of T over k so $k(y) \approx k(T)$. Let (τ, B) be the $k(y)/k$-trace of the generic fiber A_y, as in Chapter 6. By the Lang–Néron Theorem 2, we know that $A_y(k(y))/\tau B(k)$ is finitely generated. For each $t \in T^0(k^a)$ we have the specialization homomorphism

$$\sigma_t \colon A_y(k(y)) \to A_t(k^a)$$

denoted by $P \to P_t$, assuming that rational sections are morphisms.

Theorem 2.3. *Assume that $NS(T)$ is cyclic and that rational sections are morphisms. Let Γ be a finitely generated free subgroup of $A_y(k(y))$ which injects in the quotient $A_y(k(y))/\tau B(k)$. Then the set of $t \in T^0(k^a)$ such that σ_t is not injective on Γ has bounded height in $T^0(k^a)$. In particular, if k is a number field, there is only a finite number of points $t \in T^0$ of bounded degree over k such that σ_t is not injective on Γ.*

Proof. The second statement is a consequence of the first since the set of points of bounded degree and bounded height on a variety over number fields is finite. We now prove the first statement. Let $P^1, \ldots, P^r$ be a basis of Γ. Let c be an even very ample divisor class on A. Let $\langle \; . \; , \; \rangle_t$ be the positive symmetric bilinear form derived from h_{c_t}; and let $\langle \; , \; \rangle_y$ be the positive symmetric bilinear form derived from h_{c_y}. By Theorem 2.2, for two sections P, Q we get

$$\lim_{h(t) \to \infty} \langle P_t, Q_t \rangle_t / h_T(t) = \langle P_y, Q_y \rangle_y.$$

In particular,

$$\lim_{h(t) \to \infty} \det\langle P_t^i, P_t^j \rangle_t / h_T(t)^r = \det\langle P_y^i, P_y^j \rangle_y.$$

The right-hand side is $\neq 0$ by Theorem 5.4 of Chapter 6. Therefore as soon as the height $h(t)$ is sufficiently large, the determinant on the left is also $\neq 0$. This proves that the specialization homomorphism is injective on Γ when $h(t)$ is large, and concludes the proof.

Remark. Suppose that A_0 is an elliptic curve, and $T = A_0$. Then we have some obvious rational points of A_0 over $k(T)$, by corresponding generic points. Since any point of A_0 in k^a is a specialization of a generic point, this example shows that one cannot expect the specialization σ_t to be injective on all of $A_0(T)$. However, by splitting off the fixed part and considering an isogeny of A_y in the theorem with B and an abelian variety whose $k(T)/k$-trace is 0, one sees that this type of example is essentially the only possible one.

We also note that Silverman's theorem strengthens considerably what Néron obtained by the Hilbert irreducibility theorem in Chapter 9.

Silverman's theorems extend to arbitrary families results of Demjanenko [De 2] and Manin [Man 5] for the split case, which can now be recovered as a corollary.

Theorem 2.4 (Demjanenko–Manin). *Let A_0 be an abelian variety over a number field k. Let T be a projective nonsingular variety over k, and let $t_0 \in T(k)$ be a k-rational point. Let Γ be the group of morphisms $f: T \to A_0$ such that $f(t_0) = 0$. If $NS(T)$ is cyclic, and if*

$$\text{rank } \Gamma > \text{rank } A_0(k),$$

then $T(k)$ is finite.

Proof. We let $A = A_0 \times T$ and $\pi: A \to T$ the projection on T, so the algebraic family of Theorem 2.3 splits. Then Γ is a finitely generated group of sections, and since $A_0 = B$ is the $k(T)/k$-trace of the generic fiber, it follows from the condition $f(t_0) = 0$ that Γ has also the same rank mod $\tau B(k)$, or in other words mod $B(k)$ which is none other than $A_0(k)$. Theorem 2.4 is then a corollary of Theorem 2.3.

§3. Néron Heights as Intersection Multiplicities

In Chapter 11, §5 we gave Néron's interpretation of the Néron function in terms of local intersection multiplicities. We shall work here in a more global context, and develop the analogous result for the height ab ovo, on a completion of the Néron model, which however need not be viewed as such. In other words, we do not need to know about Néron models in this section.

Let k be an algebraically closed field, which need not have characteristic 0. Let T be a complete non-singular curve over k. Let A be a projective non-singular variety over k, and let

$$\pi: A \to T$$

be a surjective morphism. If y is a generic point of T over k, we let $K = k(y)$. We let $A_y = \pi^{-1}(y)$ be the generic fiber. We assume that A_y is a non-singular variety (so absolutely irreducible by our general conventions).

We define groups of divisors as follows:

$\mathrm{Div}(A, T)_k = \mathrm{Div}(A, T) =$ group of divisors in $\mathrm{Div}(A)_k$ generated by the prime divisors W such that $\pi(W) = T$. We call $\mathrm{Div}(A, T)$ the group of π-**dominant divisors**.

$\mathrm{Div}_\pi(A)_k = \mathrm{Div}_\pi(A) =$ group of divisors in $\mathrm{Div}(A)_k$ generated by irreducible fibral divisors, that is irreducible components of inverse images $\pi^{-1}(t)$ for $t \in T(k)$. We call $\mathrm{Div}_\pi(A)$ the group of **fibral divisors**.

We have a direct sum decomposition

$$\mathrm{Div}(A)_k = \mathrm{Div}(A, T) \oplus \mathrm{Div}_\pi(A).$$

Furthermore, we have a natural injection

$$\mathrm{Div}(T)_k \to \mathrm{Div}_\pi(A)$$

given by

$$Y \mapsto \pi^{-1}(Y).$$

There is also an intersection map (restriction to the generic fiber)

$$\rho_y: \mathrm{Div}(A)_k \to \mathrm{Div}(A_y)_{k(y)}$$

given by

$$\rho_y(D) = D \cdot A_y.$$

We may view $D \cdot A_y$ as the pull back of D to A_y under the inclusion map $A_y \subset A$. The homomorphism ρ_y vanishes on the group of fibral divisors $\mathrm{Div}_\pi(A)$.

Proposition 3.1. *We have an exact sequence:*

$$0 \to \mathrm{Div}_\pi(A) \to \mathrm{Div}(A)_k \to \mathrm{Div}(A_y)_{k(y)} \to 0$$

and an isomorphism $\rho_y \colon \mathrm{Div}(A, T) \to \mathrm{Div}(A_y)_{k(y)}$.

Proof. This kind of statement belongs to the foundations of algebraic geometry, and is also true in the context of schemes. In fact, the elements of $\mathrm{Div}(A, T)$ are just the thickenings as defined in Chapter 11, §5.

Let x be a generic point of A such that $\pi(x) = y$. If $z \in k(x)$ then there is a unique rational function $f \in k(A)$ such that $f(x) = z$. Then by definition of the generic fiber, the restriction of f to A_y is a rational function on A_y, denoted by f_y or $f | A_y$, and by the general properties of induced functions on subvarieties we have

$$(f_y)_{A_y} = (f)_A \cdot A_y.$$

Proposition 3.2. *Let $f \in k(A)$. If (f) is a fibral divisor then there exists a rational function $g \in k(T)$ such that $f = g \circ \pi$.*

Proof. We view $k(T)$ as contained in $k(A)$, and identify these functions fields with $k(y) \subset k(x)$. Let $z = f(x)$. If $z \notin k(y)$, then z has a zero and a pole on the generic fiber, since we assume that the generic fiber A_y is absolutely irreducible. Hence the divisor of f cannot be fibral, thus proving the proposition.

The proposition shows that we have an injection

$$0 \to \mathrm{Div}_\pi(A)/\pi^* \, \mathrm{Div}_{\mathfrak{l}}(T) \to \mathrm{Pic}(A).$$

The subgroup of divisor classes on A corresponding to the factor group on the left will be denoted by $\mathrm{Pic}_\pi(A)$ and will be called the **fibral divisor class group.** Note that the proposition implies in particular that

$$\pi^* \colon \mathrm{Pic}(T) \to \mathrm{Pic}(A)$$

is injective, and thus we have an inclusion

$$\pi^* \, \mathrm{Pic}(T) \subset \mathrm{Pic}_\pi(A).$$

If all fibers $\pi^{-1}(t)$ for all $t \in T(k)$ are irreducible, then this inclusion is an equality.

The above conditions made no further assumptions on A. We now assume that the generic fiber is an abelian variety. Let

$$P: T \to A$$

be a section. If y is a generic point of T over k then $P(y) \in A_y(k(y))$ and we have the translation

$$\tau_{P(y)}: A_y \to A_y,$$

which defines a birational map

$$\tau_P: A \to A$$

commuting with the projection $\pi: A \to T$. As in [Man 4], **we now assume**:

TR.   *For every section P, the birational map τ_P is a morphism, and therefore an automorphism since it has the inverse τ_{-P}.*

Proposition 3.3.   *The following diagram is commutative*:

and such that for x generic on A_y over $k(y)$ we have

$$\tau_P(x) = x + P(y).$$

This map τ_P maps a fiber $\pi^{-1}(t)$ into itself. If $t \in T^0(k)$, and $x \in A_t = \pi^{-1}(t)$ then

$$\tau_P(x) = x + P(t).$$

Proof. Of course, the diagram at first is only generically commutative, but once we know that τ_P is an isomorphism, then it has to be commutative at all points, so τ_P maps fibers into fibers. If the fiber is non-degenerate, then the meaning of non-degeneracy gives the compatibility of the addition on the special fiber with addition on the generic fiber, so the final formula holds. This proves the proposition.

The assumption concerning translations being morphisms is a natural one. One could also consider the following condition:

MM.   The variety A is a **minimal model over** T. This means: given any morphism

$$g: V \to T$$

of a projective non-singular variety V onto T, and a birational map $g_*: V \to A$ which makes the diagram commutative (for the rational maps involved):

then the map g_* is a morphism.

When $\dim A_y > 1$ then in general one does not expect such a minimal model to exist (so Mumford tells me). When $\dim A_y = 1$, the theorem is due to Shafarevich [Sh 3], p. 131. In any case, we shall call a family $\pi: A \to T$ satisfying these assumptions a **minimal model for a family of abelian varieties parametrized by** T, or **over** T; briefly: a **minimal model**.

If a minimal model exists, then **TR** is satisfied, in other words, **MM** implies **TR**. This follows at once from the minimality condition. Since one does not expect **MM** to be satisfied in higher dimensions, we have taken **TR** as the axiom.

We now look into applications of **TR** and Proposition 3.3. Since for each $P \in \mathrm{Sec}(T, A)$ the map τ_P is an automorphism, it follows that τ_P induces an automorphism of $\mathrm{Pic}(A)$, and also an automorphism of the fibral group $\mathrm{Pic}_\pi(A)$ by Proposition 3.3. Hence $\mathrm{Sec}(T, A)$ (or equivalently $A_y(k(y))$) operates as a group of automorphisms of $\mathrm{Pic}_\pi(A)$. We are especially interested in this representation of $\mathrm{Sec}(T, A)$.

Since T is a curve, there is only a finite number of points $t \in T(k)$ such that $\pi^{-1}(t)$ is degenerate. In any case, $\pi^{-1}(t)$ consists only of a finite number of irreducible components, which are therefore permuted by the action of τ_P. In particular, τ_P is represented in the permutation group of the components of such fibers. We define:

$A_y(k(y))_0 \approx \mathrm{Sec}(T, A)_0 =$ subgroup of sections P such that τ_P induces the identity on $\mathrm{Div}_\pi(A)$, i.e. on the group of fibral divisors.

By the above remarks, $A_y(k(y))_0$ is of finite index in $A_y(k(y))$; or in the notation of sections, $\mathrm{Sec}(T, A)_0$ is of finite index in the group of all sections $\mathrm{Sec}(T, A)$.

If every fiber $\pi^{-1}(t)$ is irreducible for all $t \in T(k)$, then

$$\mathrm{Sec}(T, A) = \mathrm{Sec}(T, A)_0,$$

in other words $\mathrm{Sec}(T, A)$ acts trivially on $\mathrm{Div}_\pi(A)$.

We now follow Manin [Man 4]. The results are less general than Néron's, but the exposition draws more on global geometric concepts which are worth while recording here.

Let P be a section. We shall write

$$\tau_P D = D_P.$$

If we restrict to the generic fiber, then $\tau_{P(y)} D_y = (D_y)_{P(y)}$ is the translation by $P(y)$. If P, Q are sections, we let

$$D_{P,Q} = D_{P+Q} - D_P - D_Q + D.$$

Inducing $D_{P,Q}$ on the generic fiber A_y gives the divisor of a rational function f_y defined over $k(y)$. This function is uniquely determined up to an element of $k(y)$, that is a non-zero function on T. There exists a function

$$f_{P,Q} = f \in k(A)$$

such that

$$f \mid A_y = f_y,$$

and such a function f is also determined up to an element of $k(T)$. By Proposition 3.1 there exists a fibral divisor $Z_{P,Q} = Z$ such that

$$D_{P,Q} = (f) + Z.$$

By Proposition 3.2 the class of Z in $\text{Pic}_\pi(A)$ is independent of the choice of f and will be denoted by $c_{P,Q}$, so that this class is characterized by the formula

$$\text{Cl}(D_{P,Q}) = c_{P,Q},$$

where $c = \text{Cl}(D)$ is the class of D in $\text{Pic}(A)$. It is obvious that $c_{P,Q}$ is symmetric in P, Q that is

$$c_{P,Q} = c_{Q,P}.$$

Proposition 3.4. *Assume that A satisfies* **TR**. *For fixed Q the map $P \mapsto c_{P,Q}$ is a 1-cocycle of $\text{Sec}(T, A)$ in $\text{Pic}_\pi(A)$. The map*

$$(P, Q) \mapsto c_{P,Q}$$

is bilinear for $P, Q \in \text{Sec}(T, A)_0$.

Proof. Let R be a third section. If

$$D_{P,Q} = (f_{P,Q}) + Z_{P,Q}$$

as previously, where $Z_{P,Q}$ is a fibral divisor, then

$$(D_{P,Q})_R + D_{R,Q} = D_{P+R,Q}$$

and

$$(D_{P,Q})_R = (\tau_R f_{P,Q}) + \tau_R Z_{P,Q}.$$

Furthermore, $\tau_R Z_{P,Q}$ is a fibral divisor by Proposition 3.3. The desired relation is then an immediate consequence of the definitions, since with the choice of function $\tau_R f_{P,Q}$ to determine $\tau_R c_{P,Q}$ we actually find

$$\tau_R Z_{P,Q} + Z_{R,Q} \sim Z_{P+R,Q},$$

which is the cocycle relation. If τ_R operates trivially on the Picard group, then the cocycle relation becomes a bilinear relation, thereby proving the proposition.

Proposition 3.5. *Assume that A satisfies* **TR**. *Let $c \in \mathrm{Pic}(A)$. Let $h_{c,y}$ be the canonical Néron height on the generic fiber A_y associated with the class c_y induced by c on A_y. If $P \in \mathrm{Sec}(T, A)_0$ then*

$$h_{c,y}(P(y)) = \deg[c.P(T) - c.O(T)]$$
$$= \deg(P^*c - O^*c),$$

where O is the zero section, and the intersection products are taken on A.

Proof. The equality

$$\deg c.P(T) = \deg P^*c$$

is a special case of the general fact that a morphism induces a contravariant ring homomorphism for the intersection product, and

$$P^*(c.P(T)) = P^*c.P^*P(T) = P^*c.T = P^*c.$$

By linearity in c we may assume that c is the linear class of a hyperplane section L of A in a given projective embedding. Let h_L denote the height with respect to this embedding. By Proposition 3.2 of Chapter 3 we have

$$h_L(P(y)) = \deg c_L.P(T).$$

It will now suffice to prove that if we subtract a constant, specifically $\deg c \cdot O(T)$, then the function

$$h: P \mapsto \deg[c \cdot P(T) - c \cdot O(T)]$$

is quadratic (this function takes the value 0 at $P = O$). As usual, we form the derived function $\Delta h(P, Q)$ which we have to verify is bilinear. But the bilinearity condition is immediately seen to be equivalent with the bilinearity of $c_{P,Q}$ which was proved in Proposition 3.4, after using the fact that τ_P is an automorphism of A, and so

$$\deg c \cdot P(T) = \deg[\tau_P(c_{-P} \cdot O(T))] = \deg c_{-P} \cdot O(T).$$

This concludes the proof of the theorem.

§4. Fibral Divisors

The goal of this section is to reach Theorem 4.9. I am indebted to Mumford for the proof via Theorems 4.5 and 4.7 concerned with numerical properties of fibral divisors.

For the beginning of this section, by a **curve** T we mean either a complete non-singular curve, or spec(o) where o is a Dedekind ring. By a **surface**, we shall mean either a non-singular projective variety of dimension 2, or a regular scheme V projective over spec(o), according to the two preceding cases, so in either case we have the proper structural morphism

$$g: V \to T.$$

For each $t \in T$ we have the fiber $V_t = F = g^{-1}(t)$. A divisor D on V is called **fibral** if its support is contained in a fiber. From the definition of intersection numbers, and the functoriality of the pull backs of divisors to curves, it is an elementary fact that if D is fibral and $E = (f)$ is the divisor of a function on V then $\deg(D.E) = 0$. Thus we may speak of the numerical equivalence class of a fibral divisor D, which we denote by c_D. We use the standard abbreviation $\deg(c.c') = (c.c')$ if c, c' are fibral.

Proposition 4.1. Let $g: V \to T$ be a projective morphism of a surface V on a curve T as above. Assume that the generic fiber is geometrically connected. Let $F = g^{-1}(t)$ be any fiber, and c any fibral class. Then $(c.c_F) = 0$. In particular, $(c_F^2) = 0$.

Proof. Without loss of generality, we may assume that o is a discrete valuation ring, because intersection numbers are defined locally. Let π be a prime element of o, and s the special point. Then the fiber V_s is defined by the equation $\pi = 0$, so V_s is a principal divisor, from which the desired assertion follows by the remarks preceding the proposition.

Keeping the same notation, let

$$F = \sum n_i C_i$$

be the expression of a fiber as a sum of absolutely irreducible components with multiplicities $n_i > 0$. Let

$$a_{ij} = (n_i C_i . n_j C_j).$$

We call (a_{ij}) the **intersection matrix** of the fiber.

Proposition 4.2. *The intersection matrix of the fiber has the following properties.*

(1) $a_{ij} = a_{ji}$ *for all i, j; that is the matrix is symmetric.*
(2) $a_{ij} \geq 0$ *if $i \neq j$.*
(3) *The row sums are 0, that is $\sum_j a_{ij} = 0$ for all i.*
(4) *For $i \neq j$ define i to be linked to j if $a_{ij} > 0$. Then any two distinct indices are connected by a chain of such linkings; that is, if $i \neq j$ there is a sequence $i = i_1, i_2, \ldots, i_m = j$ such that i_v is linked to i_{v+1}.*

Proof. The symmetry is obvious. If $i \neq j$ then $C_i \neq C_j$ so $(C_i . C_j)$ is defined and ≥ 0, which proves (2) since $n_i > 0$ for all i. The row sums are 0 because $\sum n_j C_j = F$ and $(c . c_F) = 0$ by Proposition 4.1, for any fibral class c. This proves (3). Property (4) immediately follows from the fact that a fiber F is connected (Zariski's connectedness theorem, see [Har], Chapter III, Corollary 11.3). This concludes the proof.

Lemma 4.3. *Let (a_{ij}) be a real symmetric matrix satisfying the conditions of the preceding proposition. Then its corresponding quadratic form is negative, and the null space is one-dimensional, spanned by $(1, 1, \ldots, 1)$.*

Proof. For any vector $(\ldots, x_i, \ldots)$ we have

$$\sum_{i,j} a_{ij} x_i x_j = - \sum_{i<j} a_{ij}(x_i - x_j)^2 \leq 0,$$

after using (1), (2), (3) of Proposition 4.2. If $x_i \neq x_j$ for some pair i, j then a linking chain as in (4) shows that the sum over $i < j$ has at least one non-zero term and so the sum is $\neq 0$. This proves the lemma.

Proposition 4.4. *Let c be a fibral class. Then* $(c^2) \leqq 0$; *and* $(c^2) = 0$ *if and only if c is a rational multiple of* c_F.

Proof. This is an immediate corollary of Proposition 4.2 and the lemma.

However more is true, and we shall need the following stronger statement.

Theorem 4.5. *Let X be a fibral divisor, whose support is contained in the fiber* $|F|$. *If* $(c_X^2) = 0$ *then* $X = qF$ *for some rational number q.*

Proof. Let

$$X = \sum x_i n_i C_i$$

with rational x_i. Applying the lemma to $(\dots, x_i, \dots)$ proves our proposition.

Since distinct fibers are orthogonal to each other, the proposition immediately yields that if X is a fibral divisor such that $(c_X^2) = 0$ then there exist fibers $F_1, \dots, F_m$ and rational numbers $q_1, \dots, q_m$ such that

$$X = q_1 F_1 + \cdots + q_m F_m.$$

In particular, there exists a positive integer q and a divisor Y on T such that

$$qX = g^{-1}(Y).$$

We shall now apply this result to higher dimensional fiberings, and we return to the basic assumptions of the preceding section.

For the rest of this section, let

$$\pi: A \to T$$

be a surjective morphism of projective non-singular varieties, and assume T is a curve. Also assume that the generic fiber is a variety. We suppose that π *is defined over k as before.*

In Theorem 4.5, the rational number q need not be an integer. There is a simple criterion when it is an integer, also applicable to higher dimensional families, as follows.

Proposition 4.6. *Assume that there exists a section* $P: T \to A$. *Let*

$$A_t = \pi^{-1}(t) = \sum n_i W_i$$

be the expression of a special fiber as a sum of irreducible components. Then $P(t)$ lies in exactly one component $W = W_1$ (say), and $n_1 = 1$, that is W occurs with multiplicity 1. If Z_t is a fibral divisor with support contained in $|A_t|$, and $Z_t = qA_t$ where q is rational, then in fact q is an integer.

Proof. This is immediate from the formalism

$$P^{-1}(X . Y) = P^{-1}(X) . P^{-1}(Y)$$

under conditions of proper intersection which are trivially satisfied. Here we take $X = \pi^{-1}(t)$ and $Y = P(T)$. Since $\pi \circ P = \text{id}$, the first stated result follows because $(t) . T = (t)$ with multiplicity 1. The second statement is then clear since g.c.d. $n_i = 1$.

Let dim $A = n$ and let H^{n-2} be a generic linear variety so that $A . H^{n-2} = V$ is a non-singular surface. The restriction of π to V is then a morphism

$$g: V \to T$$

of the type considered at the beginning of the section. Write

$$\pi^{-1}(t) = \sum n_i W_i$$

as above. If W is an irreducible component of this fiber, then $W . H^{n-2}$ will be called a **generic curve** on the fiber. It may have singularities.

Theorem 4.7. *Let $\pi: A \to T$ be a surjective morphism of a projective non-singular variety onto a non-singular curve, such that the generic fiber is a variety. Let Z be a fibral divisor on A such that for every generic curve C on a fiber we have*

$$(Z . C)_A = 0.$$

Then there exists a rational number q and a divisor Y on T such that $Z = q\pi^{-1}(Y)$. If π has a section, we can take $q = 1$.

Proof. Without loss of generality, we may assume that the support of Z is contained in some fiber $\pi^{-1}(t)$. Let $Z = \sum m_i W_i$. Let

$$Z_1 = Z . H^{n-2} = \sum m_i C_i$$

where $C_i = W_i . H^{n-2}$ is an irreducible curve on V. Also

$$\pi^{-1}(t) . H^{n-2} = \sum n_i C_i.$$

By assumption, letting $C = W . H^{n-2}$ for $W = W_i$, some i, we get

$$0 = (Z.C)_A = (Z_1.C)_V.$$

Therefore $(Z_1.Z_1)_V = 0$. By Theorem 4.5, $Z_1 = qg^{-1}(t)$ for some rational number q. But intersecting with a generic H^{n-2} is an injective map on the components W_i so it follows that

$$Z = q\pi^{-1}(t).$$

If π has a section then we can apply Proposition 4.6 to conclude that q is an integer, thereby proving the theorem.

As in §3, we further specialize the situation to the case when the generic fiber is an abelian variety. For our purposes, however, the axiom **TR** involving sections is not sufficient, and we shall need a stronger one. For this purpose we need the Néron model as defined in Chapter 11, §4. Of course instead of $S = \text{spec}(\mathfrak{o})$, we can take S to be covered by a finite number of such open sets, as with the curve T which has been used as our parameter curve.

Mumford has suggested the possibility of the following property being true for some completion of a Néron model **A**.

GCNM. *There exists a projective flat morphism $A \to S$ such that the Néron model is an open subset of A over S, and such that the group multiplication $A \times_S A \to A$ extends to a morphism*

$$A \times_S A \to A.$$

Observe that this axiom is stronger than **TR**, as in §3, because it gives an additional algebraic structure to the family of translations. Such a completion would not necessarily be uniquely determined up to an isomorphism over S. We shall now assume that it exists, and we consider the case when $S = T$ is a complete curve. Then $\pi: A \to T$ is the type of morphism we have been considering, but now we have the added property giving the action of the Néron model on this completion. If $\pi: A \to T$ satisfies **GCNM** we shall say that π (or A) is a **good completion of the Néron model.**

If the dimension of the generic fiber is 1, that is if A is a family of elliptic curves, then a minimal model over T as in §3 is a good completion, whose existence is therefore accepted as a fact.

From condition **GCNM** we deduce that for each fiber $\pi^{-1}(t) = A_t$ there is an action

$$A_t \times A_t \to A_t.$$

The next lemma refers to this action.

Lemma 4.8. *Let* $t \in T(k)$. *Let* $P: T \to A$ *be a section. The following conditions are equivalent:*

(1) $P(t)$ *leaves every irreducible component stable;*
(2) $P(t)$ *leaves the* $O(t)$-*component stable;*
(3) $P(t)$ *lies in the* $O(t)$-*component.*

Proof. Suppose $P(t)$ leaves the $O(t)$-component stable. By the definition of a Néron model, $O(t)$ lies in the Néron part $\mathbf{A}_t$. Then $P(t) + O(t) = P(t)$ shows that $P(t)$ lies in the $O(t)$-component. Next suppose that $P(t)$ lies in the $O(t)$-component. Let $\mathbf{A}_t^0$ be the connected component of the identity. Since $\mathbf{A}$ is a group scheme, $\mathbf{A}_t^0$ is a connected algebraic group over k, whose zero element is precisely $O(t)$. Since $\mathbf{A}_t$ is represented as a permutation group of the components of $\pi^{-1}(t)$, which form a finite set, we get a representation of $\mathbf{A}_t^0(k)$ as a group of permutations of this finite set. Since $\mathbf{A}_t^0$ is connected, all elements of $\mathbf{A}_t^0(k)$ give the identity representation, so leave every component stable. Since by assumption $P(t)$ lies in this component, it follows that $P(t)$ leaves every component stable. This proves the lemma.

Theorem 4.9. *Assume that A is a good completion of the Néron model. Let $P, Q \in \operatorname{Sec}(T, A)_0$ and let $D \in \operatorname{Div}(A)_k$ be a divisor on A. Write*

$$D_{P+Q} - D_P - D_Q + D = (f) + Z,$$

where Z is a fibral divisor and $f \in k(A)$. Then there exists a divisor Y on T such that $Z = \pi^{-1}(Y)$.

Proof. By Theorem 4.7 it will suffice to prove that $(Z.C) = 0$ for every generic curve $C = W.H^{n-2}$ on an irreducible component of a fiber A_t. From the definition of Z, it will suffice to prove that

$$(D_P.C) = (D.C).$$

As in §3, let τ_P be the automorphism of A corresponding to translation by P on the generic fiber. Then

$$\begin{aligned}
\tau_P(D).C &= \tau_P(D).(W.H^{n-2}) \\
&= (\tau_P(D).W).H^{n-2} \\
&= \tau_P(D.W).H^{n-2} \\
&= \tau_{P(t)}(D.W).H^{n-2}
\end{aligned}$$

where $P(t) \in \mathbf{A}_t^0(k)$ by assumption, the lemma, and the definition of a Néron model. But for x ranging over $\mathbf{A}_t^0(k)$, the cycles $\tau_x(D.W)$ are algebraically equivalent on A, since they are parametrized by the algebraic family arising from the operation

$$\mathbf{A}_t^0 \times A_t \to A_t.$$

Therefore the intersection numbers

$$\deg \tau_{P(t)}(D.W).H^{n-2} \quad \text{and} \quad \deg (D.W).H^{n-2}$$

are equal. This proves the theorem.

§5. The Height Determined by a Section: Tate's Theorem

This section transposes to varieties of dimension ≥ 1 a theorem of Tate [Ta 5] concerning heights on a family of elliptic curves. The theorem arose from a conjecture by Silverman. The proof follows [Ta 5] without essential change, except that we use the main result of §4 giving a criterion for a fibral divisor to be an integral multiple of the total fiber. This was harder to handle in the higher dimensional case, and depends on the existence of a good completion for the Néron model, whose existence is not known today in dimension > 1.

This section may be viewed as a continuation of §4. We work under the same assumptions and notation:

Let $\pi: A \to T$ be a surjective morphism of projective non-singular varieties such that A is a good completion of a Néron model, all defined over the algebraically closed field k. We suppose that T has dimension 1, so T is a curve.

Let $O: T \to A$ be the zero section and $P: T \to A$ any section. We have trivially

$$\tau_O \circ \tau_P = \tau_P \circ \tau_O = \tau_P \quad \text{and} \quad P = \tau_P \circ O.$$

The first relation is between automorphisms of A, and the second between maps of T into A.

To simplify the notation, let us abbreviate formal linear combinations as follows:

$$[\tau_P, \tau_Q] = \tau_{P+Q} - \tau_P - \tau_Q + \tau_O,$$
$$[P, Q] = (P + Q) - (P) - (Q) + (O).$$

Then for instance if $c \in \text{Pic}(A)$ we have

$$[P, Q]^*(c) = (P + Q)^*c - P^*c - Q^*c + O^*c,$$

and similarly for $[\tau_P, \tau_Q]^*c$ (inverse image), or $[\tau_P, \tau_Q]_* c$ (direct image).

Proposition 5.1. *Let* $c \in \text{Pic}(A)$. *Let* $P, Q \in \text{Sec}(T, A)_0$. *Then*

$$[\tau_P, \tau_Q]^*c = \pi^*[P, Q]^*c.$$

Proof. By Theorem 4.9 we know that the class $[\tau_P, \tau_Q]^*c$ on the left-hand side is a pull back π^*a for some $a \in \text{Pic}(T)$. From $P^* = O^* \circ \tau_P^*$ (and similarly with Q and $P + Q$) we get

$$\pi^*[P, Q]^*c = \pi^*O^*[\tau_{P,}, \tau_Q]^*c = \pi^*O^*\pi^*a = \pi^*a = [\tau_P, \tau_Q]^*c$$

because $\pi \circ O = \text{id}$ and $O^*\pi^* = \text{id}$. This proves the proposition.

Theorem 5.2. *Suppose now that k is the algebraic closure of a field k_0 with a proper set of absolute values satisfying the product formula. Let $P \in \text{Sec}(T, A)_0$. Let $c \in \text{Pic}(A)$ and let $c_t = c | A_t$ be the restriction of c to A_t for $t \in T^0(k)$, where T^0 is the subset of points t such that the fiber A_t is a non-degenerate specialization of the generic fiber. Let:*

$h_t = h_{c_t}$ = *canonical height on* $A_t(k)$;

q_t = *homogeneous quadratic part of* h_t.

Then for $t \in T^0(k)$ we have

$$q_t(P(t)) = \tfrac{1}{2}h_{[P, P]^*c}(t) + O_P(1).$$

Proof. We shall abbreviate $P(t)$ by P_t. By the functoriality of heights under morphisms, and Proposition 5.1, for any section $Q \in \text{Sec}(T, A)_0$ and $R \in \text{Sec}(T, A)$ we find:

$$h_{[P, Q]^*c}(t) = h_{\pi^*[P, Q]^*c}(R_t) + O_{P, Q}(1)$$

$$= h_{[\tau_P, \tau_Q]^*c}(R_t) + O_{P, Q}(1)$$

$$= h_c(P_t + Q_t + R_t) - h_c(P_t + Q_t) - h_c(Q_t + R_t) + h_c(R_t) + \beta(R_t),$$

where $\beta = \beta_{P, Q}$ is a bounded function, say $|\beta(R_t)| \leq \gamma$. We let

$$R = iP + jQ \quad \text{for } i, j \in \mathbf{Z}.$$

By iP we mean of course $\tau_P^i(O)$, and similarly for jQ. We sum for $0 \leq i$, $j \leq n - 1$. With the abbreviation $B(i, j) = h_c(iP_t + jQ_t)$, we find:

$$n^2 h_{[P, Q]^*c}(t) = \sum_{i, j} [B(i + 1, j + 1) - B(i + 1, j) - B(i, j + 1) + B(i, j)] + \beta_n$$

$$= B(n, n) - B(n, 0) - B(0, n) + B(0, 0) + \beta_n,$$

where

$$\beta_n = \sum_{i, j} \beta(iP_t + jQ_t).$$

Divide by n^2. We obtain

$$h_{[P, Q]^*c}(t) = \frac{1}{n^2} h_c(n(P_t + Q_t)) - \frac{1}{n^2} h_c(nP_t) - \frac{1}{n^2} h_c(nQ_t) + \beta'_n(t),$$

where $\beta'_n(t)$ is the average for $0 \leq i, j \leq n - 1$ of the numbers $\beta_n(iP_t + jQ_t)$, and is therefore bounded by the original bound for β, $|\beta'_n(t)| \leq \gamma$. Let n tend to infinity. We know that the quadratic form q_t can be obtained as the limit

$$q_t(x) = \lim_{n \to \infty} \frac{1}{n^2} h_c(nx)$$

for $x \in A_t(k)$. Hence

$$h_{[P, Q]^*c}(t) = q_t(P_t + Q_t) - q_t(P_t) - q_t(Q_t) + O_{P, Q}(1).$$

Putting $P = Q$ yields the theorem.

Remark. The above argument catches only the quadratic part. It is more elaborate to get both the linear and the quadratic part. When the dimension of the generic fiber is > 1, then one cannot expect $[-1]$ to be an automorphism of the total space A, and thus one cannot a priori split a divisor class on A into an even and an odd class. If the dimension of the generic fiber is 1, this can obviously be done, and one can prove an analogous result for an odd class, by exactly the same type of argument. We leave it as an exercise to the reader. At this point, I did not want to go much further into the theory, I merely wanted to illustrate Tate's argument in the simplest case.

Corollary 5.3 (Silverman's conjecture). *Suppose $T = \mathbf{P}^1$ is the projective line, and h_T is the standard height on $\mathbf{P}^1$. Let y be a generic point of T and let q_y be the quadratic form of the canonical height $h_{c, y}$ on the generic fiber. For $P \in \mathrm{Sec}(T, A)$ (and not necessarily in $\mathrm{Sec}(T, A)_0$), P fixed, and $t \in T^0(k)$, t variable, we have*

$$q_t(P(t)) = q_y(P)h_T(t) + O_P(1).$$

Proof. Both q_t and q_y satisfy the property $q(mx) = m^2 q(x)$. Hence it suffices to prove the corollary when P is replaced by mP for some positive integer m, and hence we may assume $P \in \mathrm{Sec}(T, A)_0$ after all, since $\mathrm{Sec}(T, A)_0$ is of finite index in $\mathrm{Sec}(T, A)$. Next, let a be a divisor class of degree 1 on T. Then

$$[P, P]^*c = (\deg[P, P]^*c)a + b,$$

where b has degree 0. Hence by Theorem 5.2 and Proposition 3.5,

$$q_t(P_t) = q_y(P)h_a(t) + h_b(t) + O_P(1).$$

But on the projective line, $b = 0$ so h_b can be taken to 0, and h_a can be taken to be h_T. This proves the corollary.

Corollary 5.4. *Let T be an arbitrary curve of genus ≥ 1. and let a be a divisor class of degree 1 on T. Then*

$$q_t(P_t) = q_y(P)h_a(t) + O_P(h_a(t)^{1/2}) + O_P(1).$$

Proof. The proof is the same as in the preceding corollary, taking into account Proposition 5.4 of Chapter 5.

Remark. The above results give another proof for Silverman's Theorem 2.2 when T is a curve.

CHAPTER 13

Néron Functions Over the Complex Numbers

We start by giving Néron's formula for the Néron function of an abelian variety over the complex numbers in terms of the normalized theta function associated with a divisor.

As with heights, additional things can be said on curves and on the relation of Néron functions pulled back to the curve from the Jacobian. According to Arakelov [Ar 1], [Ar 2], the Néron function over the complex numbers can be expressed in terms of the theta function on the Jacobian, in terms of a Green's function on the curve, or in terms of differentials of third kind. However, Arakelov's papers are very sketchy. Hriljac [Hri] has given complete details how to get a Green's function on the curve as a pull back of the log of the Riemann theta function. This will be carried out in the present chapter.

Throughout this chapter, we assume all varieties defined over the complex numbers.

The analytic treatment of this chapter will be self-contained from [L 18]. For related facts and more extensive treatments, see Griffiths–Harris [Gr–H] and Mumford [Mum 4]. It turns out that what we need here consists only of basic facts, which might all have been included in [L 18].

§1. The Néron Function of an Abelian Variety

Let A be an abelian variety over $\mathbf{C}$, so we have an analytic isomorphism

$$p: \mathbf{C}^n/\Lambda \to A(\mathbf{C}),$$

where Λ is a lattice in $\mathbf{C}^n$. To each divisor D on $A(\mathbf{C})$, there is a normalized theta function F_D on $\mathbf{C}^n$ whose divisor is $p^{-1}(D)$, and which is uniquely determined up to a constant factor. Cf. for instance [L 18]. By definition, a **normalized theta function** is a meromorphic function on $\mathbf{C}^n$ satisfying the condition

$$F(z + u) = F(z) \exp\left[\pi H(z, u) + \frac{\pi}{2} H(u, u) + 2\pi\sqrt{-1}\, K(u) \right],$$

where H is a hermitian form, and $K(u)$ is real valued. *Since we work analytically throughout, we identify $A(\mathbf{C})$ with $\mathbf{C}^n/\Lambda$.*

To make this chapter as self-contained as possible, we recall here the definition of a Néron function. To each divisor D, Néron associates a function $\lambda_D: A - |D| \to \mathbf{R}$ satisfying the following properties and uniquely determined by them up to an additive constant: the association $D \mapsto \lambda_D$ is additive (modulo constants); λ_D is continuous outside the support of D; we have

$$\lambda_{D_a}(z) = \lambda_D(z - a) + c(a)$$

for some constant $c(a)$ (we call this invariance under translation); if D is defined on an open set U by a meromorphic function f, then there exists a continuous function α on U such that

$$\lambda_D(z) = -\log|f(z)| + \alpha(z).$$

Theorem 1.1. *Let F_D be the normalized theta function with divisor D. Let*

$$\lambda_D(z) = -\log|F_D(z)| + \frac{\pi}{2} H(z, z).$$

Then λ_D is a Néron function whose divisor is D.

Proof. It is immediately verified that the expression on the right is periodic with respect to Λ, so defined on $\mathbf{C}^n/\Lambda$. Since $\mathbf{C}$ is locally compact, and $A(\mathbf{C})$ is compact, it follows at once that λ_D is continuous on the complement of $|D|$, in fact real analytic, and bounded on every closed subset of this complement. From the definition of the theta function representing the divisor, it follows at once that λ_D has the right logarithmic singularity on the divisor D. And finally, one sees from the transformation equation with respect to translations by lattice points that λ_D is invariant under translations, and is therefore a Néron function. This proves the theorem.

We observe that the Néron function is in fact real analytic outside the support of its divisor.

Presumably in the case of full multiplicative reduction, using Morikawa's theta function [Mor], one can give an explicit form of the Néron functions as done by Tate for elliptic curves, cf. [L 7], Chapter 3, §5.

As a matter of terminology, the hermitian form H will be said to be **associated with the divisor** D. The same H is associated with any divisor algebraically equivalent to D. Cf. [L 18] for these foundational matters concerning theta functions and the associated hermitian forms.

Let z_i $(i = 1, \ldots, n)$ be the complex coordinates on $\mathbf{C}^n$. If $z_i = x_i + \sqrt{-1} y_i$ then we have the differential operators

$$\frac{\partial}{\partial z_i} = \frac{1}{2}\left(\frac{\partial}{\partial x_i} - \sqrt{-1}\,\frac{\partial}{\partial y_i}\right) \quad \text{and} \quad \frac{\partial}{\partial \bar{z}_i} = \frac{1}{2}\left(\frac{\partial}{\partial x_i} + \sqrt{-1}\,\frac{\partial}{\partial y_i}\right).$$

We apply these operators to differential forms in the usual way. For functions f, we have the operator

$$\partial f = \sum_{i=1}^{n} \frac{\partial f}{\partial z_i} \, dz_i \quad \text{and} \quad \bar{\partial} f = \sum_{i=1}^{n} \frac{\partial f}{\partial \bar{z}_i} \, d\bar{z}_i.$$

If $\omega = g(z, \bar{z}) \, d\bar{z}$ then $\partial \omega = \dfrac{\partial g}{\partial z} \, dz \wedge d\bar{z}$, so

$$\partial \bar{\partial} f = \sum_{i,j=1}^{n} \frac{\partial^2 f}{\partial z_i \partial \bar{z}_j} \, dz_i \wedge d\bar{z}_j.$$

If $n = 1$, and $z = x + \sqrt{-1} y$, then

$$dz \wedge d\bar{z} = \sqrt{-1}\, 2\, dy \wedge dx \quad \text{and} \quad \sqrt{-1}\, dz \wedge d\bar{z} = 2\, dx \wedge dy.$$

If f is analytic then $\bar{\partial} f = 0$, so these operators ∂ and $\bar{\partial}$ do not depend on the choice of analytic coordinates.

Proposition 1.2. *Let D be a divisor on A and let H be its associated hermitian form. Then*

$$\partial \bar{\partial} \lambda_D(z) = \partial \bar{\partial} \, \frac{\pi}{2} H(z, z).$$

Proof. This is immediate because F_D is analytic, so $\partial \bar{\partial}$ kills $\log|F_D|$.

We let $M = (h_{ij})$ be the matrix representing H. Then for any column vectors z, w we have

$$H(z, w) = {}^t z M \bar{w} = \sum_{i,j} h_{ij} z_i \bar{w}_j.$$

Furthermore, the 2-form of Proposition 1.2 is given by

$$\boxed{\partial \bar{\partial} H(z, z) = \sum_{i,j} h_{ij} dz_i \wedge d\bar{z}_j.}$$

The form

$$\frac{\sqrt{-1}}{2} \, \partial \bar{\partial} H(z, z)$$

is called the **2-form associated with the divisor** D, or with the Hermitian form H. We are interested in this 2-form especially in the case of curves, their Jacobian, and the theta divisor. The purpose of the next sections is to develop the basic theory of this 2-form.

One may define a **Green's function** associated with the divisor D to be a function $G_D: A - |D| \to \mathbf{R}$ satisfying the following properties:

GF 1. G_D *is* C^∞ *except on the support of* D. *If* D *is represented by a rational function* f *on the open set* U, *then*

$$G_D(z) = -\log|f(z)| + C^\infty - function \ on \ U.$$

GF 2. $\partial\bar\partial G_D = \partial\bar\partial \dfrac{\pi}{2} H(z, z).$

GF 3. *We have*

$$\int G_D \, d\mu_A = 0$$

where $d\mu_A$ *is the flat metric on* A, *induced from the Euclidean metric on* $\mathbf{C}^n$, *and such that* $\mathbf{C}^n/\Lambda$ *has volume* 1.

Then Proposition 1.2 can be stated by saying that the Néron function is a Green's function for the divisor, when normalized to satisfy **GF 3.**

§2. The Scalar Product of Differentials of First Kind

Let C be a complete non-singular curve of genus g over the complex numbers. For this section and the next we identify C and its complex points $C(\mathbf{C})$. We fix a point P_0 on C, taken as origin, and we let

$$\psi : C \to J$$

be the map into the Jacobian such that $\psi(P_0) = 0$. As with C, we identify J with $J(\mathbf{C})$. Then $\psi(P)$ represents the divisor class of $(P) - (P_0)$. We have the summation map

$$S = S_\psi : Z(C) \to J$$

of 0-cycles on C into J. In previous chapters, we studied these maps algebraically. Here we go into their analytic theory, and [L 18] provides a sufficient background.

We suppose C (identified with its Riemann surface as stated above) represented as a quotient of a polygon with identified sides.

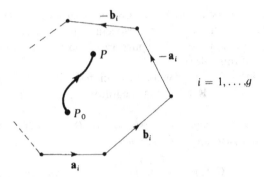

Proposition 2.1. *Relative to this representation, there exists a basis* $(\varphi_1, \ldots, \varphi_g) = \Phi$ *for the differentials of first kind (dfk), such that*

$$\int_{\mathbf{a}_i} \varphi_j = \delta_{ij}.$$

Let

$$\tau_{ij} = \int_{\mathbf{b}_i} \varphi_j,$$

and let $\tau = (\tau_{ij})$. *Then* τ *is symmetric, and* Im τ *is positive definite.*

The first statement follows by duality [L 18], Chapter IV, §5. The second is just the Riemann relations, Chapter IV, §4.

The period lattice Λ of Φ has the form $(1_g, \tau)$, with $\tau \in \mathfrak{H}_g$, where $\mathfrak{H}_g$ is the Siegel upper half space. A basis as in Proposition 2.1 will be said to be **normalized**.

A vector in $\mathbf{C}^g$ is written as a column vector,

$$z = \begin{pmatrix} z_1 \\ \vdots \\ z_g \end{pmatrix},$$

the coordinates being those of the normalized basis. The Abel–Jacobi theorem yields an analytic representation of ψ as in [L 18], Chapter IV, namely

$$\psi(P) = \int_{P_0}^{P} \Phi \mod \Lambda,$$

so $\psi \colon C \to \mathbf{C}^g/\Lambda$ gives an identification of the canonical map of C into its Jacobian J, itself identified with $\mathbf{C}^g/\Lambda$. Furthermore, for a normalized basis we have

$$\varphi_i = \psi^*(dz_i) \quad \text{for} \quad i = 1, \ldots, n.$$

If φ, η are two differentials of the first kind, then we may define their **scalar product**

$$\langle \varphi, \eta \rangle = \frac{\sqrt{-1}}{2} \int_C \varphi \wedge \bar{\eta}.$$

This defines a structure of Hilbert space on the g-dimensional space of dfk. Indeed, if z is a local analytic coordinate on some open set of C, and

$$z = x + \sqrt{-1}y,$$

then $\varphi(z) = f(z)\,dz$ for some analytic function f, $\bar{\varphi} = \overline{f(z)}\,d\bar{z}$, and

$$\varphi \wedge \bar{\varphi} = |f(z)|^2\,dz \wedge d\bar{z}$$
$$= |f(z)|^2\sqrt{-1}\,2\,dy \wedge dx,$$

so $\sqrt{-1}\varphi \wedge \bar{\varphi} = |f(z)|^2 2\,dx \wedge dy$.

In general, a differential form is said to be **positive** if when expressed locally in the form

$$h(x, y)\,dx \wedge dy,$$

the function $h(x, y)$ is real and positive. We see that $\sqrt{-1}\,\varphi \wedge \bar{\varphi}$ is a positive 2-form (except for a finite number of zeros). A positive 2-form is also called a **volume form**.

On the other hand, the pull-back

$$\psi^*: \mathrm{dfk}(J) \to \mathrm{dfk}(C)$$

from the differentials of first kind (invariant differentials) on J to those of C is an isomorphism of vector spaces. Therefore, if φ, η are dfk on J, then we can also define a hermitian product

$$\langle \varphi, \eta \rangle = \langle \psi^*\varphi, \psi^*\eta \rangle$$

by pull back to the curve.

Proposition 2.2. *Suppose that $(\varphi_1, \ldots, \varphi_g)$ is a normalized basis for the dfk on C. Let $(1_g, \tau)$ be the period matrix as in Proposition 2.1. Then*

$$\langle \varphi_i, \varphi_j \rangle = \langle dz_i, dz_j \rangle = \mathrm{Im}\ \tau_{ij}.$$

Proof. We need a lemma from the basic theory of dfk.

Lemma 2.3. *Let φ, η be differentials of first kind on C. Then*

$$\int_C \varphi \wedge \bar{\eta} = \sum_{k=1}^{g} \int_{\mathbf{b}_k} \varphi \cdot \int_{\mathbf{a}_k} \bar{\eta} - \sum_{k=1}^{g} \int_{\mathbf{a}_k} \varphi \cdot \int_{\mathbf{b}_k} \bar{\eta}.$$

Proof. The integral over C is equal to the double integral over the interior of the polygon

$$\iint_{\text{Int } \mathscr{P}} \varphi \wedge \bar{\eta},$$

using the same representation of C as a quotient of a polygon as before. Define

$$f(P) = \int_{P_0}^{P} \varphi.$$

Then f is uniquely defined inside the polygon, by taking the integral over a path lying entirely inside the polygon. On the boundary, we take paths leading to the boundary as in [L 18], Chapter IV, §1. We get functions f^+ on the segments $\mathbf{a}_i, \mathbf{b}_i$ ($i = 1, \ldots, g$) and functions f^- on the segments $-\mathbf{a}_i, -\mathbf{b}_i$.

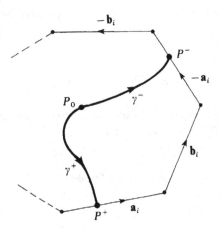

On the figure, we have shown two liftings of P to the sides $\mathbf{a}_i$ and $-\mathbf{a}_i$ before identification. The value of f is determined by integration over the paths γ^+ and γ^- as shown. Then for P on $\mathbf{a}_i$ we get

$$f^-(P) - f^+(P) = \int_{\mathbf{b}_i} \varphi,$$

and for P on $\mathbf{b}_i$ we get

$$f^-(P) - f^+(P) = \int_{-\mathbf{a}_i} \varphi.$$

We have

$$d(f\bar{\eta}) = df \wedge \bar{\eta} = \varphi \wedge \bar{\eta}.$$

Therefore by Green's theorem (Stokes' theorem in two dimensions), we get the lemma.

Now put $\varphi = \varphi_i$ and $\eta = \varphi_j$. Using the hypothesis of normalization, and the fact that $\tau_{ij} = \tau_{ji}$ the desired formula drops out from the definitions. This proves the proposition.

Let $\eta_1, \ldots, \eta_g$ be an orthogonal basis for the dfk, whether on the Jacobian or on the curve since there is a natural isometry between the spaces of dfk on both. Then the *positive form*

$$\frac{\sqrt{-1}}{2g}(\eta_1 \wedge \bar{\eta}_1 + \cdots + \eta_g \wedge \bar{\eta}_g)$$

is independent of the choice of orthonormal basis, and will be called the **canonical 2-form**, on J or on C. The canonical 2-form on C is obtained as the pull back by ψ of the canonical 2-form on J. On C, we can view the canonical 2-form as a volume form, and we denote it by $d\mu$. By definition of an orthonormal basis, we have

$$\int_C d\mu = 1;$$

in other words, the total mass of C under the canonical 2-form is 1.

Theorem 2.4. *Let H be a hermitian form, let M be its matrix, and suppose that*

$$M = (\operatorname{Im} \tau)^{-1}.$$

Then $\dfrac{\sqrt{-1}}{2g} \partial\bar{\partial}H(z, z)$ *is the canonical 2-form on J. In particular, on the curve C, we have*

$$\psi^* \frac{\sqrt{-1}}{2} \partial\bar{\partial}H(z, z) = g\, d\mu.$$

Proof. Since M is real symmetric, there exists a real symmetric $B = (b_{ik})$ such that $M = B^2$. Let

$$\beta_i = \sum_{k=1}^{g} b_{ik} dz_k.$$

From the definition of B we find at once

$$\sum_{i=1}^{g} \beta_i \wedge \bar{\beta}_i = \partial\bar{\partial}H(z, z).$$

On the other hand, by Proposition 2.2 and the hypothesis on M, we get

$$\langle \beta_i, \beta_j \rangle = \delta_{ij},$$

so the theorem follows.

In the next section, we shall find a natural theta function which has the hermitian form as in Theorem 2.4.

§3. The Canonical 2-Form and the Riemann Theta Function

Recall that the general quasi-periodicity of a theta function with respect to the lattice Λ is expressed by the equation

$$F(z + u) = F(z)e(L(z, u) + J(u)), \quad \text{all } u \in \Lambda,$$

where $L(z, u)$ is **C**-linear in z, **R**-linear in u; $J(u)$ is some number depending on u, and

$$e(w) = \exp(2\pi\sqrt{-1}w).$$

The function F is then said to be of **type** (L, J).

We let $\tau \in \mathfrak{H}_g$ so that τ is symmetric and Im τ is positive definite. The column vectors of τ are denoted by $\tau_1, \ldots, \tau_g$. Such coordinates are with respect to some basis of $\mathbf{C}^g$, to be prescribed.

A theta function F will be said to be of **Frobenius type** if there exists a basis $\{e_1, \ldots, e_g, \tau_1, \ldots, \tau_g\}$ of Λ over $\mathbf{Z}$ such that $\{e_1, \ldots, e_g\}$ is a basis of $\mathbf{C}^g$, and if coordinates are taken with respect to this basis, then:

FT 1. $F(z + e_i) = F(z)$ for $i = 1, \ldots, g$;

FT 2. $F(z + \tau_i) = F(z)e(-z_i - c_i)$ for $i = 1, \ldots, g$
                                       and some constants c_i

Recall that a **trivial theta function** is the exponential of a quadratic polynomial. Two theta functions are said to be **equivalent** if their quotient is a trivial theta function. Every theta function is equivalent to one which is of Frobenius type, cf. [L 18], Chapter VI, proof of Theorem 3.1. Theta functions of Frobenius type occur in natural ways, as in [L 18], Chapter VIII, §3.

When F is of Frobenius type, then

FT 3. $L(z, e_i) = 0$

$L(z, \tau_i) = -z_i$ for $i = 1, \ldots, g$.

The **Riemann (alternating) form** E is defined by

$$E(z, w) = L(z, w) - L(w, z).$$

For a theta function of Frobenius type, besides $E(z, z) = 0$ we have

FT 4. $E(e_i, \tau_j) = -\delta_{ij}$ for $i, j = 1, \ldots, g$.

The **associated hermitian form** H is the form whose imaginary part is E. Its real part is then $E(\sqrt{-1}\, z, w)$.

We define the **Riemann theta function** for $\tau \in \mathfrak{H}_g$ by

$$\theta(z, \tau) = \theta_\tau(z) = \sum_{m \in \mathbf{Z}^g} \mathbf{e}(\tfrac{1}{2}\,{}^t m \tau m + {}^t m z).$$

Cf. [L 18], Chapter VIII, §3 for more general functions, which satisfy easily determined transformation laws relative to translation by the lattice; the following are special cases for the Riemann theta function:

θ is even, that is $\theta(-z) = \theta(z)$.

$\theta(z + e_i) = \theta(z)$;

$\theta(z + \tau_i) = \mathbf{e}(-z_i - \tau_{ii}/2)\theta(z)$.

In other words, θ is of Frobenius type with respect to the period lattice.

Proposition 3.1. *Let H be the hermitian form associated with the Riemann theta function, and let M be its matrix. Then $M = (\mathrm{Im}\ \tau)^{-1}$.*

Proof. This is a special case of the general determination of the matrix for H as in [L 18], Chapter VIII, Lemma 1.2. However, the computation here can be carried out directly since the situation is already normalized. Since $E(e_i, e_j) = 0$ we have

$$H(e_i, e_j) = E(\sqrt{-1}e_i, e_j).$$

By **FT 4** we get

$$\delta_{ij} = E(\sqrt{-1}\, \mathrm{Im}\, \tau_j, e_i) = \sum_k (\mathrm{Im}\, \tau_{jk}) E(\sqrt{-1} e_k, e_i)$$
$$= \sum_k (\mathrm{Im}\, \tau_{jk}) H(e_k, e_i).$$

This proves the proposition.

If we now specialize to the Jacobian of a curve, we obtain:

Theorem 3.2. *Let* $(1_g, \tau)$ *be a normalized period matrix for the differentials of first kind on a curve C as in §2. Let* $\theta = \theta_\tau$ *be the Riemann theta function, and H its associated hermitian form. Then*

$$\frac{\sqrt{-1}}{2g} \partial\bar{\partial} H(z, z)$$

is the canonical 2-form, and its pull-back to the curve is the canonical volume form $d\mu$.

§4. The Divisor of the Riemann Theta Function

The object of this section is to prove Riemann's theorem describing the divisor of zeros (θ) of his theta function. We suppose the curve C embedded in its Jacobian by the mapping

$$\psi : C \to J$$

as before.

Theorem 4.1 (Riemann). *There exists a point* $w \in \mathbf{C}^g$ *such that*

$$(\theta) = \Theta_{-w},$$

where Θ *is the usual sum of* $\psi(C)$ *taken* $g - 1$ *times in J. Furthermore,* $2w = S(\mathfrak{k})$ *where* $\mathfrak{k}$ *is the canonical class.*

The proof will require some lemmas. We shall prove first that for almost all translations of θ, the intersection with the curve consists of g points (taken with multiplicity). Then we prove that the sum of these g points is what it should be. This establishes the connection with the algebraic theory of Chapter 5, Proposition 5.8 (originally due to Weil). We may even state this connection in a statement characterizing the theta divisor. As a matter of notation, if D is a divisor on J, we denote

$$D \cdot C = \psi^*(D).$$

Proposition 4.2. *Let D be a positive divisor on J having the following properties:*

Th 1. D *is even, that is* $D^- = D$.

Th 2.   *There is some element $w \in J$ such that for u generic,*

$$D_u \cdot C = \sum_{i=1}^{g} (P_i) \qquad \text{and} \qquad \sum_{i=1}^{g} \psi(P_i) = u + w.$$

Then $D_w = \Theta$, and $2w = S(\mathfrak{f})$ where $\mathfrak{f}$ is the canonical class.

Proof. Let $P_1, \ldots, P_g$ be sufficiently general points on the curve (say generic independent over a field of definition for C, J and w, finitely generated over the rational numbers, or anything else you want in that line). Let $u = \sum \psi(P_i) - w$. Then

$$\psi(P_1) + \cdots + \psi(P_{g-1}) - w = \sum_{i=1}^{g} \psi(P_i) - w - \psi(P_g)$$
$$= u - \psi(P_g).$$

This element lies in D if and only if $\psi(P_g) - u$ lies in D, by **Th 1**. But **Th 2** implies that $\psi(P_g) \in D_u$. So we conclude that

$$\psi(P_1) + \cdots + \psi(P_{g-1}) \in D_w.$$

This proves that $\Theta \subset D_w$. But Θ is irreducible, so

$$D_w = m\Theta + X,$$

where m is a positive integer, and X is a positive divisor not containing Θ. On the Jacobian, the curve C is numerically effective in the sense that for any non-zero positive divisor X, we have $\deg(C.X) > 0$. This comes from the fact that C is the intersection $g - 1$ times of an ample divisor (Θ itself), with a positive rational coefficient up to numerical equivalence. Since D_w and Θ have the same intersection degree with C, it follows that $X = 0$. This proves that $D_w = \Theta$.

As to the second statement of the proposition, we know from the standard theory (Theorem 5.8 of Chapter 5) that

$$\Theta^-_u \cdot C = \sum (P_i) \qquad \text{with} \qquad \sum \psi(P_i) = u,$$

and $\Theta^- = \Theta_{-S(\mathfrak{t})}$. The assertion $2w = S(\mathfrak{t})$ follows at once. This proves the proposition.

In the application, we let $D = (\theta)$ be the divisor of the theta function. Then D is even since $\theta(-z) = \theta(z)$, so **Th 1** is satisfied. The next results prove that D satisfies **Th 2**, and will therefore prove Theorem 4.1.

Assume first for simplicity that θ does not vanish identically on $\psi(C)$. The polygonal representation can then be taken so that the boundary of the polygon does not meet the divisor of zeros of θ. Then the number of points of intersection of $\psi(C)$ and (θ) is the number of zeros of θ on $\psi(C)$ taken with multiplicities, and is given by:

$$\text{number of zeros of } \theta \circ \psi = \frac{1}{2\pi\sqrt{-1}} \int_{\text{Bd }\mathscr{P}} d \log \theta \circ \psi.$$

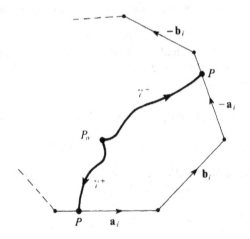

For P on $\mathbf{a}_i$ we let

$$\psi_i^+(P) = \int_{\gamma^+} \varphi_i \quad \text{and} \quad \psi_i^-(P) = \int_{\gamma^-} \varphi_i,$$

where γ^+ and γ^- are paths as shown on the figure. We make a similar definition for P on $\mathbf{b}_i$. The integral

$$\psi(P) = \int_{P_0}^{P} (\varphi_1, \ldots, \varphi_g)$$

is well-defined for P inside the polygon, and on the boundary, gives rise to the two values $\psi^+(P)$ and $\psi^-(P)$ such that for P on $\mathbf{a}_i$ we have

$$\psi^-(P) - \psi^+(P) = \tau_i.$$

Then

$$\theta(\psi^+(P)) = \theta(\psi^-(P))\mathbf{e}(\psi_i^+(P) + \tau_{ii}/2),$$

where ψ_i^+ is the i-th component of ψ^+. Then:

$$\frac{1}{2\pi\sqrt{-1}} \int_{\mathbf{a}_i} d\log\theta \circ \psi^+ + \frac{1}{2\pi\sqrt{-1}} \int_{-\mathbf{a}_i} d\log\theta \circ \psi^-$$

$$= \frac{1}{2\pi\sqrt{-1}} \int_{\mathbf{a}_i} d\log\theta \circ \psi^+ - d\log\theta \circ \psi^-$$

$$= \int_{\mathbf{a}_i} \varphi_i \quad (\text{because } d\psi_i = \varphi_i)$$

$$= 1.$$

Similarly, for P on $\mathbf{b}_i$, we have

$$\psi^+(P) - \psi^-(P) = e_i.$$

The same argument using the transformation law for the theta function shows that the corresponding integral is 0.

Thus for the pairs of integrals taken over $\mathbf{a}_i$, $-\mathbf{a}_i$, $\mathbf{b}_i$, $-\mathbf{b}_i$ we see that g of them will give a value of 1, and the other g have value 0. The sum is therefore equal to g, which proves that the intersection of (θ) with $\psi(C)$ consists of exactly g points.

If θ vanishes identically on $\psi(C)$, then we let $F = \theta_c$ be some translate of θ which does not vanish identically on $\psi(C)$. The integrals are computed in the same way, using the transformation formula, to yield the same result.

We now prove that the divisor (θ) satisfies **Th 2**, and formulate the result as a lemma.

Lemma 4.3. *There exists some $w \in \mathbb{C}^g$ having the following property. Let $u \in \mathbb{C}^g$ be such that θ_u does not vanish identically on $\psi(C)$, where*

$$\theta_u(z) = \theta(z - u).$$

Let P_i $(i = 1,\ldots,g)$ be the g zeros of θ_u in C. Then

$$\sum_{i=1}^{g} \psi(P_i) = u + w.$$

Proof. Cauchy's residue formula yields

$$\sum_{i=1}^{g} \psi(P_i) = \frac{1}{2\pi\sqrt{-1}} \int_{\mathrm{Bd}\,\mathscr{P}} \psi\, d\log\theta_u \circ \psi.$$

The values of ψ on the boundary are of course the values ψ^+ and ψ^- as previously described. For P on $\mathbf{a}_i$, the transformation equation for the theta function gives

$$\theta_u(\psi^+(P)) = \theta_u(\psi^-(P))e(\psi_i^+(P) + \tfrac{1}{2}\tau_{ii} + u_i).$$

Hence

$$d \log \theta_u(\psi^+(P)) - d \log \theta_u(\psi^-(P)) = 2\pi\sqrt{-1}\,\varphi_i.$$

Therefore

$$\frac{1}{2\pi\sqrt{-1}}\left[\int_{\mathbf{a}_i} \psi_j^+ \, d \log \theta_u \circ \psi^+ + \int_{-\mathbf{a}_i} \psi_j^- \, d \log \theta_u \circ \psi^-\right]$$

$$= -\frac{1}{2\pi\sqrt{-1}}\,\tau_{ij}\int_{\mathbf{a}_i} d \log \theta_u \circ \psi^+ + \tau_{ij}\int_{\mathbf{a}_i}\varphi_i + \int_{\mathbf{a}_i}\psi_j^+ \varphi_j.$$

The last two terms are constant. The image of the path $\mathbf{a}_i$ under ψ^+ is a path whose end points differ by e_i, and since

$$\theta_u(z + e_i) = \theta_u(z),$$

it follows that the image of $\mathbf{a}_i$ under $\theta_u \circ \psi^+$ is a closed path. Hence

$$\int_{\mathbf{a}_i} d \log \theta_u \circ \psi^+ \in \mathbf{Z}.$$

Therefore this last integral, which depends continuously on u, is also constant. This shows that the contribution of the integral over the boundary of the polygon given by the pair of integrals over $\mathbf{a}_i$ and $-\mathbf{a}_i$ can be absorbed into the constant w.

We now consider the pair of integrals corresponding to a side $\mathbf{b}_i$.

Let P be a point on $\mathbf{b}_i$. Then

$$\psi^+(P) - \psi^-(P) = e_i \qquad \text{so} \qquad \theta_u(\psi^+(P)) = \theta_u(\psi^-(P)).$$

Hence

$$\frac{1}{2\pi\sqrt{-1}}\left[\int_{\mathbf{b}_i} \psi_j^+ \, d \log \theta_u \circ \psi^+ + \int_{-\mathbf{b}_i} \psi_j^- \, d \log \theta_u \circ \psi^-\right]$$

$$= \frac{1}{2\pi\sqrt{-1}}\,\delta_{ij}\int_{\mathbf{b}_i} d \log \theta_u \circ \psi^+.$$

We shall prove that

$$\int_{\mathbf{b}_i} d \log \theta_u \circ \psi^+ = u_i + \text{constant}.$$

This will complete the determination of the desired integral over the pairs of sides $\mathbf{b}_i$, $-\mathbf{b}_i$, and will give the desired value

$$\frac{1}{2\pi\sqrt{-1}} \int_{\mathrm{Bd}\,\mathscr{P}} \psi \, d \log \theta_u \circ \psi = u + \text{constant},$$

thus finishing the proof of the theorem.

So we compute the integral of $d \log \theta_u \circ \psi^+$ over $\mathbf{b}_i$. The map ψ^+ gives a map

$$\psi^+ : \mathbf{b}_i \to \text{ path from a point } v \in \mathbf{C}^g \text{ to } v + \tau_i.$$

Then $\theta_u \circ \psi^+$ maps $\mathbf{b}_i$ to a path from $\theta_u(v)$ to $\theta_u(v + \tau_i)$, and we know that

$$\theta_u(v + \tau_i) = \theta_u(v)\mathbf{e}(-v_i + u_i - \tau_{ii}/2).$$

So we get

$$\int_{\mathbf{b}_i} d \log \theta_u \circ \psi^+ \equiv \log \theta_u(v + \tau_i) - \log \theta_u(v) \bmod 2\pi\sqrt{-1}\mathbf{Z},$$

and therefore

$$\frac{1}{2\pi\sqrt{-1}} \int_{\mathbf{b}_i} d \log \theta_u \circ \psi^+ - u_i \equiv -v_i - \tau_{ii}/2 \quad \bmod \mathbf{Z}.$$

The right-hand side of the congruence is constant mod $\mathbf{Z}$, and the left-hand side depends continuously on u, so is constant. This proves that

$$\frac{1}{2\pi\sqrt{-1}} \int_{\mathbf{b}_i} d \log \theta_u \circ \psi^+ = u_i + \text{constant}.$$

As already stated, this concludes the proof of Lemma 4.3, and also concludes the proof of Theorem 4.1.

§5. Green, Néron, and Theta Functions

Let D be a divisor on C. By a **Green's function** associated with D one means a real valued function $G = G_D$ on C satisfying the following properties as in Arakelov [Ar 1].

GF 1. *The function G is C^∞ except on the support of D. If P occurs with multiplicity n in D, then in terms of a local analytic coordinate z at P, the function G has the form*

$$G(z) = -n \cdot \log|z| + C^\infty \text{ function.}$$

GF 2.   *We have* $\dfrac{\sqrt{-1}}{\pi}\,\partial\bar{\partial}G = (\deg D)\,d\mu.$

GF 3. $\displaystyle\int_C G\,d\mu = 0.$

Remark 1.   If $\deg D = 0$, then $\partial\bar{\partial}G = 0$, so G is harmonic except on the support of D.

Remark 2.   The uniqueness of a Green's function is immediate. Indeed, the first two conditions determine G up to an additive constant, because the difference between two functions satisfying these two conditions must be everywhere harmonic, so constant. The third condition then eliminates the remaining ambiguity.

Theorem 5.1.   *A Green's function exists for any divisor on the curve.*

Proof. We shall prove that it suffices to show the theorem in the case when D is a non-special divisor, positive of degree g. Assume the theorem proved in this case. Let P be any point on the curve. We assume throughout that $g \geq 1$, so $l(P) = 1$. Let $P = P_1$, and let $P_2, \ldots, P_g$ be independent generic points such that the divisor

$$\mathfrak{a} = \sum_{i=1}^{g} (P_i)$$

is non-special. Suppose we have proved the theorem for $\mathfrak{a}$. Note that each divisor $g(P_i)$ for $i = 2, \ldots, g$ is non-special. By **Q**-linearity, the theorem then follows for P_1, as desired.

Now for a non-special divisor, we can give the Green's function more explicitly. The following theorem is suggested in [Ar 2], and details are given in [Hri].

Theorem 5.2.   *Let D be a positive divisor of degree g on C, and non-special. Let $u = S(D)$. Let λ_u be the Néron function associated with $\Theta^{-}{}_u$. Then*

$$\lambda_u \circ \psi$$

satisfies the two properties **GF 1** *and* **GF 2** *of a Green's function. In other words, it has the correct logarithmic singularities on the support of D, and furthermore*

$$\frac{\sqrt{-1}}{\pi}\,\partial\bar{\partial}(\lambda_u \circ \psi) = g\,d\mu.$$

Proof. The intersection of $\Theta^-{}_u$ with C is precisely the divisor D. The value for $\partial\bar{\partial}(\lambda_u \circ \psi)$ drops out of previous computations, namely Theorem 1.2 and Theorem 2.4. Of course, we have used Riemann's theorem identifying the divisor of the theta function with the usual theta divisor.

§6. The Law of Interchange of Argument and Parameter

We deal here with another relation between periods of differentials of third kind, classically called the law of **interchange of argument and parameter**, or a **reciprocity law**.

Let (P, Q) be a pair of points on the Riemann surface R. Then there exists a unique differential of third kind $\omega = \omega_{P,Q}$ having poles only at P, Q, with residues $+1$ at P and -1 at Q, and such that the periods of ω are pure imaginary. We call ω the dtk **associated** with (P, Q). As in [L 18], Chapter IV, we view the Riemann surface represented as a quotient of a polygon $\mathscr{P}$, whose sides are denoted $\mathbf{a}_i$, $\tilde{\mathbf{a}}_i$ $(i = 1,\ldots,g)$ where g is the genus. We also write $\tilde{\mathbf{a}}_i = \mathbf{a}_{i+g}$.

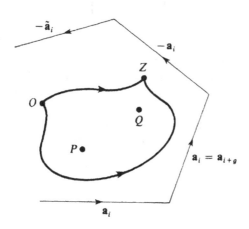

Let O be a point in the interior, $O \neq P, Q$. Then taking the integral along any path,

$$\operatorname{Re} \int_O^Z \omega_{P,Q}$$

is well defined as a real analytic function on the Riemann surface, except for logarithmic singularities at P and Q,

Let

$$f(Z) = f_{P,Q}(Z) = \int_O^Z \omega_{P,Q} \quad \text{along some path.}$$

Then f is defined only modulo the periods, but we can describe how f varies inside the polygon somewhat more precisely. Suppose we restrict ourselves to paths beginning at O, and lying entirely inside the polygon. If we take the above integral along any path which does not go through P, Q and has the same winding number with respect to P and to Q, then the value of f is independent of the choice of such paths. This is the value we take in the sequel, and specifically in the statement of the next theorem.

Theorem 6.1. *Let (P_1, Q_1) and (P_2, Q_2) be two pairs of points such that all four points are distinct and inside the polygon. Let ω_1, ω_2 be the dtk associated with these pairs. Then we have the relation:*

$$2\pi\sqrt{-1}\left[\int_{P_2}^{Q_2}\omega_1 - \int_{P_1}^{Q_1}\omega_2\right] = -\sum_{i=1}^{2g}\int_{\mathbf{a}_i}\omega_2\int_{\bar{\mathbf{a}}_i}\omega_1.$$

In particular, we have equality of real parts:

$$\mathrm{Re}\int_{P_2}^{Q_2}\omega_1 = \mathrm{Re}\int_{P_1}^{Q_1}\omega_2.$$

Proof. The second relation follows from the first because the periods are pure imaginary, so the right-hand side in the first relation is real. Therefore the pure imaginary part of the left-hand side must be 0, and this amounts to the relation between the real parts of the integrals as stated.

The proof of Theorem 6.1 is carried out by integrating around the polygon. As in [L 18], this integral is taken as the limit of integrals of polygons $\mathscr{P}'$ lying entirely inside $\mathscr{P}$ as shown on the figure.

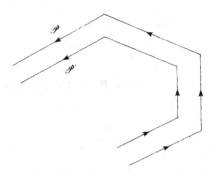

But we have to take care of the change of values of the integral

$$f_1(Z) = \int_0^Z \omega_1$$

when the path has different winding number around P_1 and Q_1. We shall find the value of the integral

$$\int_{Bd\,\mathscr{P}} f_1\omega_2$$

in two ways, First, taking the integral over the polygon following the representation by sides a_i, $\hat{a}_i$ we find the following value, referring to the steps preceding Theorem 1.1 of [L 18]:

6.1.1
$$\int_{Bd\,\mathscr{P}} f_1\omega_2 = -\sum_{i=1}^{2g} \int_{a_i} \omega_2 \int_{\hat{a}_i} \omega_1.$$

This value is the limit of the integrals taken around polygons $\mathscr{P}'$. On the other hand, the integral over $\mathscr{P}'$

$$\int_{Bd\,\mathscr{P}'} f_1\omega_2$$

can be computed as the integral over the following indicated path, minus the integral over the closed path γ encircling the segment L from P_1 to Q_1.

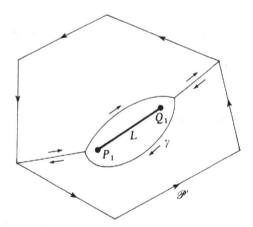

Let Z be a point on L. We define $f_1^+(Z)$ to be the value of the integral

$$f_1^+(Z) = \int_0^Z \omega_1$$

taken over a path approaching Z from one side of the segment as shown on fig. (a); and $f_1^-(Z)$ to be the value obtained by the sum of this integral, and the integral taken on a closed curve with winding number -1 around Q_1 and 0 around P_1, as shown on fig. (b).

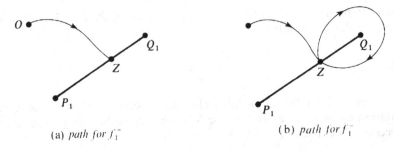

(a) *path for f_1^-*   (b) *path for f_1^-*

Then

$$f_1^- - f_1^+ = \text{integral of } \omega_1 \text{ around a loop around } Q_1 \text{ with}$$
$$\text{negative orientation}$$
$$= -2\pi\sqrt{-1}.$$

We obtain

$$\int_{\text{Bd }\mathscr{D}'} f_1\omega_2 = 2\pi\sqrt{-1} \sum \text{res}(f_1\omega_2) - \int f_1\omega_2$$

$$= 2\pi\sqrt{-1}[f_1(Q_2) - f_1(P_2)] - \int_{P_1, l.}^{Q_1} (f_1^+ - f_1^-)\omega_2$$

6.1.2. $= 2\pi\sqrt{-1}[f_1(Q_2) - f_1(P_2)] - 2\pi\sqrt{-1}[f_2(Q_1) - f_2(P_1)].$

Putting **6.1.1** and **6.1.2** together proves the theorem.

§7. Differentials of Third Kind and Green's Function

Arakelov [Ar 1] refers to classical Riemann surface theory as in Schiffer–Spencer [S–S] for the basic properties of differentials of third kind, which he uses in connection with Green functions on curves. In this section, we only give a simple application.

The reciprocity law of the previous section allows us to define a symmetric pairing $\langle a, b \rangle$ between divisors of degree 0 on a curve C, when the divisors are of type $(P) - (Q)$, and their support is disjoint, by means of the real part of the integral in Theorem 6.1. We can then extend this pairing by linearity to all pairs of divisors of degree 0 with disjoint supports, since any divisor of degree 0 can be expressed as a sum of terms of the above type.

Theorem 7.1. *The pairing $\langle a, b \rangle$, so defined satisfies the properties of a Néron symbol on curves, as in Theorem 3.6 of Chapter 11.*

Proof. All four properties are immediately verified.

Actually, only one of the divisors need be of degree 0 to obtain a Green function on the curve. Let $\mathfrak{a}$ be of degree 0, and let $\omega_\mathfrak{a}$ be the differential of third kind associated with $\mathfrak{a}$, so that:

$\omega_\mathfrak{a}$ has poles only at the points occurring in $\mathfrak{a}$, with residue equal to the multiplicity of the point;

the periods of $\omega_\mathfrak{a}$ are pure imaginary.

For P not in the support of $\mathfrak{a}$, we define

$$G_\mathfrak{a}(P) = \mathrm{Re} \int_{P_0}^{P} \omega_\mathfrak{a}.$$

Theorem 7.2. *The function $G_\mathfrak{a}$ satisfies the first two properties of a Green function, namely it has the right logarithmic singularities on the support of $\mathfrak{a}$, and is harmonic outside the support of $\mathfrak{a}$.*

Proof. The singularities come from the real part of the integral. On the other hand, consider the function

$$f(P) = \int_{P_0}^{P} \omega_\mathfrak{a},$$

well defined modulo the imaginary periods and modulo $2\pi\sqrt{-1}\mathbf{Z}$. Then f is analytic in the neighborhood of any given point P, and hence

$$\partial\bar\partial(f + \bar f) = 0.$$

But $f + \bar f$ is essentially $G_\mathfrak{a}$, so $\partial\bar\partial G_\mathfrak{a} = 0$. This proves the theorem.

Appendix

Review of S. Lang's *Diophantine Geometry*, by L. J. Mordell, Bull. Amer. Math. Soc. **70** (1964), 491–498.*

Review of L. J. Mordell's *Diophantine Equations*, by S. Lang, Bull. Amer. Math. Soc. **76** (1970), 1230–1234.*

* Reprinted with permission from the American Mathematical Society.

Diophantine geometry. By Serge Lang. Interscience Tracts in Pure and Applied Mathematics, No. 11. John Wiley and Sons, Inc., New York, 1962. 1 + 170 pp. $7.45.

Let $(x) = (x_1, x_2, \ldots, x_n)$ be n variables in a given ring R. Let $f(x) = f(x_1, x_2, \ldots, x_n)$ be a polynomial with coefficients in a given field F. We suppose to begin with that R and F are sets of algebraic numbers. The term "Diophantine Analysis" may be applied to the two following topics:

(1) Diophantine Equations, and here there is the question of discussing the solution in R of $f(x) = 0$.

(2) Diophantine Approximation, and now the problem is to investigate the lower bound of $|f(x)|$ for all (x) in R.

This book is concerned with probably the three most important developments in the period 1909–1963. There is to begin with the so-called Mordell–Weil Theorem. In 1922, the reviewer showed that if $n = 2$, and both R and F are the rational field, and if $f(x_1, x_2)$ is a cubic polynomial such that the curve $f(x_1, x_2) = 0$ is of genus 1, so that the curve has no double point, then a composition process can be given whereby an Abelian group can be associated with the rational points of the cubic, and this group has a finite basis. In simpler geometric terms, this meant that all the rational points can be derived from a finite number by the familiar tangent and chord process. A simpler and more perspicuous proof was given by Weil. He then extended the result to systems of p points on curves of genus p where R and F are now arbitrary algebraic number fields.

Next, the Thue–Siegel–Roth Theorem. In 1909, Thue proved that if θ is an algebraic number of degree n, then the inequality $|q\theta - p| < 1/q^{n/2 + \delta}$ has only a finite number of rational integer solutions in p, q for given $\delta > 0$. This result was improved by Siegel, who showed that the exponent $n/2 + \delta$ can be replaced by $2\sqrt{n}$. Finally Roth showed in 1955 that the $n/2$ could be replaced by 1, a best possible result.

Lastly, there is Siegel's Theorem. In 1929, he proved that when R is the ring of algebraic integers in an algebraic number field F, then $f(x_1, x_2) = 0$ has only a finite number of solutions except in some simple well-defined cases.

For many problems and for a long time, the chief centre of interest was in the cases when R is either the rational field Q or the rational integers Z, and F is Q. Obvious generalizations suggested themselves to more general rings and fields, for example, finite fields, function fields, to the new absolute values associated with valuation theory, and also to several polynomials instead of one.

Various methods have been applied in the past to the problems; arithmetical, algebraical, analytical and geometrical. The necessary geometric ideas were on the whole of such a character that they could be considered algebraic in nature. In recent times, powerful new geometric ideas and methods have been developed by means of which important new arithmetical theorems and related results have been found and proved and some of these are not easily proved otherwise. Further, there has been a tendency to clothe the old results, their extensions, and proofs in the new geometrical language. Sometimes, however, the full implications of results are best described in a geometrical setting. Lang has these aspects very much in mind in this book, and seems to miss no opportunity for geometric presentation. This accounts for his title "Diophantine Geometry." As indicated above, the three main developments are essentially of an arithmetic or algebraic nature. This also applies to the fourth topic treated by Lang, namely, Hilbert's Irreducibility Theorems, of which the simplest case is as follows. Let $f(x, y)$ be a polynomial with coefficients in a field F, and suppose that $f(x, y)$ is irreducible in F. Are there sets of values of y in F for which the polynomial remains irreducible in x? This is so, for example, if F is a rational field, but is not so if F is the field of complex numbers.

A general question that immediately suggests itself to a reader is what object an author has in mind when writing a book. Some have the true teacher's spirit or even a missionary spirit, wishing to introduce their subject to a wide circle of readers in the most attractive way. Such an author's treatment is essentially self-contained, the presentation is made as simple and complete as possible, and there is no undue generalization that would tend to make unnecessarily difficult the comprehension of the simpler and really fundamental cases; he is painstaking in his efforts to save the reader unnecessary and troublesome effort. When the subject makes undue demands on the reader, the author tries to give the reader some idea of the proof in easily understood language.

Lang is not such an author.[1] Much of the book is practically unreadable unless one is familiar with, among others, Bourbaki, the author's books on Algebraic Geometry and Abelian Varieties, and Weil's Foundations of Algebraic Geometry, and is prepared occasionally to go to the original sources for proofs of some theorems needed in the present volume. Lang may take the point of view that he is only interested in such readers and caters for no others. In fact, he says in his Foreword, "Diophantine problems represent some of the strongest attractions to algebraic geometry." However, in his pages on prerequisites, he refers to the elementary nature of a number of his chapters and their self-containment. Many readers will not accept either of these statements. The topics brought together in this volume are of the greatest interest to a far wider class of readers than those he seems to have in mind. It is unfortunate that it will be exceedingly difficult for them to learn something about most of these topics from the presentation given in this book.

The author's style and exposition leave a great deal to be desired. The results in the book appear as theorems, propositions, properties, lemmas, and even a criterion. The logical distinction between these is not at all clear. When a reference is made to one of them, the reader must turn over the pages

---

[1] In his recent book on Calculus, he states, ". . . One writes an advanced monograph for one's self, because one wants to give permanent form to one's vision of some beautiful part of mathematics, not otherwise accessible . . ." Compare with Chaucer's clerk (the scholar and teacher of those days) of whom Chaucer says "and gladly would he learn and gladly teach."

of the relevant chapter to find it. Occasionally it is non-existent, as when on p. 126, Lang refers to Corollary of Theorem 6 of Chapter VI, §5, and on p. 72 to Proposition 7 of Chapter IV, §4, and on p. 148 to Proposition 1 of §1 when he means Corollary 1 of Theorem 1.

His presentation of proofs often seems unattractive. He frequently starts off with various definitions and then gives some arguments and finally says that we have now proved Theorem X, repeating again many of his definitions, sometimes adding a few lines to finish the proof of Theorem X. How much better it would be if he started by stating his theorem and then gave the proof! He would avoid much unnecessary repetition that also makes the enunciation of many results of inordinate length. He would save a great deal of space which could be used most advantageously otherwise.

For example, in Chapter IV, he starts off by saying, "Throughout this chapter, F is a field with a proper set of absolute values M_F satisfying the product formula." At the end of the page, he repeats, "Let F be a field, etc.," and does it again in Lemma 1. It would have been very helpful if he always stated theorems explicitly instead of referring to them by the name of their originators, and he could easily have found room for this.

There is often an impression of vagueness about some of his proofs. They seem to lack the clarity one associated with demonstrations. He deals unsatisfactorily with details. His attitude to them is given elsewhere where he says that in matters of this kind, it is customary to omit details and that he proposes to follow this custom. This will suffice if a writer is giving a sketch of a proof. However, if a result is supposed to have been proved, then it seems desirable that the proof should be complete and presented in a logical manner, e.g., follow the pattern, if A implies B and B implies C, then A implies C. Too often do we see in this book remarks such as something is obvious, or clear, or that it follows, or that it can be proved in the well-known manner, or that the reader will immediately verify, or that it is left as an exercise, or that the proof can be found elsewhere.

Reading is made more difficult because the author gives a far greater number of definitions than is usual with other writers, and it is not easy to retain these in mind. In some chapters, he introduces without explanation technical terms and concepts which are not really necessary for the demonstration and which could be easily dispensed with. He sometimes takes it for granted that when he refers to a theorem by name, the reader will be familiar with what he has in mind. Lang is a very learned mathematician—he oozes mathematics—and assumes that his readers are au fait with all the entities hurled at them. Whenever possible all the resources of algebraic geometry are brought into the proofs of theorems and their generalizations. He is often not content with the ordinary language used in expressing simple ideas, but uses a language of his own which many readers may find difficult to understand. Further he seems to use a method of infinite ascent in expounding his proofs, that is, simple ideas are often developed by using more complicated ones.

Let us now look at the reaction of that would-be reader not too familiar with algebraic geometry, whom Lang possibly did not have in mind but who nevertheless would like to learn something about "Diophantine Geometry," and the important theorems brought together in the present book. The title of the book suggests to him that here is an opportunity for learning something really worth while. He will soon be disillusioned and be faced with a titanic struggle. He will require the patience of Job, the courage of Achilles, and the stength of Hercules to understand the proofs of some of the essential theorems. He will realize that some of the proofs will be above his head, but at any rate he may hope to get some idea of what is being done. He would have been helped enormously if Lang had given a glossary containing definitions of many terms that he uses. Many of these are independent of algebraic geometry. One may begin by mentioning terms which will be familiar to many readers, but not to others, such as: "injective," "bijective" and "surjective"—because in his "Algebraic Geometry," written a few years before, he thinks it necessary to define "surjective." Then there are "finitely generated field" "morphism," "regular extensions of a field," "extension of a field k free from K over k_1," "extension of k which is independent of K over k." Naturally there are a large number of geometric terms, and their definitions may sometimes be found in his other two books from their indexes, but not always. Among these are "transport of structure," "supp ϕ," "isogeny," "fibering," "compatibility of intersections and reduction," "inclusion," "ample." Some of the terms can be defined quite simply, and what a blessing for many readers it would have been if this had been done in the present volume!

Though he gives a table of notation at the end of the book, there are many other symbols which the reader is supposed to know from the author's geometry or to guess at. Though he defines the sets Q, Z of rational numbers and of integers, he does not deem it necessary to define R as the real number field when he first uses the symbol. The meaning of symbols, such as $v(nd)$, and the associated $A \cdot B$ is taken for granted or is obscure, and some of his geometric formulae might be puzzling. It would be helpful to define a commutative diagram.

Even when he does give definitions or details, he often gives them in the wrong place and at the wrong time, and when they are not really necessary. On p. 65, he refers to "two linearly equivalent divisors" and he first defines the term on p. 131. He reminds his readers of the "meaning of the simple symbol A_K as the group of rational points in K in the commutative group variety A," something one could hardly forget, and also that "P and P' are congruent mod TB_K" means "$P - P' \in TB_K$." He finds it necessary to inform his readers of the definition of a discrete lattice. Distrusting their mathematical powers, he finds it desirable to prove that $\sqrt{(1 + x)} < 1 + x/2$ since in a bracket, he says "square both sides."

Let us now examine the contents of the book. The author states that Chapters I, II give a good part of classical number theory. Chapter I deals with absolute values on a field. He states that his treatment is much influenced by Artin. The reader would be advised to read Artin since Lang's treatment is not self-contained, and is sometimes lacking in precision. Thus in §§2 on "completions," he states that a field K is complete if every Cauchy sequence converges, omitting the condition that the limit must be in K. He is also under the impression that if $x_n \nrightarrow 0$, then there exists a number $a > 0$ such that $|x_n| > a$ for all sufficiently large n. Chapter II deals with the product formula for absolute values, ideals, divisors, units and the class number. He calls these two chapters "elementary." This does not mean that they are easy to read. Here the author begins with his numerous definitions. We have "well-behaved absolute value," 'proper sets of absolute values," definitions which many other writers do not find necessary; and then references to schemes, fiber and various learned references e.g. Noether's normalization theorem, Krull's principal ideal theorem. His examples 2 and 3 cannot be called elementary.

Chapters III and IV deal with the heights of points and also with what have been called the heights of polynomials. The simplest properties are given in Chapter III. The extension to heights on abelian groups leads to the main idea involved in the method of infinite descent, a process first applied by Fermat and of fundamental importance not only in the study of Diophantine Equations but which also has other applications. Chapter IV is concerned with inequalities connecting heights of points and their mappings—in particular by means of linear systems—and the results are of great importance for the main theorems.

Let us now turn to the main theorems of the book—following the sequence given in the book—first, the so-called Mordell-Weil Theorem. Its early history has been already mentioned and there has been a great deal of work done since on the subject. It should be stated again that the only part played by Mordell was in his theorem of some 40 years ago. The Mordell-Weil theorem states, "Let K be a finitely generated field over the prime field. Let A be an abelian variety defined over K. Then the points of K lying in A are finitely generated." In this form the theorem is of considerable generality and so makes great demands upon the technical knowledge of the reader. These are increased since he also shows that the ideas in the proof suffice to prove "the Theorem of the Base."

Let us note some of the concepts required in the chapter. There are a "$K|k$ trace of A," a "Theorem of Chow," "Chow's Regularity Theorem," "Chow Coordinates," "compatibility of projections with specialization," "blowing up a point," "Albanese Variety," "Picard Variety," "Jacobian of a curve," "Chow's theory of the $k(u)|k$-trace." When proof of an extension makes it exceedingly difficult to understand the simpler cases, it might sometimes be better if the generalizations were left in the Journals. The reviewer is reminded of Rip Van Winkle, who went to sleep for a hundred years and woke up to a state of affairs and a civilization (and perhaps a language) completely different from that to which he had been accustomed. There were, however, some things still familiar to him—which is more than can be said by the reviewer about the presentation of the present treatment.

Next, the Thue–Siegel–Roth Theorem. This is presented with the utmost generality. The variables are now elements of an algebraic number field K, and the absolute values are those associated with the field. The author claims to follow Roth's proof. The reader might prefer to read this which requires only a knowledge of elementary algebra and then he need not be troubled with axioms which are very weak forms of the Riemann–Roch Theorem.

Thirdly, Siegel's Theorem. There is given an account of the modern version of Siegel's proof in which θ-functions are replaced by the Jacobian, and Roth's Theorem is used instead of Thue's Theorem. It also gives the extension of the theorem to the case in which the solution is extended to subrings of a field K of finite type over Q.

The final section is on Hilbert's Irreducibility Theorem. Siegel, in his great memoir on Diophantine Approximation, had given an application of his results to this theorem. Lang gives not only the usual elementary proof for the irreducibility over the rational numbers, but also considerable extensions and generalizations. He finds it necessary to introduce into his account Zariski open sets without defining them.

So much for the detailed criticism. This may tend to blind us to some merits of the book. One must mention that the topics have been dealt with systematically so as to show and emphasize algebraic structure and the significance of algebraic geometry in some of the deepest and most important aspects of number theory. One will not be surprised if some of the geometers who read this book will seek further acquaintance with the queen of mathematics.

In conclusion, the reader will need no convincing that Lang, as has already been said, is a very learned mathematician, thoroughly familiar with every aspect of the topics he deals with, and their developments. His interesting and valuable historical notes give futher evidence of this. Lang assumes that his readers are as knowledgeable as he is, and can grapple with the subject with the same ease that he does. Even if they could, Lang's style is not such as to make matters easy for them. Lang in writing is not a follower of Gauss, whose motto was "pauca sed matura." Further thought and care about his book, before publication, would have been well worth while. Those who can understand the book will be indebted to him for having brought together in one volume the important results contained in it. How much greater thanks would he have earned if the book had been written in such a way that more of it could have been more easily comprehended by a larger class of readers! It is to be hoped that some one will undertake the task of writing such a book.

L. J. Mordell

Diophantine equations, by L. J. Mordell, Academic Press, New York and London, 1969.

The theory of diophantine equations is one of the oldest in mathematics, one of its most attractive, and also at the moment one which is still fairly undeveloped as being exceptionally hard. One reason for this is perhaps that in the full generality of the Hilbert problem, it cannot be effectively dealt with. Nevertheless, I personally would expect a wide class of diophantine problems to be effectively solvable (e.g. those on curves or abelian varieties), and in any case, many special cases are solvable.

Because of difficulties which have been encountered historically, a portion of the subject has developed as an accumulation of special diophantine equations, mostly in two variables, i.e. curves. It was well understood in the nineteenth century that nonsingular cubic curves have a group law on them, parametrized by the elliptic functions from a complex torus, but Poincaré was the first to draw attention to the special group of rational points when this curve is defined by an equation with rational coefficients, and he guessed that this group might be finitely generated. Mordell proved this fact in 1922, and thereby provided the first opportunity to behold the beginnings of a much broader approach to this type of equation. He also conjectured that a curve of genus ≥ 2 has only a finite number of rational points, and this magnificent conjecture remains unproved today. These matters, which are perhaps Mordell's greatest contributions to the subject, are treated in Chapters 16 and 17 of the present book.

The other parts of the book are roughly distributed as follows. A number of concrete special equations of degrees 2, 3 and 4 are discussed at the beginning, mostly with the method of congruences. Chapter 7 gives a discussion of the fundamental theorem concerning quadratic forms over the rationals (solvability globally is equivalent to solvability locally everywhere).

Chapter 8 deals with Pell's equation, which essentially solves effectively for the units of a real quadratic field. The treatment is classical. Next comes a sequence of chapters on surfaces, mostly cubic and quartic, dealing with special cases when rational or integral points can be found. A brief chapter mentions the role of units as affecting certain equations in number fields, and examples are worked out. After the general discussion already mentioned on curves of genus 1 or > 1, we return to special cases which can be handled without the general theory, somewhat more effectively using Minkowski's theorem on convex bodies and congruence methods, applied to the representation of numbers by quadratic and cubic forms. Next we have Thue's theorem on diophantine approximations (the weak version, not Roth's version), and its application to equations of type $f(x, y) = m$ where f is a homogeneous polynomial of higher degree. The next chapter mentions Skolem's method by p-adic analysis, but does not go into details of proofs.

We then return to cubic and quartic forms, or rather special cases, involving the explicit determination of integral solutions. The discriminant and other covariants of these forms are discussed (indispensable means to get at the solutions effectively). The scene shifts back once more to cubic and quartic curves of the elliptic type, like $y^2 = x^3 + k$ and $y^2 = f(x)$, where f is a cubic polynomial with no multiple roots, looking for integral points rather than rational points. Mordell's original proof that the number of these is finite is given, using Thue's theorem. The next chapter indicates the extension of this result to the case when f has arbitrary degree (proved by Siegel in 1926), but refers to an earlier Siegel paper for the stronger version of the analogue of Thue's theorem needed to make the proof go through. Mordell also states Siegel's general theorem of 1929, that a curve of genus at least 1 has only a finite number of integral points, giving explicitly the exceptional cases of genus 0 when infinitely many such points may occur. The book concludes with other special equations of higher degree, for instance special results on $ax^n - by^n = c$ and the Fermat curve, proving the nonexistence of integral solutions in a few simple cases, while assuming assorted facts of algebraic number theory, both of the standard variety, but also more specialized, like those involving regular primes.

As can be seen from this sketch, the contents of the book are jumpy, and some comments are now in order concerning the broader implications of Mordell's style, his point of view, and the context in which he writes.

Special concrete cases like cubic curves have provided much of the testing ground for experimentation, methods, theorems, and conjectures in diophantine analysis, and hence it is very welcome to have some of these cases brought together, as Mordell has done. I emphasize: He collects together special cases, without particular unifying order, or any design that I could make out, that might tie them together or make their succession in 30 chapters more than what appears to be an arbitrary succession. That is Mordell's taste, and I cannot quarrel with it. The book, as it is, will be very useful to those interested in diophantine equations, and wishing to work out special cases with essentially elementary techniques. I personally had bought a copy of the book before being sent the review copy, now given to the library.

But the reader must be aware of the limitations of Mordell's exposition. For one thing, Mordell clings systematically to the chronological development of the subject throughout the book. Even when an important development has taken place, e.g. Roth's theorem on diophantine approximations, subsuming previous results in the subject (by Thue and Siegel, say), Mordell gives the earliest theorem, namely Thue's, and only briefly refers to Roth's paper for the extension, when only a few additional pages at an equally elementary level would have been needed to get the full result. When a stronger version is needed to handle the equation $y^2 = f(x)$ with f of higher degree, Mordell refers to an earlier paper of Siegel (Math. Zeitschrift, 1921, a misprint in the reference gives the date erroneously as 1961), which also treats the number field case needed for this particular application. However, the inexperienced reader will have to figure out for himself that a single formulation of Roth's theorem in number fields can be used effectively for all these applications.

Even though I find the succession of equations treated somewhat arbitrary, there seems to be one thread which runs through them, suggested by the "List of Equations and Congruences" appearing at the end of the book in lieu of an index. This list is ordered according to degree (degree 1, degree 2, degree 3, degree 4, degree > 4) and then according as to whether the equation is homogeneous or not. Of course, one's first attempt in dealing with diophantine equations is to experiment with equations of low degree and small coefficients. But it soon becomes apparent that the degree is not a good invariant for the behavior of these equations, whether searching for rational points or integral points, and the classification by degree is to a large extent misleading. However, Mordell's taste when faced with a theorem like Siegel's on curves of higher genus is just to say: "The proof is of a very advanced character." And leave it at that.

Nor does Mordell tell us of Weil's generalization in 1929 concerning the finite generation of the group of rational points in the higher dimensional case; which is a pity, because this is one of the approaches which gives a method of attack for the Mordell conjecture on curves of genus ≥ 2: We embed them in their Jacobians, and look at their intersection with the finitely generated group of rational points on this Jacobian. By this method, and the positive definite quadratic form of Néron–Tate on this group, Mumford was for instance able to show that the gaps between rational points of ascending height become exponentially large.

It is also possible to connect both results and methods of diophantine analysis with algebraic geometry, and I found it interesting in my book *Diophantine Geometry* to present the known results which allowed us to make this connection coherently (e.g. as it applies to Severi's theorem of the base, following work of Néron, and with Néron). The intense dislike which Mordell has for this kind of exposition is clearly evidenced by his famous review of the book (Bull. Amer. Math. Soc. **70** (1964), 491–498). (If the review is not famous, it should be.) In this connection, I can do no better than to reproduce an exchange of letters with him in November 1966, shortly after I had sent him a copy of some of my other books. He kindly wrote me:

Dear Professor Lang, Thank you very much for the textbooks which I shall be glad to read. I hope I shall not have to struggle with them as I did with D. G. You may be interested to know that I found your *Algebra* quite readable and very useful. It was obviously meant to be understood. (You may quote this if you wish to.) . . ."

And after a few other kind remarks, he expressed the hope to meet me at a talk of his on diophantine equations to be given shortly in New York. (We met, and I enjoyed it.) Still, I answered Mordell on the substantial points raised both by his review and his letter:

> Dear Professor Mordell, Thanks for your letter. What you write there prompts me to clarify some points about book writing.
>
> I see no reason why it should be prohibited to write very advanced monographs, presupposing substantial knowledge in some fields, and thus allowing certain expositions at a level which may be appreciated only by a few, but achieves a certain coherence which would not otherwise be possible.
>
> This of course does not preclude the writing of elementary monographs. For instance, I could rewrite Diophantine Geometry by working entirely on elliptic curves, and thus make the book understandable to any first year graduate student (not mentioning you · · ·). Both books would then coexist amicably, and neither would be better than the other. Each would achieve different ends.
>
> When you write of any book that it is "obviously meant to be understood", whether as a compliment for one book or blame for another, you are still missing the point: I never meant Diophantine Geometry to be understood specifically by you, or anyone who did not have the rather vast background required for its reading. All my books are meant to be understood by readers having the prerequisites for the level at which the books are written. These prerequisites vary from book to book, depending on the subject matter, my mood, and other aesthetic feelings which I have at the moment of writing. When I write a standard text in Algebra, I attempt something very different from writing a book which for the first time gives a systematic point of view on the relations of diophantine equations and the advanced contexts of algebraic geometry. The purpose of the latter *is* to jazz things up as much as possible. The purpose of the former is to educate someone in the first steps which might eventually culminate in his knowing the jazz too, if his tastes allow him that path. And if his tastes don't, then my blessings to him also. This is known as aesthetic tolerance. But just as a composer of music (be it Bach or the Beatles), I have to take my responsibility as to what *I* consider to be beautiful, and write my books accordingly, not just with the intent of pleasing one segment of the population. Let pleasure then fall where it may. With best regards, Serge Lang.
>
> SERGE LANG

Note: Mordell in his review quotes from the preface of the first edition of my calculus book, but with an elision. The full text of my sentence runs as follows:

> One writes an advanced monograph for oneself, because one wants to give permanent form to one's vision of some beautiful part of mathematics, not otherwise accessible, somewhat in the manner of a composer setting down his symphony in musical notation.

I stand by the text as written, not as quoted. The musical analogy is an essential part of what I meant.

Bibliography

[A1]   A. ALTMAN, The size function on abelian varieties, *Trans. Amer. Math. Soc.* 164 (1972) pp. 153–161

[Ar 1] S. J. ARAKELOV, Families of algebraic curves with fixed degeneracies, *Izv. Akad. Nauk SSSR Ser. Mat.* 35 No. 6 (1971) = *Math. USSR Izv.* 5 No. 6 (1971) pp. 1277–1302

[Ar 2] S. J. ARAKELOV, Intersection theory of divisors on an arithmetic surface, *Izv. Akad. Nauk SSSR Ser. Mat.* 38 No. 6 (1974) = *Math. USSR Izv.* 8 No. 6 (1974) pp. 1167–1180.

[Ar 3] S. J. ARAKELOV, *Theory of Intersections on the Arithmetic Surface*, Proc. Int. Congress Math. Vancouver (1974) pp. 405–408

[Ar]   E. ARTIN, *Algebraic Numbers and Algebraic Functions*, Notes from Princeton and New York Universities, 1950–1951, Gordon & Breach, 1967

[A–W]   E. ARTIN and G. WHAPLES, Axiomatic characterization of fields by the product formula, *Bull. Amer. Math. Soc.* 51 No. 7 (1945) pp. 469–492

[Ba]   A. BAKER, *Transcendental Number Theory*, Cambridge University Press, 1975

[Bl]   S. BLOCH, A note on height pairings, Tamagawa numbers, and the Birch–Swinnerton-Dyer Conjecture, *Invent. Math.* 58 (1980) pp. 65–76

[Bo 1] F. BOGOMOLOV, Sur l'algébricité des représentations l-adiques, *C. R. Acad. Sci.* Paris 290 (1980) pp. 701–704

[Bo 2] F. BOGOMOLOV, Points of finite order on an abelian variety, *Izv. Akad. Nauk SSSR Ser. Mat.* 44 (1980), *AMS translation Math. USSR Izvestija* 17 No. 1 (1981) pp. 55–72

[Bom]   E. BOMBIERI, On the Thue-Siegel-Dyson theorem, *Acta Math.* 148 (1982) pp. 255–296

[Ca 1] J. W. CASSELS, Arithmetic on curves of genus 1, *J. reine angew. Math.* 202 (1959) pp. 52–99

[Ca 2] J. W. CASSELS, The rational solutions of the Diophantine equation $Y^2 = X^3 - D$, *Acta Math.* 82 (1950) pp. 243–273; Addenda and corrigenda to the above, *Acta Math.* 84 (1951) p. 299

[C–F]   J. W. CASSELS and A. FROHLICH, Editors, *Algebraic Number Theory*, Proceedings of a Conference, Academic Press, 1967

[Cha 1] C. CHABAUTY, Sur les équations diophantiennes liées aux unités d'un corps de nombres algébriques fini, thèse, *Annali di Math.* 17 (1938) pp. 127–168

[Cha 2]   C. CHABAUTY, Sur les points rationnels des variétés algébriques dont l'irrégularité est supérieure à la dimension, *C. R. Acad. Sci. Paris* **212** (1941) pp. 1022–1024

[Chow]   W. L. CHOW, Abelian varieties over function fields, *Trans. Amer. Math. Soc.* **78** (1955) pp. 253–275

[Ch–L]   W. L. CHOW and S. LANG, On the birational equivalence of curves under specialization, *Amer. J. Math.* **79** (1957) pp. 649–652

[Ch–W]   C. CHEVALLEY and A. WEIL, Un théorème d'arithmétique sur les courbes algébriques, *C. R. Acad. Sci. Paris* **195** (1932) pp. 570–572

[Co–T]   J. L. COLLIOT-THÉLÈNE, Variation de la hauteur sur une famille de courbes élliptiques particulière, *Acta. Arith.* **XXXI** (1976) pp. 1–16

[Da]   H. DAVENPORT and K. F. ROTH, Rational approximations to algebraic numbers, *Mathematika* **2** (1955) pp. 160–167

[De 1]   V. A. DEMJANENKO, An estimate of the remainder term in Tate's formula, *Mat. Zametki* **3** (1968) pp. 271–278 = *Math. Notes* **3** (1968) pp. 173–177

[De 2]   V. A. DEMJANENKO, Rational points of a class of algebraic curves, *Izv. Akad. Nauk SSSR Ser. Mat.* **30** (1966) pp. 1373–1396; English transl., *AMS translation* (2) **66** (1968) pp. 246–272.

[Do]   K. DÖRGE, Einfacher Beweis des Hilbertschen Irreduzibilitätssatzes, *Math. Ann.* **96** (1927) pp. 176–182

[Fa 1]   G. FALTINGS, Calculus on arithmetic surfaces, to appear

[Fa 2]   G. FALTINGS, Arakelov's theorem for abelian varieties, to appear

[Fr]   W. FRANZ, Untersuchungen zum Hilbertschen Irreduzibilitätssatz, *Math. Ann.* **33** (1931) pp. 275–293

[Gra]   H. GRAUERT, Mordell's Vermutung über rationale Punkte auf algebraischen Kurven und Funktionenkörper, *Pub. IHES* **25** (1965) pp. 131–149

[Gre]   M. GREEN, Holomorphic maps to complex tori, *Amer. J. Math.* **100** (1978) pp. 615–620

[Gro]   A. GROTHENDIECK and J. DIEUDONNE, Eléments de géometrie algébrique, *Pub. Math. IHES*, Paris 1960, 1961

[Hall]   M. HALL, The diophantine equation $x^3 - y^2 = k$, Oxford conference, Academic Press, 1971 pp. 173–198

[Har]   R. HARTSHORNE, *Algebraic Geometry*, Springer-Verlag, 1977

[Hi]   D. HILBERT, Über die Irreduzibilität ganzer rationaler Funktionen mit ganzähligen Koeffizienten, *J. reine angew. Math.* **110** (1892) pp. 104–129

[Hri]   P. HRILJAC, The Néron–Tate height and intersection theory on arithmetic surfaces, Thesis, MIT, 1982.

[In]   E. INABA, Über den Hilbertschen Irreduzibilitätssatz, *Jap. J. Math.* **19** (1944) pp. 1–25

[Ig]   J. IGUSA, Betti and Picard numbers of algebraic surfaces, *Proc. Nat. Acad. Sci. USA* **45** (1960) pp. 724–726

[K–O]   S. KOBAYASHI and T. OCHAI, Meromorphic mappings onto compact complex spaces of general type. *Invent. Math.* **31** (1975) pp. 7–16

[K–L]   D. KUBERT and S. LANG, *Modular Units*, Springer-Verlag, 1981

[L 1]   S. LANG, Unramified class field theory over function fields in several variables, *Ann. of Math.* **64** No. 2 (1956) pp. 285–325

[L 2]   S. LANG, *Introduction to Algebraic Geometry*, Interscience, New York, 1959

[L 3]   S. LANG, *Abelian Varieties*, Interscience, New York, 1959

[L 4]   S. LANG, Integral points on curves, *Pub. Math. IHES*, Paris 1960

[L 5]   S. LANG, *Algebraic Number Theory*, Addison-Wesley, 1970

[L 6]   S. LANG, Les formes bilinéaires de Néron et Tate, *Sém. Bourbaki* No. 274, 1963–1964

[L 7]   S. LANG, *Elliptic Curves: Diophantine Analysis*, Springer-Verlag, 1978

[L 8]   S. LANG, On the zeta function of number fields. *Invent. Math.* **12** (1971) pp. 337–345

[L 9]   S. LANG, Division points on curves, *Annali di matematica pura ed applicata* **(IV) LXX** (1965) pp. 229–234

[L 10]   S. LANG, *Introduction to Diophantine Approximation*, Addison-Wesley, 1966

[L 11]   S. LANG, *Introduction to Transcendental Numbers*, Addison-Wesley, 1966

[L 12]   S. LANG, Diophantine approximation on toruses, *Amer. J. Math.* **86** (1964) pp. 521–533

[L 13]   S. LANG, Transcendental numbers and diophantine approximations, *Bull. Amer. Math. Soc.* **77** (1971) pp. 635–677

[L 14]   S. LANG, Higher dimensional diophantine problems, *Bull. Amer. Math. Soc.* **80** No. 5 (1974) pp. 779–787

[L 15]   S. LANG, Conjectured diophantine estimates on elliptic curves. Volume dedicated to Shafarevich, Birkhauser, 1983

[L 16]   S. LANG, Report on diophantine approximations, *Bull. Soc. Math. France* **93** (1965) pp. 177–192

[L 17]   S. LANG, *Complex Multiplication*, Springer-Verlag, 1983

[L 18]   S. LANG, *Introduction to Algebraic and Abelian Functions*, Springer-Verlag, Graduate Texts in Mathematics, 1983

[L–N]   S. LANG and A. NÉRON, Rational points of abelian varieties over function fields, *Amer. J. Math* **81** No. 1 (1959) pp. 95–118

[L–T]   S. LANG and J. TATE, Principal homogeneous spaces over abelian varieties, *Amer. J. Math.* **80** (1958) pp. 659–684

[Lap]   A. I. LAPIN, Subfields of hyperelliptic fields, *Izv. Akad. Nauk SSSR Ser. Mat.* **28** (1964) pp. 953–988

[Li 1]   P. LIARDET, Sur une conjecture de Serge Lang, *C. R. Acad. Sci. Paris* **279** (1974) pp. 435–437

[Li 2]   P. LIARDET, Sur une conjecture de Serge Lang, *Astérisque*, **24–25**, Soc. Math. France, 1975

[Mah 1]   K. MAHLER, Über die rationalen Punkte auf Kurven vom Geschlecht Eins, *J. reine angew. Math.* **170** (1934) pp. 168–178

[Mah 2]   K. MAHLER, On the fractional parts of the powers of a rational number, *Mathematika* **4** (1957) pp. 122–124

[Man 1]   J. MANIN, Rational points of algebraic curves over function fields (in Russian), *Izvestia Akad. Nauk* **27** (1963) pp. 1395–1440. See also

Manin's announcement: Proof of the analog of Mordell's conjecture for algebraic curves over function fields, *Dokl. Acad. Nauk SSSR* **152** (1963) pp. 1061–1063 = *Soviet Mathematics* **4** (1963) pp. 1505–1507

[Man 2]   J. MANIN, Cyclotomic fields and modular curves, *Russian Math. Surveys* **26** No. 6 (November–December 1971) pp. 7–78

[Man 3]   J. MANIN, The refined structure of the Néron–Tate height, *Math. Sbornik.* (1970), *AMS Transl.* pp. 325–342

[Man 4]   J. MANIN, The Tate height of points on an abelian variety, its variants and applications, *Izv. Akad. Nauk SSSR Ser. Mat.* **28** (1964), *AMS Transl.* (2) **59** (1966) pp. 82–110

[Man 5]   J. MANIN, The p-torsion of elliptic curves is uniformly bounded, *Izv. Akad. Nauk SSSR Ser. Mat.* **33** (1969) pp. 433–438

[Man 6]   J. MANIN, Algebraic curves over fields with differentiation, *Izv. Akad. Nauk SSSR Ser. Mat.* **22** (1958) pp. 737–756; *AMS translations* Series 2 **37** (1964) pp. 59–78

[M–Z]   J. MANIN and Y. ZARHIN, Heights on families of abelian varieties, *Math. Sbornik* **89** (1972) pp. 171–181 (*Math. USSR Sbornik* **18** (1972) No. 2, pp. 169–179

[MD–LM 1]   M. MARTIN-DESCHAMPS and R. LEWIN-MENEGAUX, Applications rationnelles séparables dominantes sur une variété de type general, *Bull. Soc. Math. France* **106** (1978) pp. 279–287

[MD–LM 2]   M. MARTIN-DESCHAMPS and R. LEWIN-MENEGAUX, Surfaces de type général dominées par une variété fixe, *Bull. Soc. Math. France* **110** (1982) pp. 127–146

[Maz 1]   B. MAZUR, Modular curves and the Eisenstein ideal, *Pub. IHES* **47** (1977) pp. 129–162

[M–M]   B. MAZUR and W. MESSING, Universal extensions and one-dimensional crystalline cohomology, Springer *Lecture Notes* **370**, 1974

[Ma–T]   B. MAZUR and J. TATE, *Canonical Height Pairings Via Biextensions*, Volume dedicated to Shafarevich, Birkhauser, 1983

[Mo 1]   L. J. MORDELL, On the rational solutions of the indeterminate equation of the third and fourth degrees, *Proc. Cambridge Philos. Soc.* **21** (1922) pp. 179–192

[Mo 2]   L. J. MORDELL, *Diophantine Equations*, Academic Press, 1969

[Mor]   H. MORIKAWA, On theta functions and abelian varieties over valuation fields of rank one, I and II, *Nagoya Math. J.* **20** (1962) pp. 1–27 and 231–250

[Mum 1]   D. MUMFORD, A remark on Mordell's conjecture, *Amer. J. Math.* **LXXXVII** No. 4 (1965) pp. 1007–1016

[Mum 2]   D. MUMFORD, *Abelian Varieties*, Oxford University Press, 1974

[Mum 3]   D. MUMFORD, On the equations defining abelian varieties I, *Invent. Math.* **1** (1966) pp. 287–354

[Mu 4]   D. MUMFORD, Bi-extensions of formal groups, *Proc. Bombay Colloquium on Algebraic Geometry*, Tata Institute of Fundamental Research, Bombay, 1966

[Mur]   J. P. Murre, On generalized Picard varieties, *Math. Ann.* **145** (1962) pp. 334–353

[Ne 1]   A. Néron, Problèmes arithmétiques et géométriques rattachés à la notion de rang d'une courbe algébrique dans un corps, *Bull. Soc. Math. France* **80** (1952) pp. 101–166

[Ne 2]   A. Néron, Arithmétique et classes de diviseurs sur les variétés algébriques, *Proc. International Symposium on Algebraic Number Theory*, Tokyo-Nikko, Tokyo 1955, pp. 139–154

[Ne 3]   A. Néron, Quasi-fonctions et hauteurs sur les variétés abéliennes, *Ann. of Math.* **82** No. 2 (1965) pp. 249–331

[Ne 4]   A. Néron, Modèles minimaux des variétés abéliennes sur les corps locaux et globaux, *Pub. IHES* **21** (1964) pp. 361–482

[Ne 5]   A. Néron, Propriétés arithmétiques de certaines familles de courbes algébriques, *Proc. Int. Congress Amsterday III* (1954) pp. 481–488

[Ne 6]   A. Néron, Hauteurs et fonctions theta, *Rend. Sem. Mat. Fis. Milano*, **XLVI** (1976) pp. 111–135

[No 1]   J. Noguchi, A higher dimensional analogue of Mordell's conjecture over function fields, *Math. Ann.* **158** (1981) pp. 207–212

[No 2]   J. Noguchi and T. Sunaba, Finiteness of the family of rational and meromorphic mappings into algebraic varieties, *Amer. J. Math.* **104** (1982) pp. 887–900

[No 1]   D. G. Northcott, An inequality in the theory of arithmetic on algebraic varieties, *Proc. Cambridge Philos. Soc.* **45** (1949) pp. 502–509

[No 2]   D. G. Northcott, A further inequality in the theory of arithmetic on algebraic varieties, *Proc. Camb. Philos. Soc.* **45** (1949) pp. 510–518

[Oe]   J. Oesterle, Construction de hauteurs archimédiennes et p-adiques suivant la méthode de Bloch, *Séminaire de Théorie des Nombres 1980–81*, Birhauser–Boston

[Par 1]   A. N. Parsin, Algebraic curves over function fields I, *Izv. Akad. Nauk SSSR Ser. Mat.* **32** No. 5 (1968) = *Math. USSR Izv.* **2** No. 5 (1968) pp. 1145–1170

[Par 2]   A. N. Parsin, Minimal models of curves of genus 2 and homomorphisms of abelian varieties defined over a field of finite characteristic, *Izv. Akad. Nauk SSSR, Ser. Mat.* **36** No. 1 (1972) = *Math. USSR Izv.* **6** No. 1 (1972) pp. 65–108

[Po]   H. Poincaré, Sur les propriétés arithmétiques des courbes algébriques, *J. de Liouville* (V) **7** (1901) pp. 161–233

[Ra]   M. Raynaud, *Courbes sur une variété abélienne et points de torsion*,

[Ri 1]   D. Ridout, The p-adic generalization of the Thue–Siegel–Roth theorem, *Mathematika* **5** (1958) pp. 40–48

[Ri 2]   D. Ridout, Rational approximations to algebraic numbers, *Mathematka* **4** (1957) pp. 125–131

[Roq 1]   P. Roquette, Über das Hassesche Klassenkörperzerlegungsgesetz und seine Verallgemeinerung f¨r beliebige abelsche Functionenkörper, *J. reine angew. Math.* **197** (1957) pp. 49–67

[Roq 2]   P. Roquette, Einheiten und Divisorenklassen in endlich erzeugbaren Körpern, *J. Deutsch. Math. Verin.* **60** (1958) pp. 1–27

[Roth]   K. F. Roth, Rational approximations to algebraic numbers, *Mathematika* **2** (1955) pp. 1–20

[Scha]   S. Schanuel, Heights in number fields, *Bull. Soc. Math. France* **107** (1979) pp. 433–449

[Schm 1]　　W. SCHMIDT, Simultaneous approximation to algebraic numbers by rationals, *Acta Math.* **125** (1970) pp. 189–201

[Schm 2]　　W. SCHMIDT, *Diophantine Approximations*, Springer *Lecture Notes* **785**, 1980

[Schm 3]　　W. SCHMIDT, A metrical theorem in diophantine approximation, *Can. J. Math.* **12** (1960) pp. 619–631

[Schn]　　T. SCHNEIDER, *Einführung in die Theorie der Transzendenten Zahlen*, Springer-Verlag, Berlin, 1957

[Sel 1]　　E. SELMER. The diophantine equation $ax^3 + by^3 + cz^3 = 0$, *Acta Math.* **85** (1951) pp. 203–362

[Sel 2]　　E. SELMER, Ditto, Completion of the tables, *Acta Math.* **92** (1954) pp. 191–197

[Ser 1]　　J. P. SERRE, Groupes Algébriques et Théorie du Corps de Classes, Hermann, Paris, 1958

[Ser 2]　　J. P. SERRE, Quelques propriétés des groupes algébriques commutatifs, Appendix to M. Waldschmidt, Nombres transcendants et groupes algébriques, *Asterisque* **69–70**, 1979

[Ser 3]　　J. P. SERRE, Proprietes galoisiennes des points d'ordre fini des courbes elliptiques, *Invent. Math.* **15** (1972) pp. 259–331

[Sh 1]　　I. SHAFAREVICH, The group of principal homogeneous algebraic manifolds. *Doklady Akad. Nauk SSSR* **124** (1959) pp. 42–43

[Sh 2]　　I. SHAFA. 'VICH, Exponents of elliptic curves. *Dokl. Acad. Nauk SSSR* **114** (1957) pp. 714–716

[Sh 3]　　I. SHAFAREVICH, *Lectures on Minimal Models and Birational Transformations of Two-Dimensional Schemes*, Tate Institute, Bombay, 1966

[Sh-T]　　I. SHAFAREVICH and J. TATE, The rank of elliptic curves, *Dokl. Akad. Nauk. USSR* **175** (1967) pp. 770–773; *AMS Translation* **8** (1967) pp. 917–920

[Shi]　　G. SHIMURA, Reduction of algebraic varieties with respect to a discrete valuation of the basic field, *Amer. J. Math.* **77** (1955) pp. 134–176

[Sie 1]　　C. L. SIEGEL. Über einege Anwendungen Diophantischer Approximationen, *Abh. Preuss. Akad. Wiss. Phys. Math. Kl.* (1929) pp. 41–69

[Sie 2]　　C. L. SIEGEL, Abschätzung von Einheiten, *Nachr. Akad. Wiss. Göttingen* (1969) pp. 71–86

[Sil 1]　　J. SILVERMAN, *The Néron–Tate height on elliptic curves*, Thesis, Harvard, 1981

[Sil 2]　　J. SILVERMAN, Lower bound for the canonical height on elliptic curves, *Duke Math. J.* **48** No. 3 (1981) pp. 633–648

[Sil 3]　　J. SILVERMAN, Integer points and rank of Thue elliptic curves, *Invent. Math.* **66** (1982) pp. 395–404

[Sil 4]　　J. SILVERMAN, Heights and the specialization map for families of abelian varieties, to appear *J. Reine angew. Math.*

[Sil 5]　　J. SILVERMAN, Lower bounds for height functions, to appear

[Sil 6]　　J. SILVERMAN, Representations of integers by binary forms and the rank of the Mordell-Weil group, to appear.

[Ta 1]　　J. TATE, WC groups over p-adic fields, *Séminaire Bourbaki* 1958

[Ta 2]   J. TATE, On the conjectures of Birch and Swinnerton-Dyer and a geometric analog, *Séminaire Bourbaki* 1965–1966

[Ta 3]   J. TATE, The arithmetic of elliptic curves, *Invent. Math.* (1974) pp. 179–206

[Ta 4]   J. TATE, Algebraic cycles and poles of zeta functions, in *Arithmetic Algebraic Geometry*, Conference held at Purdue University, 1963, Harper & Row, New York, 1965

[Ta 5]   J. TATE, Variation of the canonical height of a point depending on a parameter, *Amer. J. Math.* **105** (1983) pp. 287–294

[Ta 6]   J. TATE, letter to Serre, 21 June 1968 and letter to Serre, 12 October 1979 (Appendix to the first letter)

[Vo]   P. VOJTA, Thesis, Harvard, 1983.

[Wa]   M. WALDSCHMIDT, Nombres transcendants et groupes algébriques, *Astérisque* **69–70**, Soc. Math. France, 1979

[We 1]   A. WEIL, Arithmetic on algebraic varieties, *Ann. of Math.* 53 **3** (1951) pp. 412–444

[We 2]   A. WEIL, L'arithmétique sur les courbes algébriques, *Acta Math.* **52** (1928) pp. 281–315

[We 3]   A. WEIL, *Foundations of Algebraic Geometry*, New York Second edition

[We 4]   A. WEIL, Sur les critères d'equivalences en géometrie algébrique, *Math. Ann.* **128** (1954) pp. 95–127

[We 5]   A. WEIL, *Variétés Abéliennes et Courbes Algébriques*, Hermann, Paris 1948, reprinted 1951

[We 6]   A. WEIL, Arithmétique et géometrie sur les variétés algébriques, *Act. Sc. et Ind. no. 206*, Hermann, Paris 1935

[We 7]   A. WEIL, Variétés abéliènnes, Colloque d'Algèbre et théorie des nombres, Paris 1949, pp. 125–128

[Za]   Ju. ZARHIN, Néron Pairing and Quasicharacters, *Izv. Akad. Nauk SSSR Ser. Mat.* **36** No. 3 (1972) *Math. USSR Izv.* **6** No. 3 (1972) pp. 491–503

[Za]   O. ZARISKI, The concept of a simple point on an abstract algebraic variety, *Trans. Math. Soc.* **62** (1947) pp. 1–52

[Zi]   H. ZIMMER, On the difference of the Weil height and the Néron–Tate height, *Math. Z.* **147** (1976) pp. 35–51

Index